205 Home Plans

MULTI-LEVEL DESIGNS

HOME PLANNERS, INC.®

23761 RESEARCH DRIVE
FARMINGTON HILLS, MICHIGAN 48024
TELEPHONE: (313) 477-1854

Contents

Edited by: Net Gingras
Cover design by: D. M. Naidus

Published by Home Planners, Inc., 23761 Research Drive, Farmington Hills, Michigan 48024. Printed in the United States of America. International Standard Book Number (ISBN): 0-918894-29-8.

Index to Designs

On the Cover: Cover Designs can be found on the following pages: Front cover - Design 42254, page 16. Back cover - top, Design 42580, page 109; middle, Design 42583, page 165; bottom, Design 42608, page 18.

How to read floor plans and blueprints

Selecting the most suitable house plan for your family is a matter of matching your needs, tastes, and life-style against the many designs we offer. When you study the floor plans in this issue, and the blueprints that you may subsequently order, remember that they are simply a two-dimensional representation of what will eventually be a three-dimensional reality.

Floor plans are easy to read. Rooms are clearly labeled, with dimensions given in feet and inches. Most symbols are logical and self-explanatory: The location of bathroom fixtures, planters, fireplaces, tile floors, cabinets and counters, sinks, appliances, closets, sloped or beamed ceilings will be obvious.

A blueprint, although much more detailed, is also easy to read; all it demands is concentration. The blueprints that we offer come in many large sheets, each one of which contains a different kind of information. One sheet contains foundation and excavation drawings, another has a precise plot plan. An elevations sheet deals with the exterior walls of the house; section drawings show precise dimensions, fittings, doors, windows, and roof structures. Our detailed floor plans give the construction information needed by your contractor. And each set of blueprints contains a lengthy materials list with size and quantities of all necessary components. Using this list, a contractor and suppliers can make a start at calculating costs for you.

When you first study a floor plan or blueprint, imagine that you are walking through the house. By mentally visualizing each room in three dimensions, you can transform the technical data and symbols into something more real.

Start at the front door. It's preferable to have a foyer or entrance hall in which to receive guests. A closet here is desirable; a powder room is a plus.

Look for good traffic circulation as you study the floor plan. You should not have to pass all the way through one main room to reach another. From the entrance area you should have direct access to the three principal areas of a house—the living, work, and sleeping zones. For example, a foyer might provide separate entrances to the living room, kitchen, patio, and a hallway or staircase leading to the bedrooms.

Study the layout of each zone. Most people expect the living room to be protected from cross traffic. The kitchen, on the other hand, should connect with the dining room—and perhaps also the utility room, basement, garage, patio or deck, or a secondary entrance. A homemaker whose workday centers in the kitchen may have special requirements: a window that faces the backyard; a clear view of the family room where children play; a garage or driveway entrance that allows for a short trip with groceries; laundry facilities close at hand. Check for efficient placement of kitchen cabinets, counters, and appliances. Is there enough room in the kitchen for additional appliances, for eating in? Is there a dining nook?

Perhaps this part of the house contains a family room or a den/bedroom/office. It's advantageous to have a bathroom or powder room in this section.

As you study the plan, you may encounter a staircase, indicated by a group of parallel lines, the number of lines equaling the number of steps. Arrows labeled "up" mean that the staircase leads to a higher level, and those pointing down mean it leads to a lower one. Staircases in a split-level will have both up and down arrows on one staircase because two levels are depicted in one drawing and an extra level in another.

Notice the location of the stairways. Is too much floor space lost to them? Will you find yourself making too many trips?

Study the sleeping quarters. Are the bedrooms situated as you like? You may want the master bedroom near the kids, or you may want it as far away as possible. Is there at least one closet per person in each bedroom or a double one for a couple? Bathrooms should be convenient to each bedroom—if not adjoining, then with hallway access and on the same floor.

Once you are familiar with the relative positions of the rooms, look for such structural details as:

- Sufficient uninterrupted wall space for furniture arrangement.
- Adequate room dimensions.
- Potential heating or cooling problems—i.e., a room over a garage or next to the laundry.
- Window and door placement for good ventilation and natural light.
- Location of doorways—avoid having a basement staircase or a bathroom in view of the dining room.
- Adequate auxiliary space—closets, storage, bathrooms, countertops.
- Separation of activity areas. (Will noise from the recreation room disturb sleeping children or a parent at work?)

As you complete your mental walk through the house, bear in mind your family's long-range needs. A good house plan will allow for some adjustments now and additions in the future.

Each member of your family may find the listing of his, or her, favorite features a most helpful exercise. Why not try it?

How to choose a contractor

A contractor is part craftsman, part businessman, and part magician. As the person who will transform your dreams and drawings into a finished house, he will be responsible for the final cost of the structure, for the quality of the workmanship, and for the solving of all problems that occur quite naturally in the course of construction. Choose him as carefully as you would a business partner, because for the next several months that will be his role in your life.

As soon as you have a building site and house plans, start looking for a contractor, even if you do not plan to break ground for several months. Finding one suitable to build your house can take time, and once you have found him, you will have to be worked into his schedule. Those who are good are in demand and, where the season is short, they are often scheduling work up to a year in advance.

There are two types of residential contractors: the construction company and the carpenter-builder, often called a general contractor. Each of these has its advantages and disadvantages.

The carpenter-builder works directly on the job as the field foreman. Because his background is that of a craftsman, his workmanship is probably good—but his paperwork may be slow or sloppy. His overhead—which you pay for—is less than that of a large construction company. However, if the job drags on for any reason, his interest may flag because your project is overlapping his next job and eroding his profits.

Construction companies handle several projects concurrently. They have an office staff to keep the paperwork moving and an army of subcontractors they know they can count on. Though you can be confident that they will meet deadlines, they may sacrifice workmanship in order to do so. Because they emphasize efficiency, they are less personal to work with than a general contractor. Many will not work with an individual unless he is represented by an architect. The company and the architect speak the same language; it requires far more time to deal directly with a homeowner.

To find a reliable contractor, start by asking friends who have built homes for recommendations. Check with local lumber yards and building supply outlets for names of possible candidates.

Once you have several names in hand, ask the Chamber of Commerce, Better Business Bureau, or local department of consumer affairs for any information they might have on each of them. Keep in mind that these watchdog organizations can give only the number of complaints filed; they cannot tell you what percent of those claims were valid. Remember, too, that a large-volume operation is logically going to have more complaints against it than will an independent contractor.

Set up an interview with each of the potential candidates. Find out what his specialty is—custom houses, development houses, remodeling, or office buildings. Ask each to take you into—not just to the site of—houses he has built. Ask to see projects that are complete as well as work in progress, emphasizing that you are interested in projects comparable to yours. A $300,000 dentist's office will give you little insight into a contractor's craftsmanship.

Ask each contractor for bank references from both his commercial bank and any other lender he has worked with. If he is in good financial standing, he should have no qualms about giving you this information. Also ask if he offers a warranty on his work. Most will give you a one-year warranty on the structure; some offer as much as a ten-year warranty.

Ask for references, even though no contractor will give you the name of a dissatisfied customer. While previous clients may be pleased with a contractor's work overall, they may, for example, have had to wait three months after they moved in before they had any closet doors. Ask about his follow-through. Did he clean up the building site, or did the owner have to dispose of the refuse? Ask about his business organization. Did the paperwork go smoothly, or was there a delay in hooking up the sewer because he forgot to apply for a permit?

Talk to each of the candidates about fees. Most work on a "cost plus" basis; that is, the basic cost of the project—materials, subcontractors' services, wages of those working directly on the project, but not office help—plus his fee. Some have a fixed fee; others work on a percentage of the basic cost. A fixed fee is usually better for you if you can get one. If a contractor works on a percentage, ask for a cost breakdown of his best estimate and keep very careful track as the work progresses. A crafty contractor can always use a cost overrun to his advantage when working on a percentage.

Do not be overly suspicious of a contractor who won't work on a fixed fee. One who is very good and in great demand may not be willing to do so. He may also refuse to submit a competitive bid.

If the top two or three candidates are willing to submit competitive bids, give each a copy of the plans and your specifications for materials. If they are not each working from the same guidelines, the competitive bids will be of little value. Give each the same deadline for turning in a bid; two or three weeks is a reasonable period of time. If you are willing to go with the lowest bid, make an appointment with all of them and open the envelopes in front of them.

If one bid is remarkably low, the contractor may have made an honest error in his estimate. Do not try to hold him to it if he wants to withdraw his bid. Forcing him to build at too low a price could be disastrous for both you and him.

Though the above method sounds very fair and orderly, it is not always the best approach, especially if you are inexperienced. You may want to review the bids with your architect, if you have one, or with your lender to discuss which to accept. They may not recommend the lowest. A low bid does not necessarily mean that you will get quality with economy.

If the bids are relatively close, the most important consideration may not be money at all. How easily you can talk with a contractor and whether or not he inspires confidence are very important considerations. Any sign of a personality conflict between you and a contractor should be weighed when making a decision.

Once you have financing, you can sign a contract with the builder. Most have their own contract forms, but it is advisable to have a lawyer draw one up or, at the very least, review the standard contract. This usually costs a small flat fee.

A good contract should include the following:

- Plans and sketches of the work to be done, subject to your approval.
- A list of materials, including quantity, brand names, style or serial numbers. (Do not permit any "or equal" clause that will allow the contractor to make substitutions.)
- The terms—who (you or the lender) pays whom and when.
- A production schedule.
- The contractor's certification of insurance for workmen's compensation, damage, and liability.
- A rider stating that all changes, whether or not they increase the cost, must be submitted and approved in writing.

Of course, this list represents the least a contract should include. Once you have signed it, your plans are on the way to becoming a home.

A frequently asked question is: "Should I become my own general contractor?" Unless you have knowledge of construction, material purchasing, and experience supervising subcontractors, we do not recommend this route.

How to shop for mortgage money

Most people who are in the market for a new home spend months searching for the right house plan and the ideal building site. Ironically, these same people often invest very little time shopping for the money to finance their new home, though the majority will have to live with the terms of their mortgage for as long as they live in the house.

The fact is that all banks are not alike, nor are the loans that they offer—and banks are not the only financial institutions that lend money for housing. The amount of down payment, interest rate, and period of the mortgage are all, to some extent, negotiable.

• Lending practices vary from one city and state to another. If you are a first-time builder or are new to an area, it is wise to hire a real estate (not divorce or general practice) attorney to help you unravel the maze of your specific area's laws, ordinances, and customs.

• Before talking with lenders, write down all your questions. Take notes during the conversation so you can make accurate comparisons.

• Do not be intimidated by financial officers. Keep in mind that *you are not begging for money*, you are buying it. Do not hesitate to reveal what other institutions are offering; they may be challenged to meet or better the terms.

• Use whatever clout you have. If you or your family have been banking with the same firm for years, let them know that they could lose your business if you can get a better deal elsewhere.

• Know your credit rights. The law prohibits lenders from considering only the husband's income when determining eligibility, a practice that previously kept many people out of the housing market. If you are turned down for a loan, you have a right to see a summary of the credit report and change any errors in it.

A GUIDE TO LENDERS

Where can you turn for home financing? Here is a list of sources for you to approach:

Savings and loan associations are the best place to start because they write well over half the mortgages in the United States on dwellings that house from one to four families. They generally offer favorable interest rates, require lower down payments, and allow more time to pay off loans than do other banks.

Savings banks, sometimes called mutual savings banks, are your next best bet. Like savings and loan associations, much of their business is concentrated in home mortgages.

Commercial banks write mortgages as a sideline, and when money is tight many will not write mortgages at all. They do hold about 15 percent of the mortgages in the country, however, and when the market is right, they can be very competitive.

Mortgage banking companies use the money of private investors to write home loans. They do a brisk business in government-backed loans, which other banks are reluctant to handle because of the time and paperwork required.

Some credit unions are now allowed to grant mortgages. A few insurance companies, pension funds, unions, and fraternal organizations also offer mortgage money to their membership, often at terms more favorable than those available in the commercial marketplace.

A GUIDE TO MORTGAGES

The types of mortgages available are far more various than most potential home buyers realize.

Traditional Loans

Conventional home loans have a fixed interest rate and fixed monthly payments. About 80 percent of the mortgage money in the United States is lent in this manner. Made by private lending institutions, these fixed rate loans are available to anyone whom the bank officials consider a good credit risk. The interest rate depends on the prevailing market for money and is slightly negotiable if you are willing to put down a large down payment. Most down payments range from 15 to 33 percent.

You can borrow as much money as the lender believes you can afford to pay off over the negotiated period of time—usually 20 to 30 years.

The FHA does not write loans; it insures them against default in order to encourage lenders to write loans for first-time buyers and people with limited incomes. The terms of these loans make them very attractive. The interest rate is fixed by FHA at 13½ percent, and you may be allowed to take as long as 25 to 30 years to pay it off.

The down payment also is substantially lower with an FHA-backed loan. At present it is set at 3 percent of the first $25,000 and 5 percent of the remainder, up to the $60,000 limit. This means that a loan on a $60,000 house would require a $750 down payment on the first $25,000 plus $1,750 on the remainder, for a total down payment of $2,500. In contrast, the down payment for the same house financed with a conventional loan could run as high as $20,000.

Anyone may apply for an FHA-insured loan, but both the borrower and the house must qualify.

The VA guarantees loans for eligible veterans, and the husbands and wives of those who died while in the service or from a service-related disability. The VA guarantees up to 60 percent of the loan or $27,500, whichever is less. Like the FHA, the VA determines the appraised value of the house, though with a VA loan, you can borrow any amount up to the appraised value.

The Farmers Home Administration offers the only loans made directly by the government. Families with limited incomes in rural areas can qualify if the house is in a community of less than 20,000 people and is outside of a large metropolitan area; if their income is less than $15,600; and if they can prove that they do not qualify for a conventional loan.

For more information, write Farmers Home Administration, Department of Agriculture, Washington, D.C. 20250, or contact your local office.

New loan instruments

If you think that the escalating cost of housing has squeezed you out of the market, take a look at the following new types of mortgages.

The graduated payment mortgage features a monthly obligation that gradually increases over a negotiated period of time—usually five to ten years. Though the payments begin lower, they stabilize at a higher monthly rate than a standard fixed rate mortgage. Little or no equity is built in the first years, a disadvantage if you decide to sell early in the mortgage period.

These loans are aimed at young people who can anticipate income increases that will enable them to meet the escalating payments. The size of the down payment is about the same or slightly higher than for a conventional loan, but you can qualify with a lower income. As of last year, savings and loan associations can write these loans, and the FHA now insures five different types.

The flexible loan insurance program (FLIP) requires that part of the down payment, which is about the same as a conventional loan, be placed in a pledged savings account. During the first five years of the mortgage, funds are drawn from this account to supplement the lower monthly payments.

The deferred interest mortgage, another graduated program, allows you to pay a lower rate of interest during the first few years and a higher rate in the later years of the mortgage. If the house is sold, the borrower must pay back all the interest, often with a prepayment penalty. Both the FLIP and deferred interest loans are very new and not yet widely available.

The variable rate mortgage is most widely available in California, but its popularity is growing. This instrument features a fluctuating interest rate that is linked to an economic indicator—usually the lender's cost of obtaining funds for lending. To protect the consumer against a sudden and disastrous increase, regulations limit the amount that the interest rate can increase over a given period of time.

To make these loans attractive, lenders offer them without prepayment penalties and with "assumption" clauses that allow another buyer to assume your mortgage should you sell.

Flexible payment mortgages allow young people who can anticipate rising incomes to enter the housing market sooner. They pay only the interest during the first few years; then the mortgage is amortized and the payments go up. This is a valuable option only for those people who intend to keep their home for several years because no equity is built in the lower payment period.

The reverse annuity mortgage is targeted for older people who have fixed incomes. This very new loan instrument allows those who qualify to tap into the equity on their houses. The lender pays them each month and collects the loan when the house is sold or the owner dies.

Traditional
Split-Level Homes

Design 42624

904 Sq. Ft. – Main Level; 1,120 Sq. Ft. – Upper Level
404 Sq. Ft. – Lower Level; 39,885 Cu. Ft.

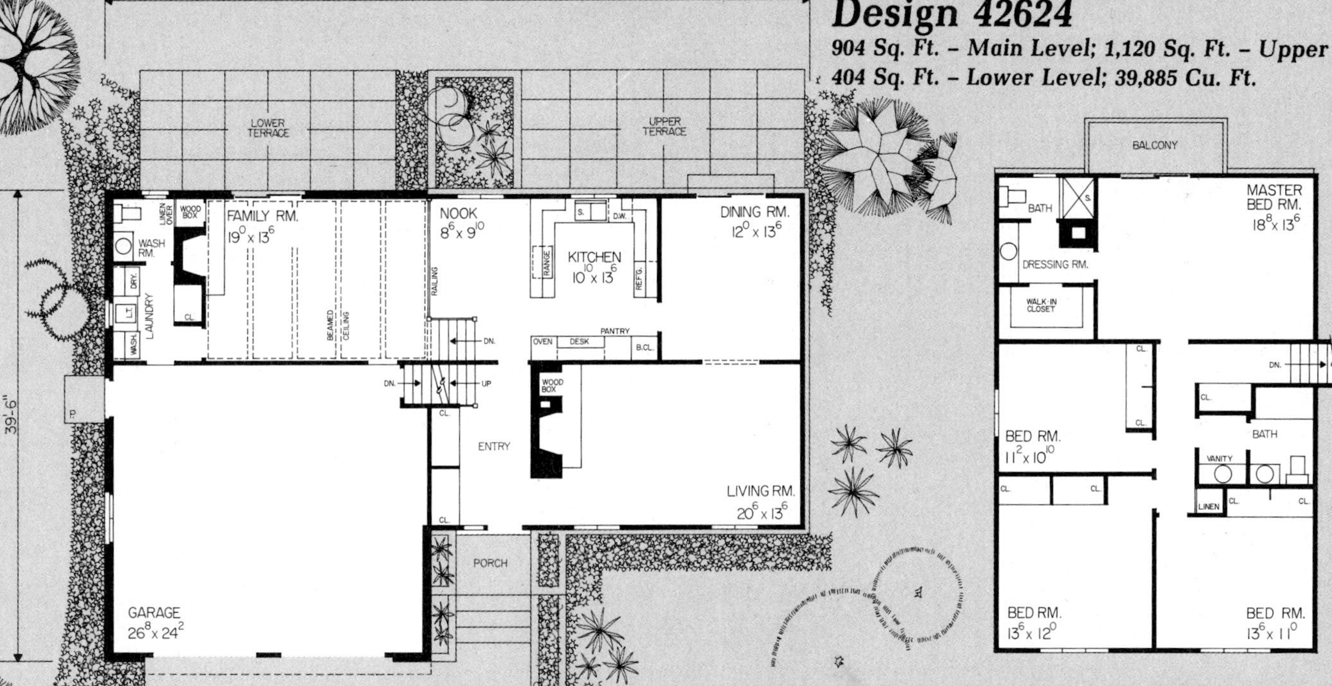

● This is tri-level living at its best. The exterior is that of the most popular Tudor styling. A facade which will hold its own for many a year to come. Livability will be achieved to its maximum on the four (including basement) levels. The occupants of the master bedroom can enjoy the outdoors on their private balcony. Additional outdoor enjoyment can be gained on the two terraces. That family room is more than 19' x 13' and includes a beamed ceiling and fireplace with wood box. Its formal companion, the living room, is similar in size and also will have the added warmth of a fireplace.

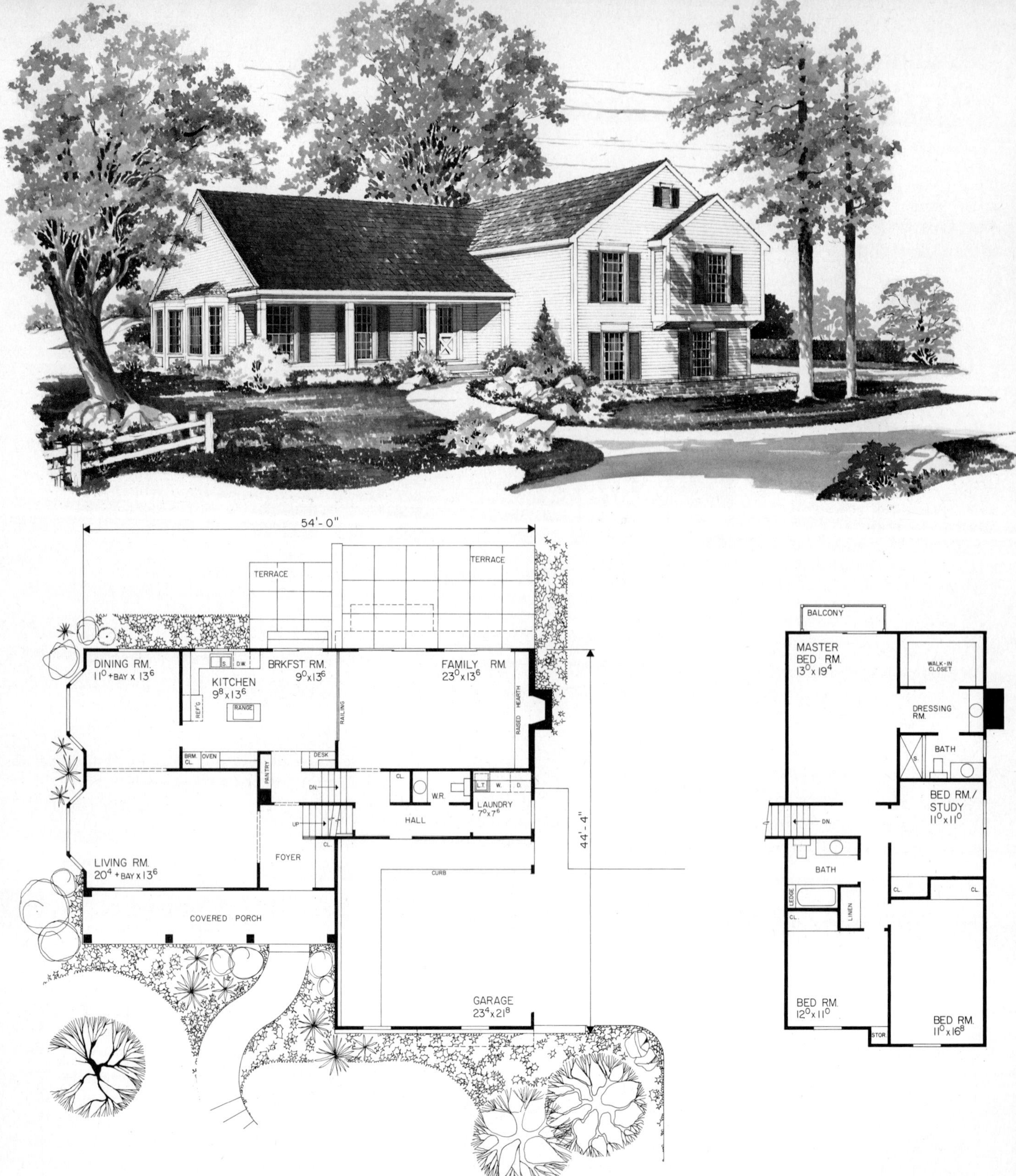

Design 42786 *871 Sq. Ft. – Main Level; 1,132 Sq. Ft. – Upper Level; 528 Sq. Ft. – Lower Level; 44,000 Cu. Ft.*

● A bay window in each the formal living room and dining room. A great interior and exterior design feature to attract attention to this tri-level home. The exterior also is enhanced by a covered front porch to further the Colonial charm. The interior livability is outstanding, too. An abundance of built-ins in the kitchen create an efficient work center. Features include an island range, pantry, broom closet, desk and breakfast room with sliding glass doors to the rear terrace. The lower level houses the informal family room, wash room and laundry. Further access is available to the outdoors by the family room to the terrace and laundry room to the side yard.

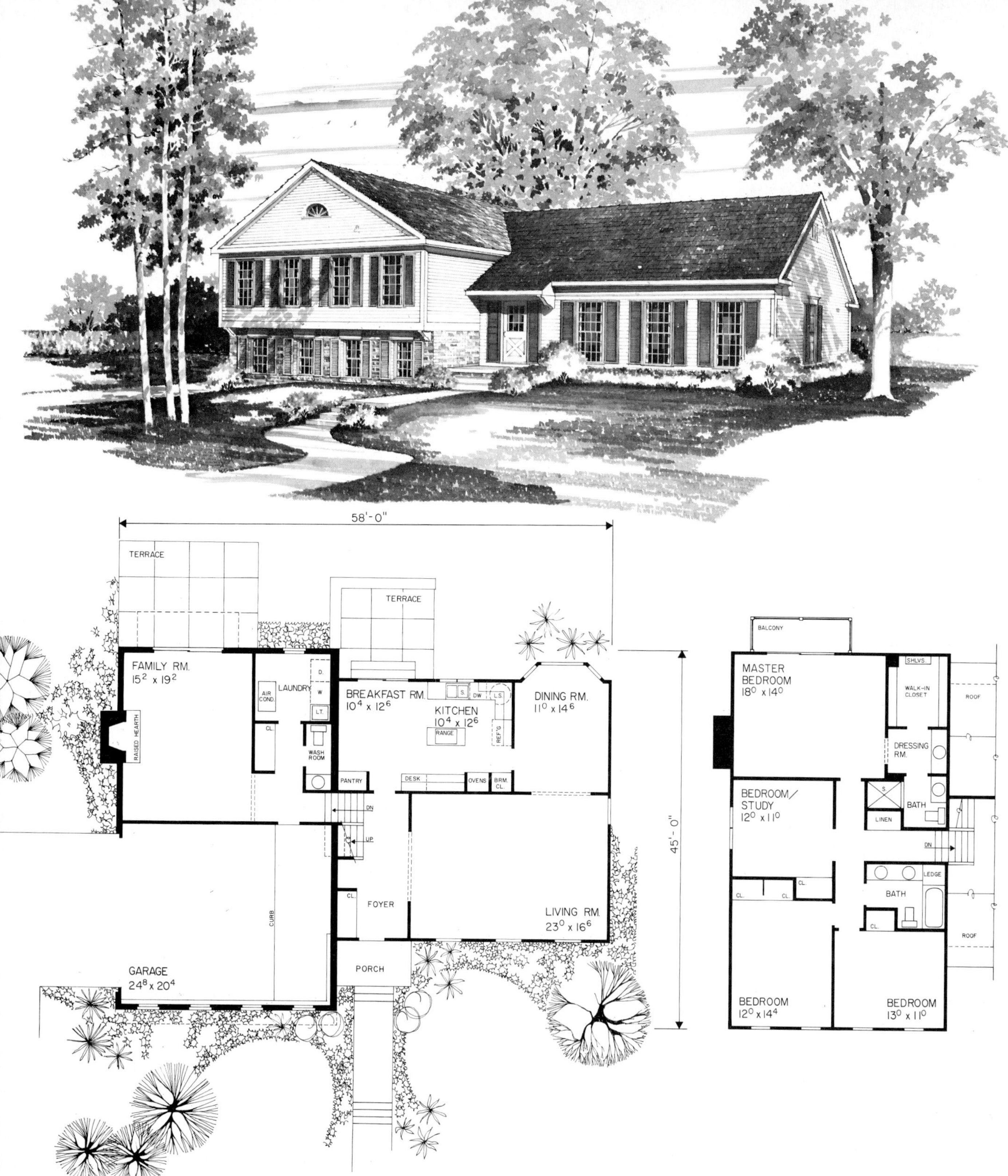

Design 42787 *976 Sq. Ft. – Main Level; 1,118 Sq. Ft. – Upper Level; 524 Sq. Ft. – Lower Level; 36,110 Cu. Ft.*

● Three level living! Main, upper and lower levels to serve you and your family with great ease. Start from the bottom and work your way up. Family room with raised hearth fireplace, laundry and wash room on the lower level. Formal living and dining rooms, kitchen and breakfast room on the main level. Stop and take note at the efficiency of the kitchen with its many outstanding extras. The upper level houses the three bedrooms, study (or fourth bedroom if you prefer) and two baths. This design has really stacked up its livability to serve its occupants to their best advantage. This design has great interior livability and exterior charm.

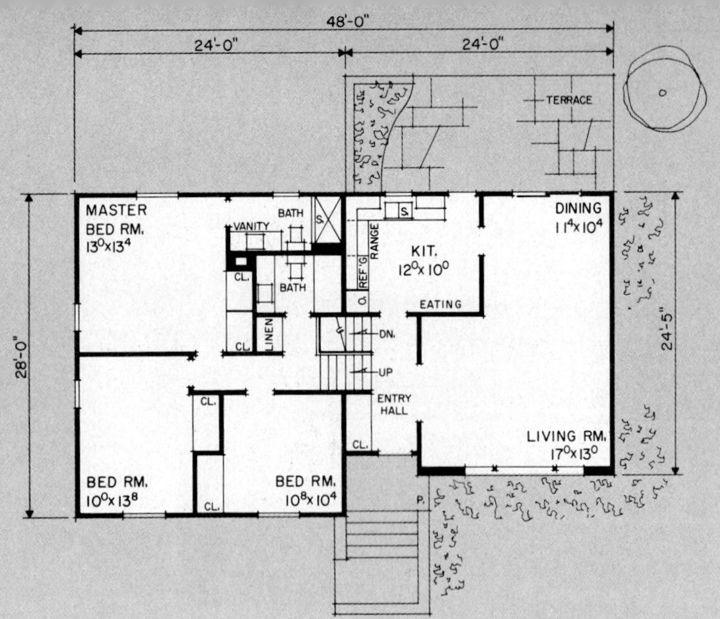

● Here are three charming split-levels designed for the modest budget. They will not require a large, expensive piece of property. Nevertheless, each is long on livability and offers all the features necessary to guarantee years of convenient living.

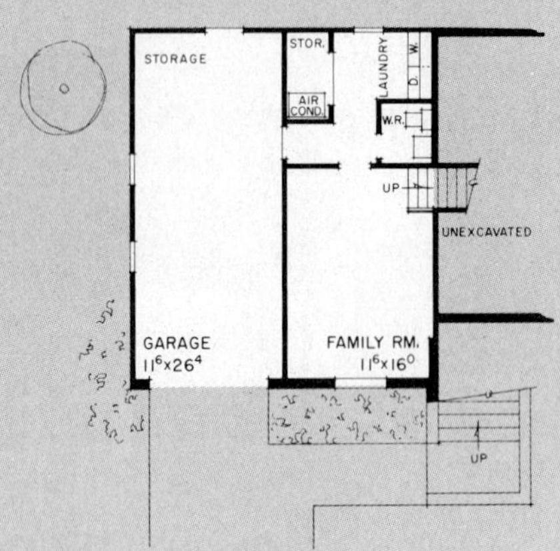

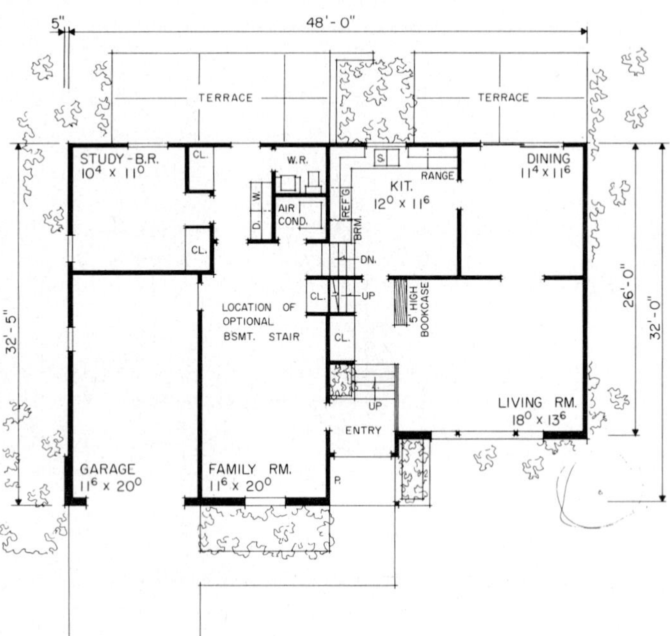

● Charming? It certainly is. And with good reason, too. This delightfully proportioned split level is highlighted by fine window treatment, interesting roof lines, an attractive use of materials and an inviting front entrance with double doors.

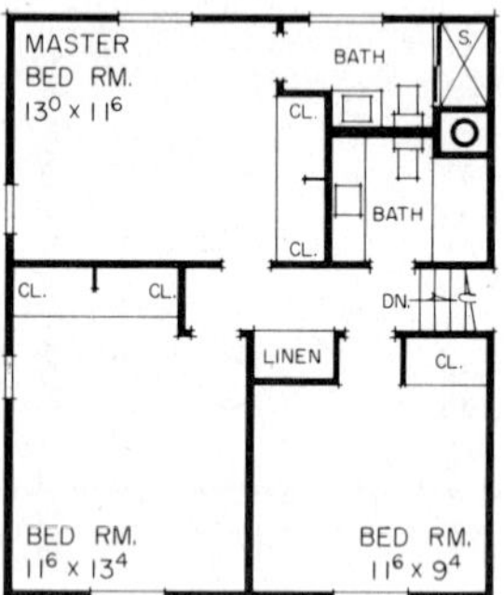

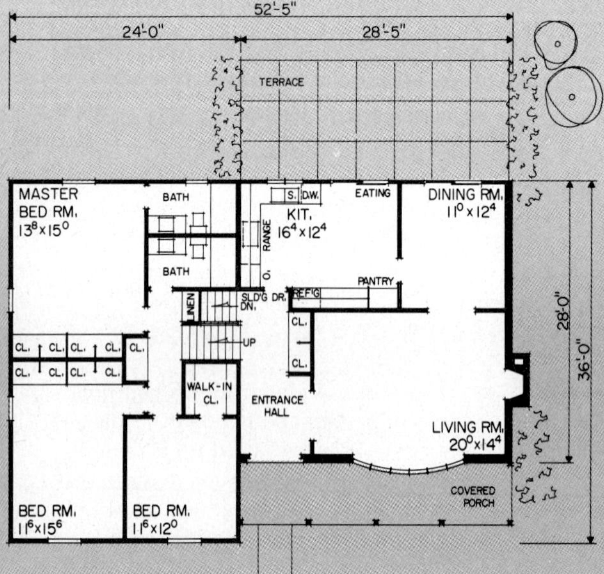

● Four level livability. And what livability it will be! This home will be most economical to build. As the house begins to take form you'll appreciate even more all the livable space you and your family will enjoy. List features that appeal to you.

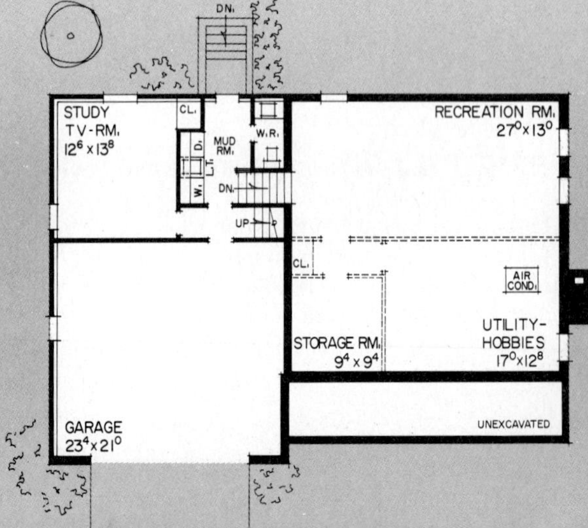

Design 41358 *576 Sq. Ft. – Main Level; 672 Sq. Ft. – Upper Level: 328 Sq. Ft. – Lower Level; 20,784 Cu. Ft.*

Design 41770 *636 Sq. Ft. – Main Level; 672 Sq. Ft. – Upper Level; 528 Sq. Ft. – Lower Level; 19,980 Cu. Ft.*

Design 41882 *800 Sq. Ft. – Main Level; 864 Sq. Ft. – Upper Level; 344 Sq. Ft. – Lower Level; 28,600 Cu. Ft.*

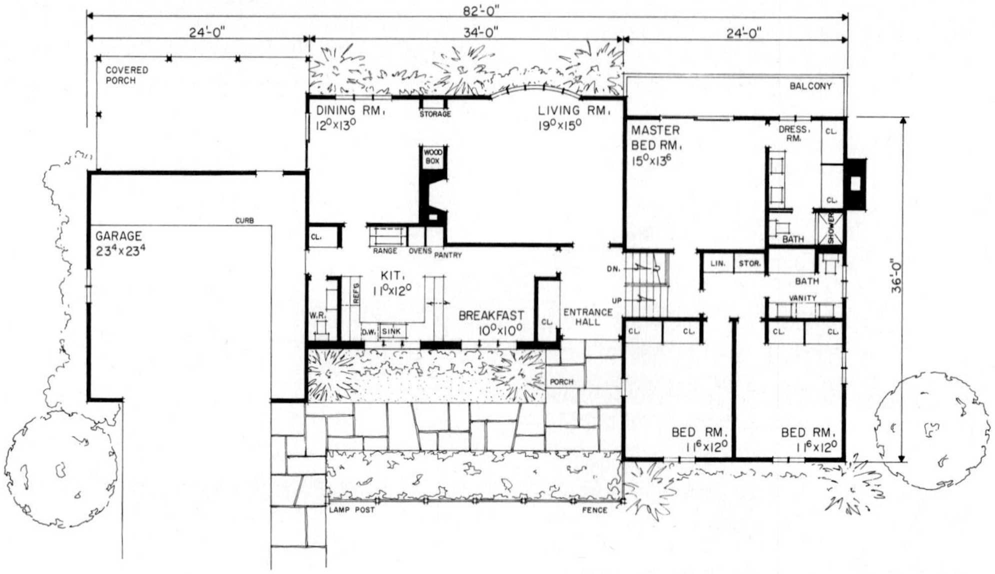

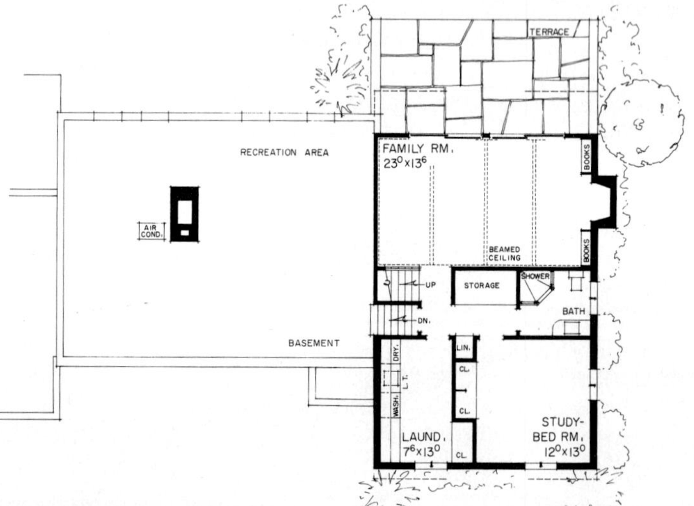

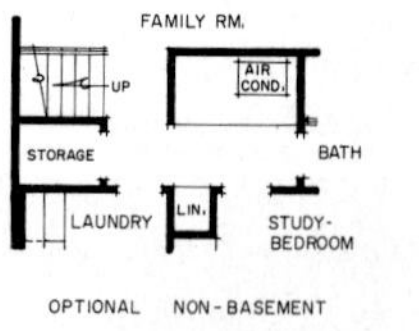

82'-0"
24'-0" 34'-0" 24'-0"
COVERED PORCH
DINING RM. $12^{0} \times 13^{0}$
STORAGE
LIVING RM. $19^{0} \times 15^{0}$
BALCONY
MASTER BED RM. $15^{0} \times 13^{6}$
DRESS. RM.
CL.
WOOD BOX
GARAGE $23^{4} \times 23^{4}$
CURB
CL.
RANGE OVENS PANTRY
BATH
SHOWER
DN.
LIN. STOR.
BATH
KIT. $11^{0} \times 12^{0}$
REFG.
W.R.
D.W. SINK
BREAKFAST $10^{0} \times 10^{0}$
CL.
ENTRANCE HALL
UP
VANITY
CL. CL. CL. CL.
36'-0"
PORCH
BED RM. $11^{6} \times 12^{0}$
BED RM. $11^{6} \times 12^{0}$
LAMP POST
FENCE

TERRACE
RECREATION AREA
FAMILY RM. $23^{0} \times 13^{6}$
BOOKS
AIR COND.
BEAMED CEILING
BOOKS
UP
STORAGE
SHOWER
BATH
DN.
BASEMENT
LIN.
CL.
CL.
DRY.
WASH
L.T.
LAUND. $7^{6} \times 13^{0}$
CL.
STUDY-BED RM. $12^{0} \times 13^{0}$

FAMILY RM.
UP
AIR COND.
STORAGE
BATH
LAUNDRY
LIN.
STUDY-BEDROOM
OPTIONAL NON-BASEMENT

Design 41935

904 Sq. Ft. – Main Level; 864 Sq. Ft. – Upper Level; 840 Sq. Ft. – Lower Level; 26,745 Cu. Ft.

● If there was ever a design that looked a part of the ground it was built on, this particular multi-level looks just that. This design will adapt equally well to a flat or sloping site. There would be no question about the family's ability to adapt to what the interior has to offer. Everything is present to satisfy the family's desire to "live a little". There are features such as the covered porch, the balcony, the two fireplaces, the extra study, the beamed ceiling family room, the complete laundry and the basement level for added recreational and storage space. Be sure not to miss the three full baths, plus the extra wash room. Blueprints for this design include optional non-basement details for construction for those not in need of the extra space.

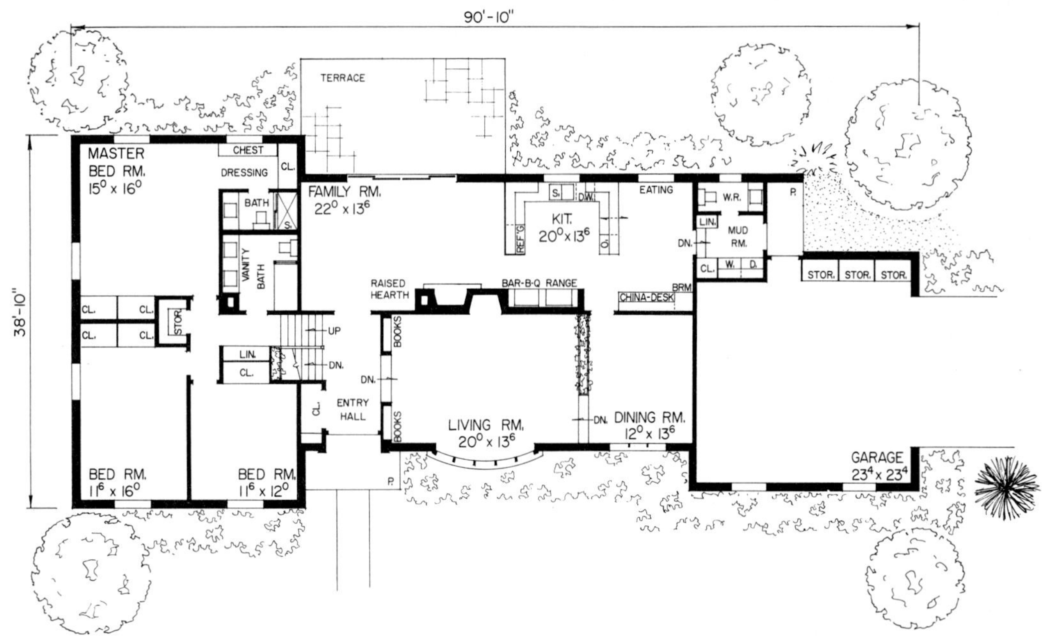

Design 41265

1,298 Sq. Ft. - Main Level; 964 Sq. Ft. - Upper Level 964 Sq. Ft. - Lower Level; 48,588 Cu. Ft.

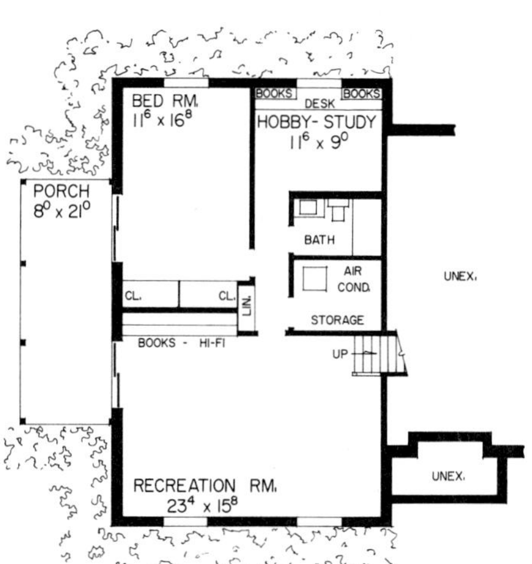

● Impressive, may be just the word to describe this appealingly formal, traditional tri-level. Changes in level add interest to its plan and make this already spacious house seem even larger. Big, glamorous living room two steps below entry level features a handsome fireplace and opposite it a broad bay window. Beyond, and raised two additional steps to emphasize the sunken area, is the formal dining room, partitioned from the living room by a built-in planter. Back of the garage, handy to the rear yard, the service porch and laundry/wash room double as a mud room area. Up a few stairs from the entry hall is the quiet sleeping level with two baths. Down a few stairs is the lower level with areas which will lend themselves to flexible living patterns. A covered porch is accessible from this level.

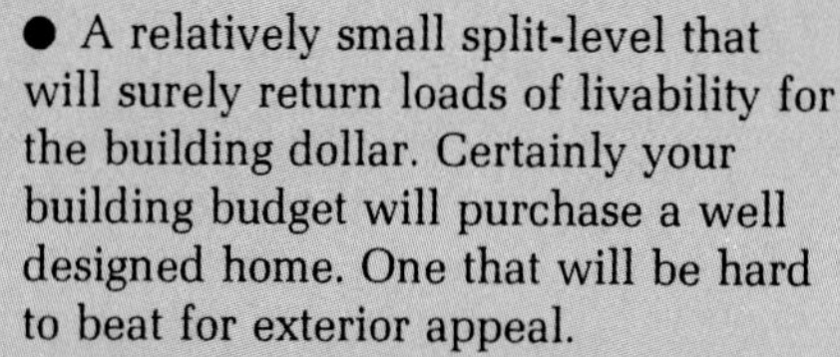

● A relatively small split-level that will surely return loads of livability for the building dollar. Certainly your building budget will purchase a well designed home. One that will be hard to beat for exterior appeal.

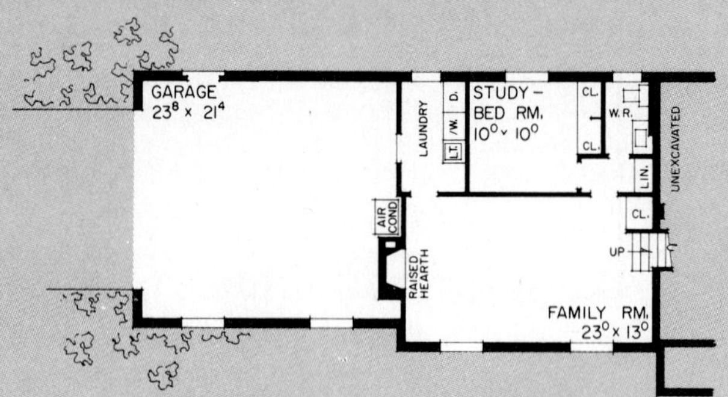

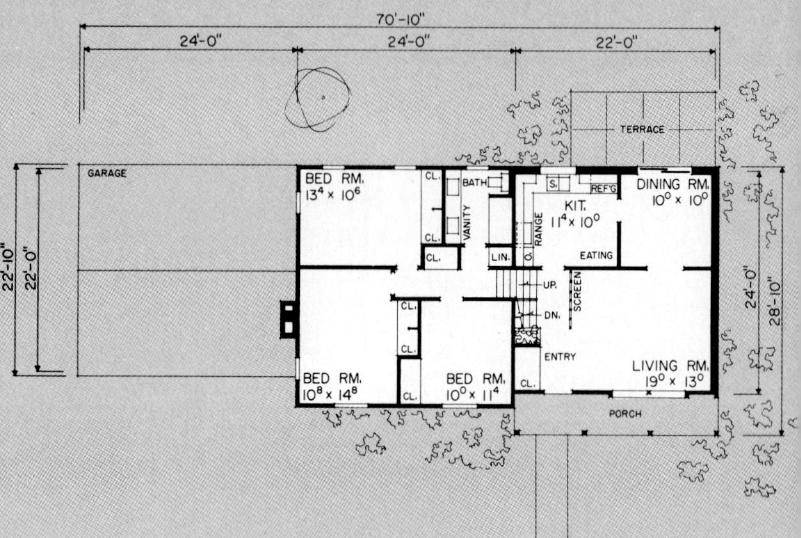

● Impressive, indeed, wherever located. Gabled roofs, muntined windows with shutters, a covered front porch, paneled double doors and a cupola over the garage give a Colonial touch to the exterior of this smartly, planned split-level.

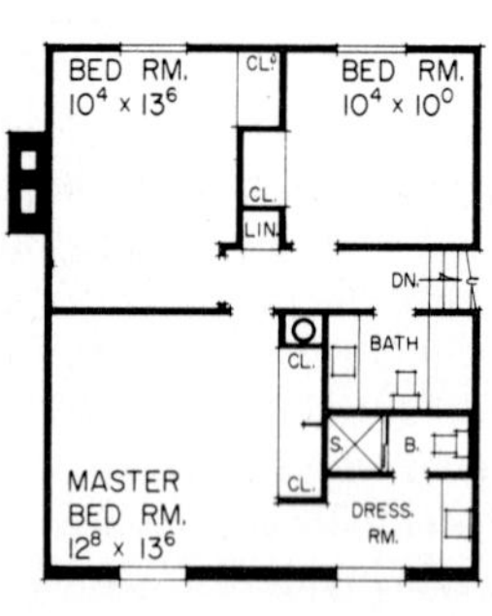

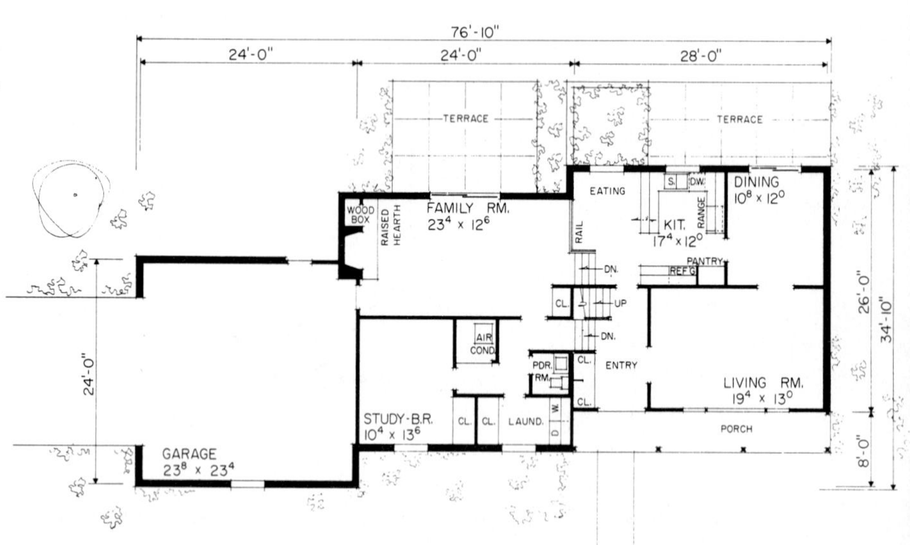

● You'll have fun living in this home. There will be four bedrooms, three baths, large formal living and dining rooms, an efficient kitchen with breakfast nook, a covered porch and much more to serve your family.

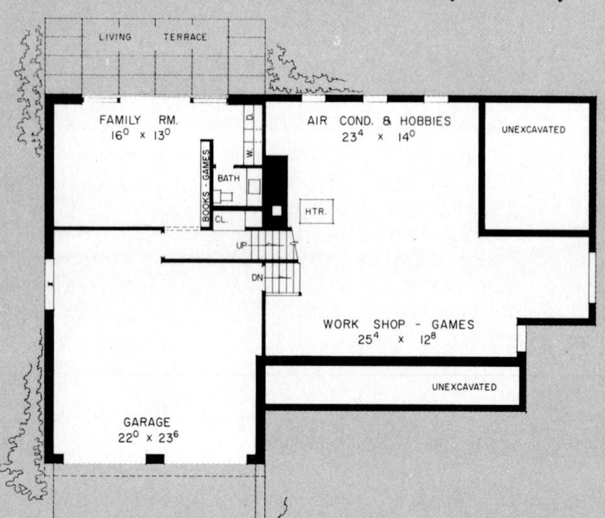

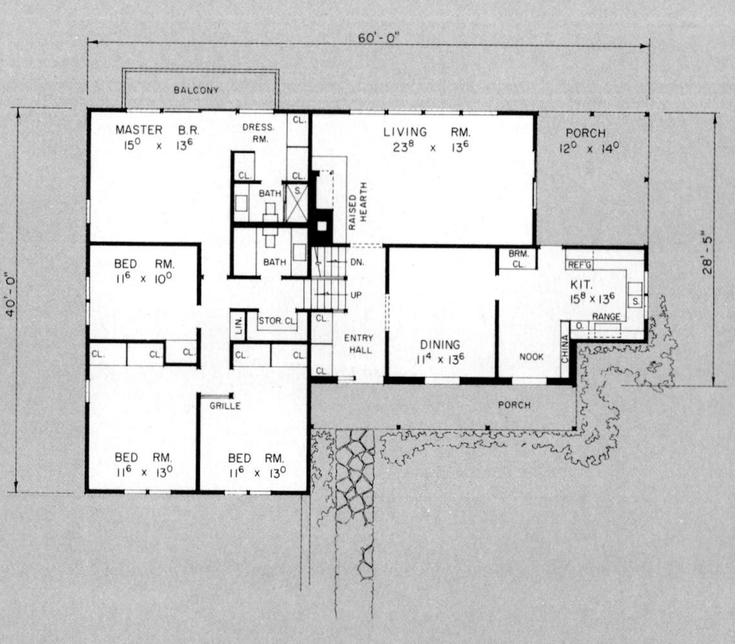

Design 41717 556 Sq. Ft. – Main Level; 624 Sq. Ft. – Upper Level; 596 Sq. Ft. – Lower Level; 17,975 Cu. Ft.

Design 41347 750 Sq. Ft. – Main Level; 672 Sq. Ft. – Upper Level; 664 Sq. Ft. – Lower Level; 21,646 Cu. Ft.

Design 43148 808 Sq. Ft. – Main Level; 960 Sq. Ft. – Upper Level; 374 Sq. Ft. – Lower Level; 33,931 Cu. Ft.

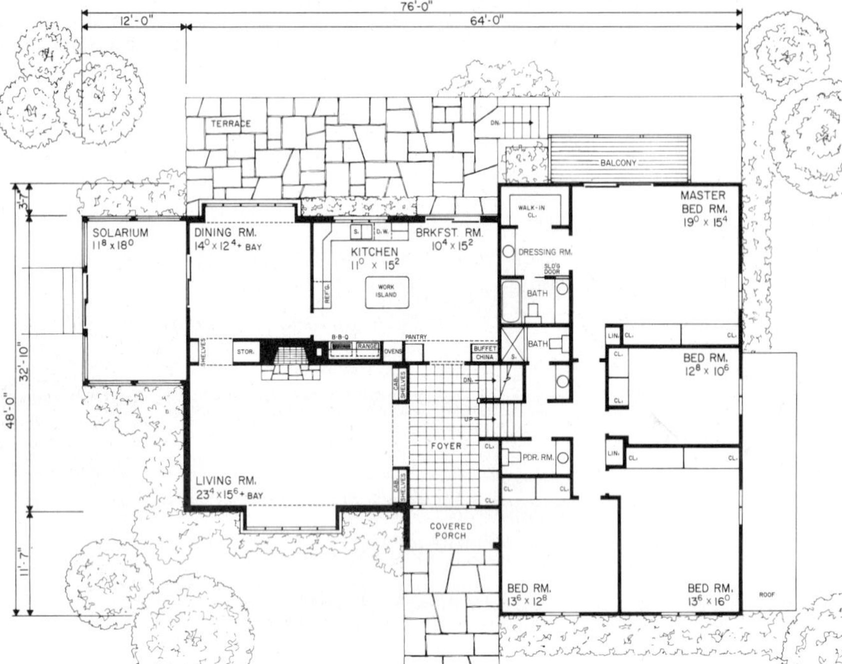

Design 42254

1,220 Sq. Ft. – Main Level
1,344 Sq. Ft. – Upper Level
659 Sq. Ft. – Lower Level
56,706 Cu. Ft.

● Tudor charm is deftly exemplified by this outstanding four level design. The window treatment, the heavy timber work and the chimney pots help set the character of this home. Contributing an extra measure of appeal is the detailing of the delightful solarium. The garden view of this home is equally appealing. The upper level balcony looks down onto the two terraces. The covered front entry leads to the spacious formal entrance hall with its slate floor. . .

Design 42243

1,274 Sq. Ft. – Main Level; 960 Sq. Ft. – Upper Level
936 Sq. Ft. – Lower Level; 42,478 Cu. Ft.

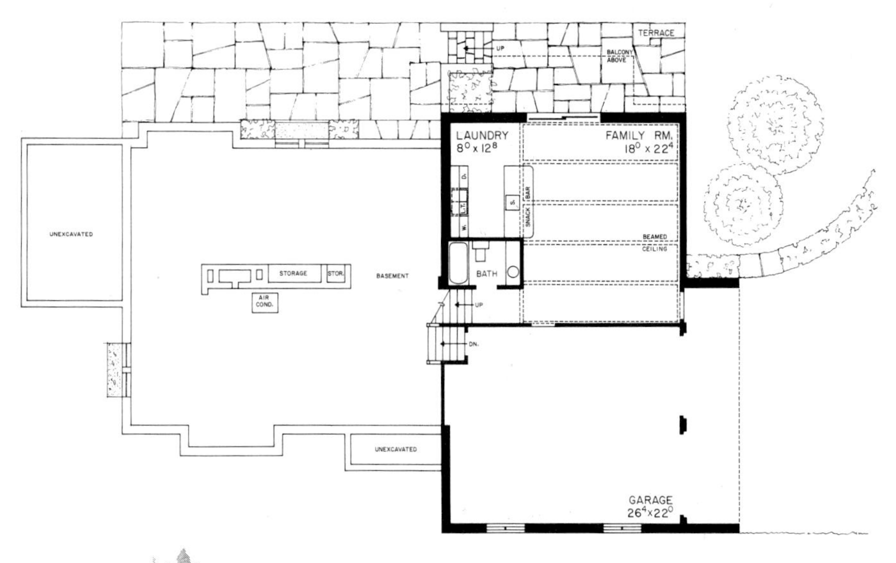

. . . Straight ahead is the kitchen and nook. The open planning of this area results in a fine feeling of spaciousness. Both living and dining rooms are wonderfully large. Each room highlights a big bay window. Notice the built-in units. Upstairs there are four bedrooms, two full baths and a powder room. Count the closets. The lower level is reserved for the all-purpose room, the separate laundry and a third full bath. The garage is adjacent. A fourth level is a basement with an abundance of space for storage and hobbies.

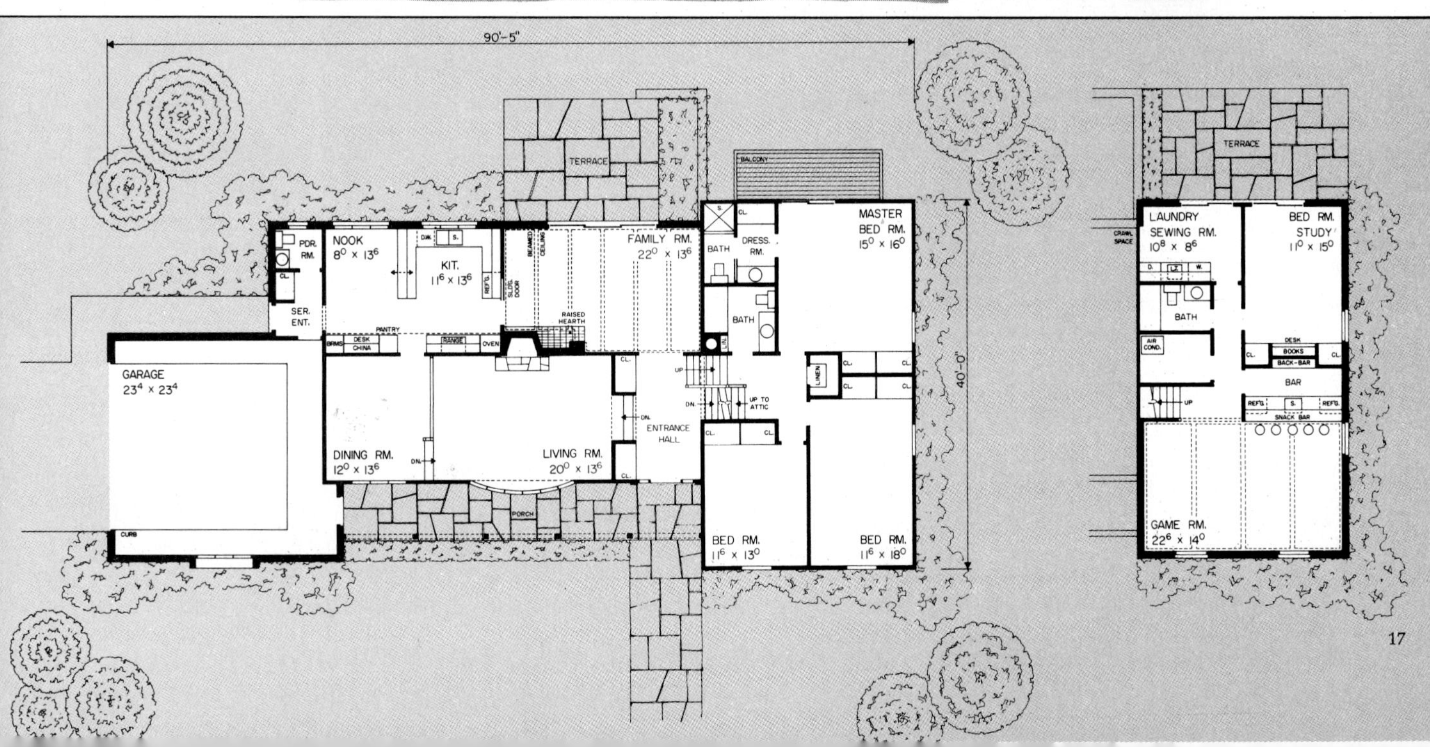

● Projecting over the lower level in Garrison Colonial style is the upper level containing three bedrooms a compartmented bath with twin lavatories and two handy linen closets. The main level consists of an L-shaped kitchen with convenient eating space, a formal dining room with sliding glass doors to the terrace and a sizable living room. On the lower level there is access to the outdoors, a spacious family room and a laundry-wash room area.

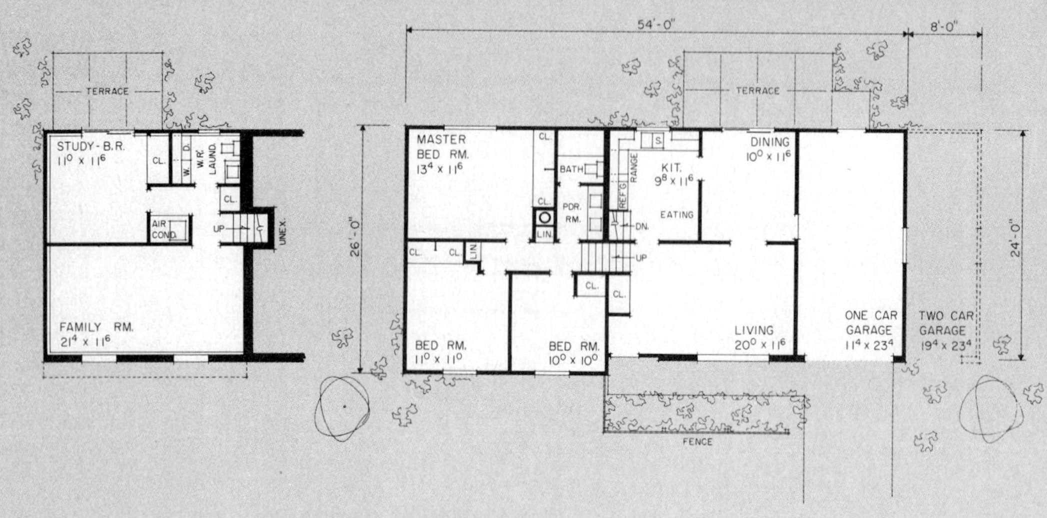

● Here are four levels just waiting for the opportunity to serve the living requirements of the active family. The traditional appeal of the exterior will be difficult to beat. Observe the window treatment, the double front doors, the covered front porch and the wrought iron work.

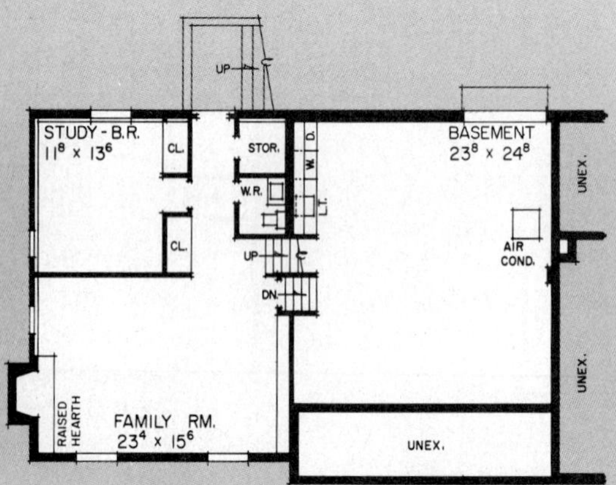

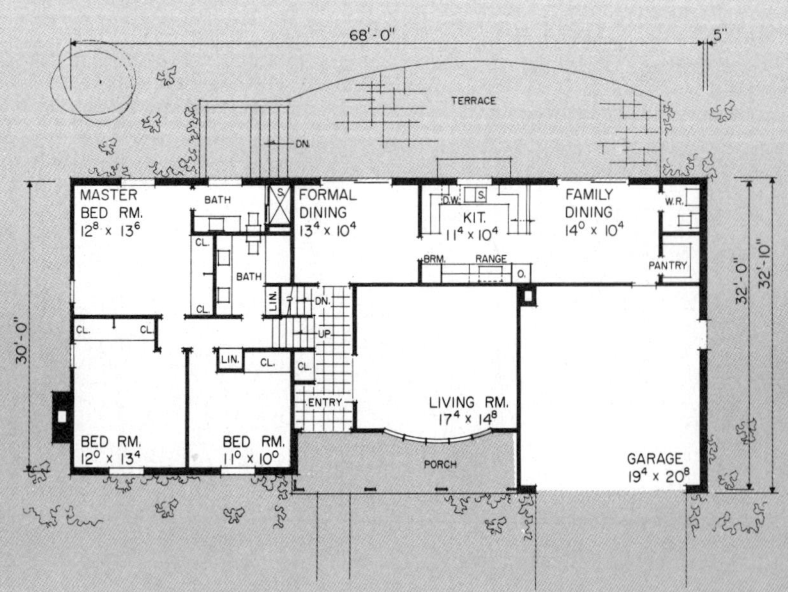

Design 41308 *496 Sq. Ft. – Main Level; 572 Sq. Ft. – Upper Level; 537 Sq. Ft. – Lower Level; 16,024 Cu. Ft.*

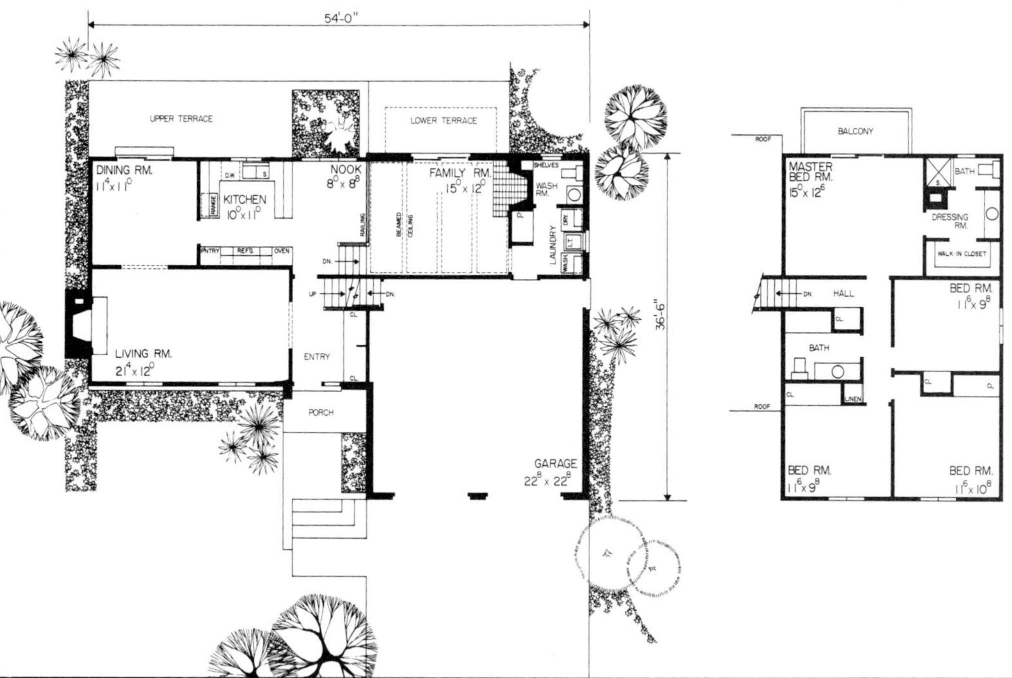

Design 42608

728 Sq. Ft. – Main Level
874 Sq. Ft. – Upper Level
310 Sq. Ft. – Lower Level
27,705 Cu. Ft.

● Here is tri-level livability with a fourth basement level for bulk storage and, perhaps, a shop area. There are four bedrooms, a handy laundry, two eating areas, formal and informal living areas and two fireplaces. Sliding glass doors in the formal dining room and the family room open to a terrace. The U-shaped kitchen has a built-in range/oven and pantry.

Design 41768 *844 Sq. Ft. – Main Level; 740 Sq. Ft. – Upper Level; 740 Sq. Ft. – Lower Level; 29,455 Cu. Ft.*

Design 41356

528 Sq. Ft. – Main Level; 576 Sq. Ft. – Upper Level
576 Sq. Ft. – Lower Level; 16,700 Cu. Ft.

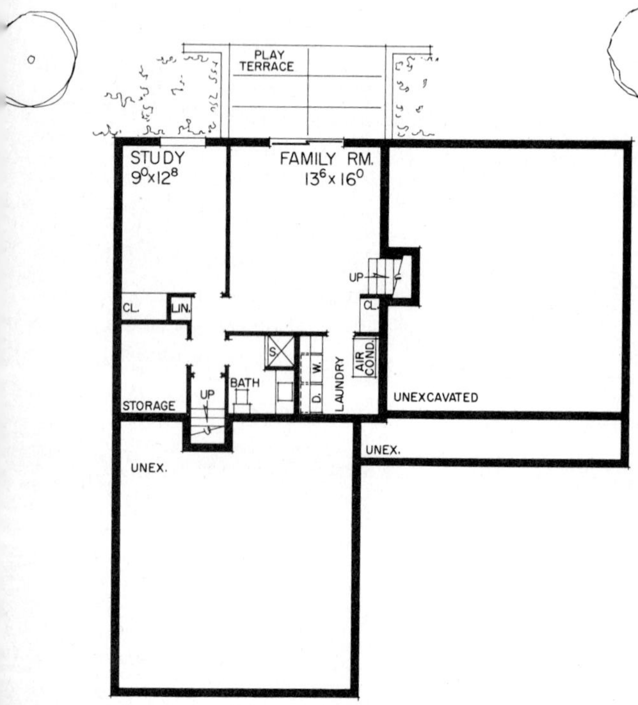

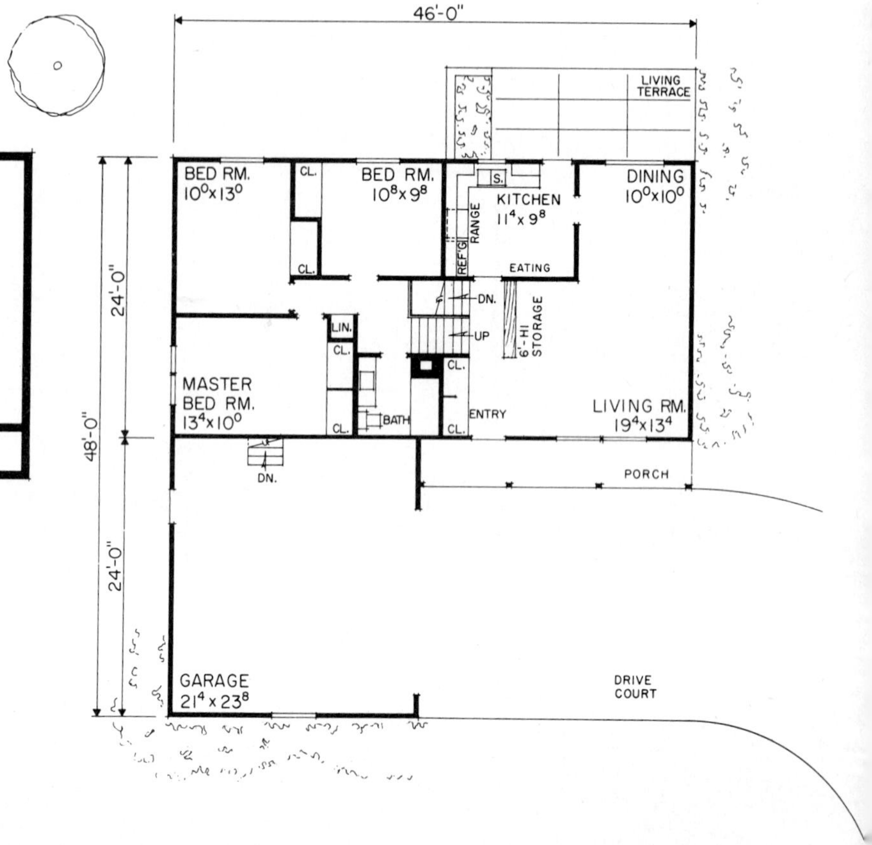

PLAY TERRACE
STUDY $9^{0} \times 12^{8}$
FAMILY RM. $13^{6} \times 16^{0}$
UP
CL.
LIN.
CL.
S.
D.
W.
LAUNDRY
AIR COND.
BATH
STORAGE
UP
UNEXCAVATED
UNEX.
UNEX.

46'-0"
LIVING TERRACE
BED RM. $10^{0} \times 13^{0}$
CL.
BED RM. $10^{8} \times 9^{8}$
CL.
S.
KITCHEN $11^{4} \times 9^{8}$
RANGE
REF'G.
EATING
DINING $10^{0} \times 10^{0}$
DN.
UP
6'-HI STORAGE
LIN.
CL.
CL.
MASTER BED RM. $13^{4} \times 10^{0}$
CL.
BATH
CL.
ENTRY
LIVING RM. $19^{4} \times 13^{4}$
DN.
PORCH
24'-0"
48'-0"
24'-0"
GARAGE $21^{4} \times 23^{8}$
DRIVE COURT

● An attractive L-shaped traditional split-level design for a narrow building site. There is space galore in which the young and growing family may move around in this modest sized design. Three distinct levels offer fine livabilty. The main level features a spacious, open planned formal living and dining area. The efficient kitchen offers an area for informal eating. A living terrace is but a few steps from the work center. A built-in unit, six feet in height, is a nice feature of the living room. The upper level permits the three bedrooms to enjoy their full measure of privacy. The lower level represents an area the family can expand for informal living. There is a family room with adjacent play terrace, a study, a full bath, a storage room and a laundry.

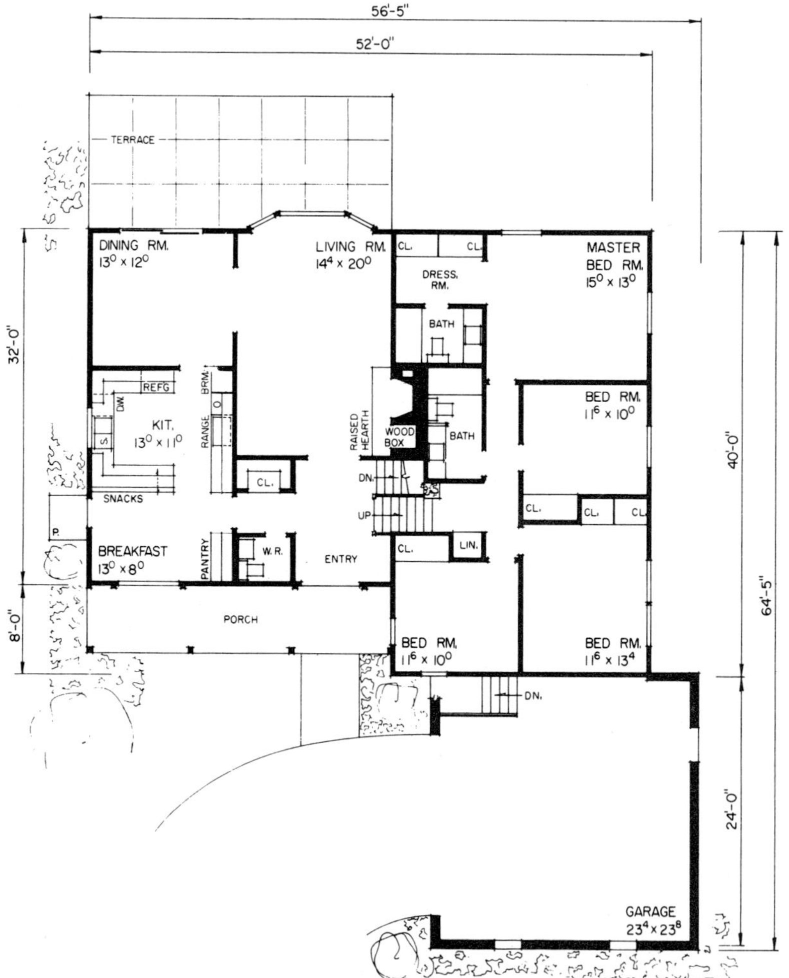

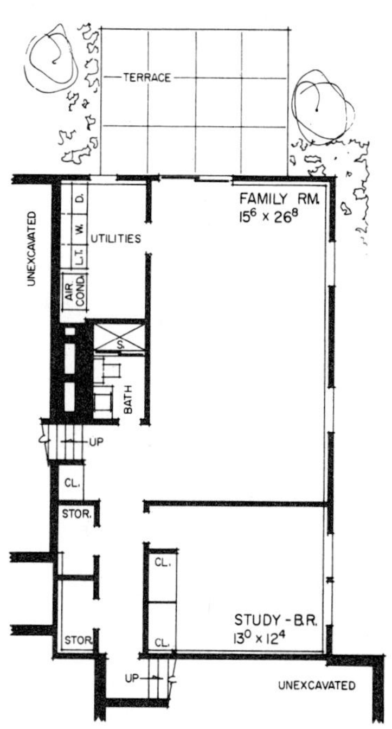

Design 41721

896 Sq. Ft. - Main Level
960 Sq. Ft. - Upper Level
960 Sq. Ft. - Lower Level
32,595 Cu. Ft.

● Ideal for a relatively narrow site, this L-shaped tri-level will serve the large family wonderfully. The double doors of the front entry are sheltered by the long covered porch and lead to the spacious hall which routes traffic efficiently to all areas. The kitchen has an abundance of counter and cupboard space flanked by the informal breakfast room and the separate dining room. The bay window of the living room overlooks the outdoor terrace. Four bedrooms and two baths comprise the quiet upper sleeping level. An abundance of dual-use space is found in the lower level, accessible from garage. Note the utility room, the bath with stall shower, the study/bedroom, the large family room and two storage closets which are all features of the lower level.

Design 41112

686 Sq. Ft. – Main Level
672 Sq. Ft. – Upper Level
336 Sq. Ft. – Lower Level
19,132 Cu. Ft.

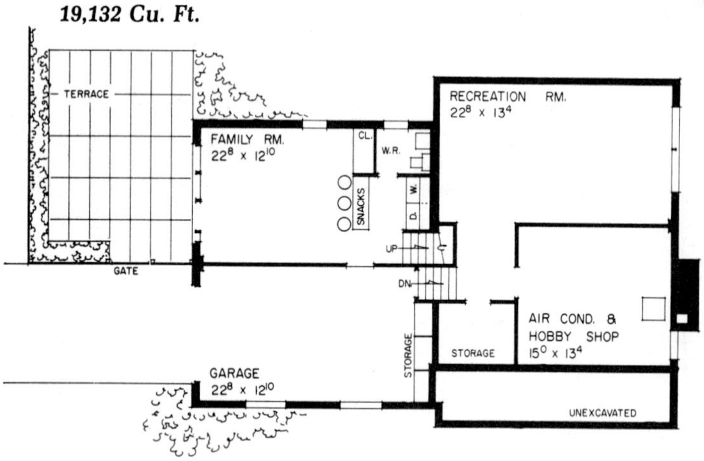

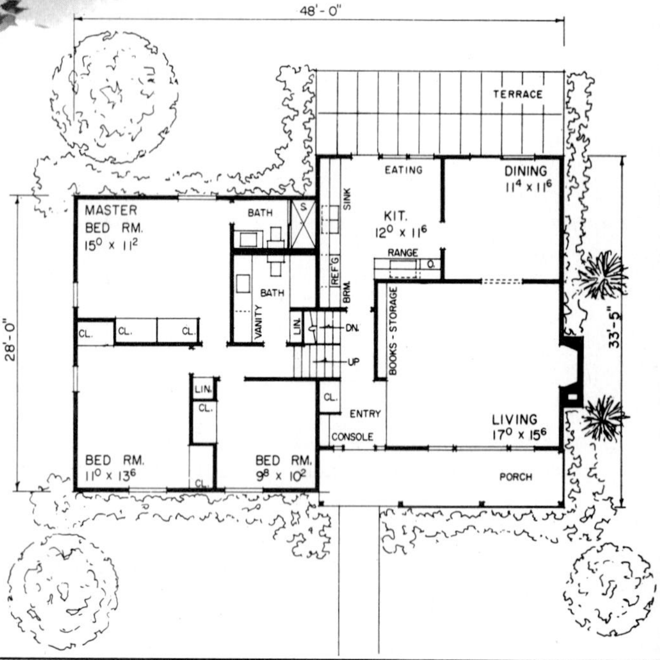

● Traditional tri-level living is guaranteed from this three bedroom design. Three floors of outstanding livability with a fourth level serving as additional recreational space or just use it for storage.

Design 41391

413 Sq. Ft. – Main Level
483 Sq. Ft. – Upper Level; 495 Sq. Ft. – Lower Level
13,715 Cu. Ft.

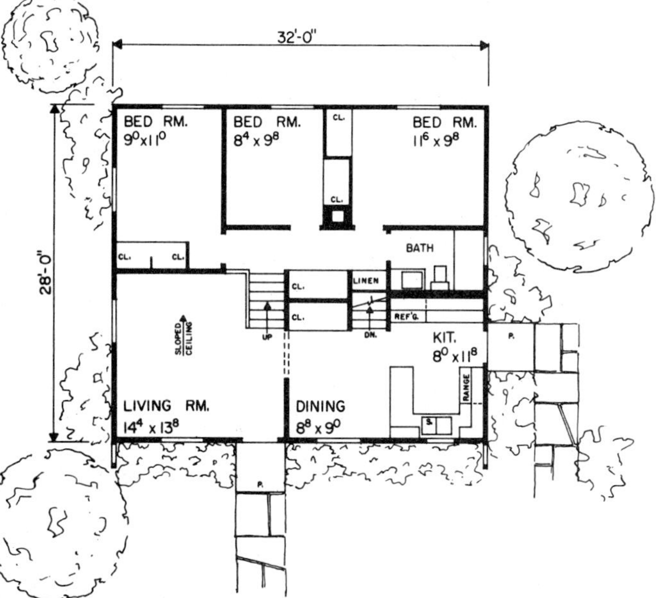

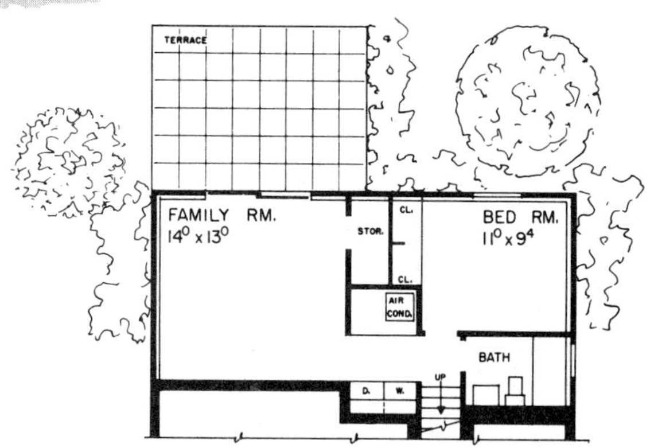

● This compact tri-level will build most economically. Its charming traditional facade will never fail to elicit enthusiastic comment. Designed to assure the most economical use of lumber, this house is a perfect rectangle. When built on a site which slopes to the rear, the lower level can become exposed. This permits the family room to function with the outdoor terrace. The extra bedroom becomes completely livable. This house then features four bedrooms and two full baths! How would the other two designs on these pages suit your family's requirements?

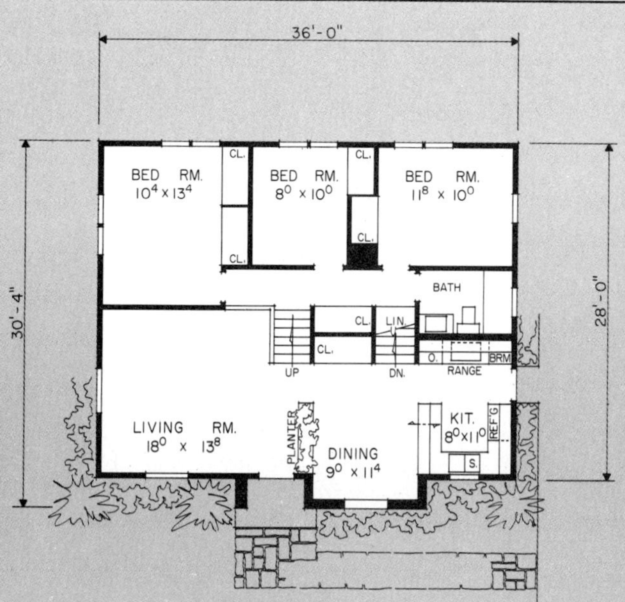

Design 43166 *487 Sq. Ft. – Main Level*

545 Sq. Ft. – Upper Level; 504 Sq. Ft. – Lower Level
15,066 Cu. Ft.

● A front kitchen is ideal to view approaching callers. Plus it has a door to the side yard. A built-in planter divides the dining room from the living room. Three bedrooms and bath are across the rear of the house on the upper level. Family room, study and bath occupy the lower level.

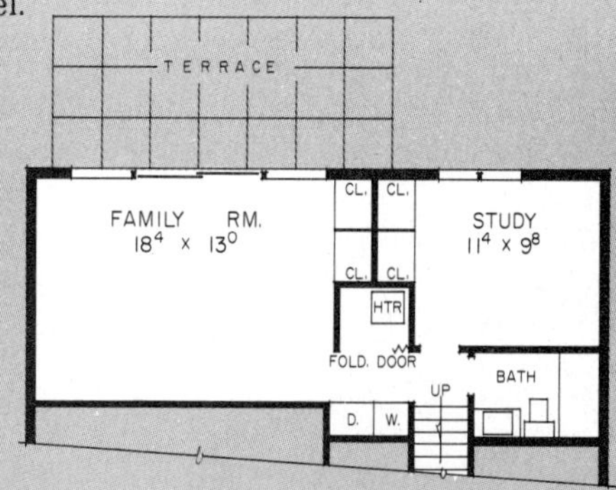

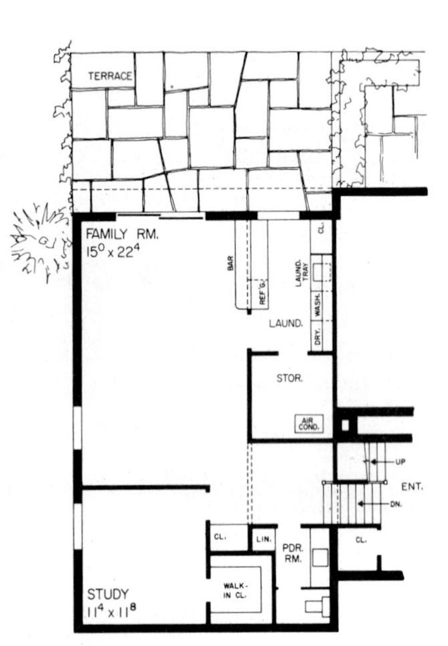

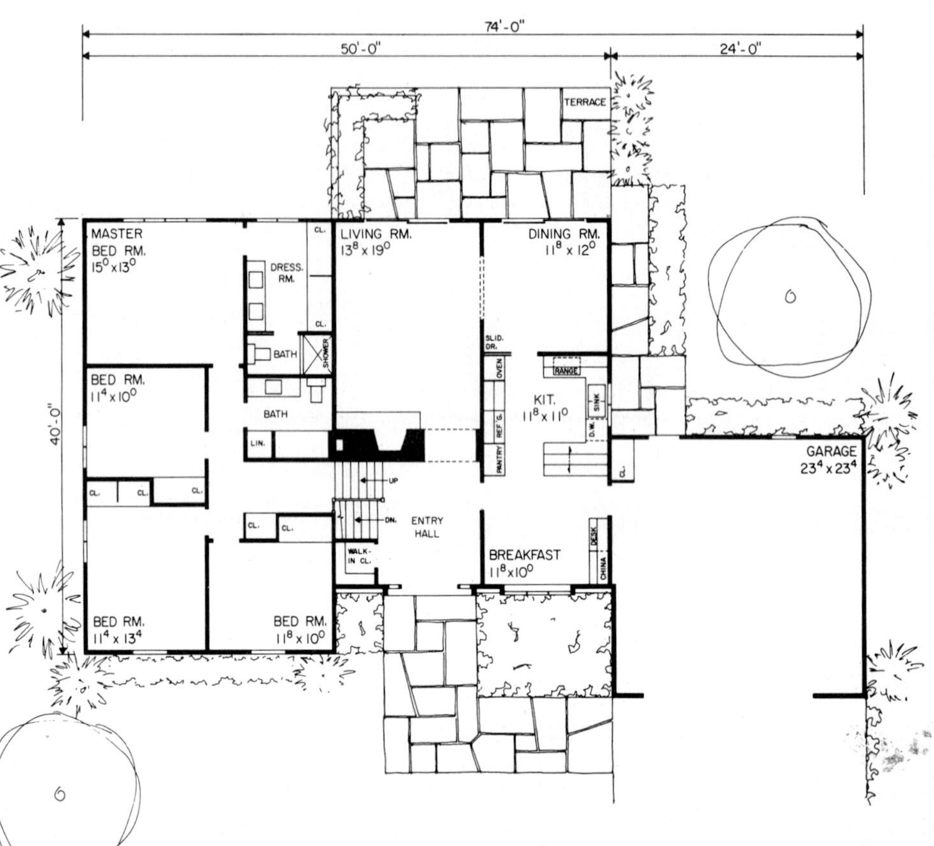

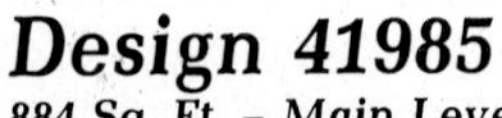

Design 41985

884 Sq. Ft. – Main Level
960 Sq. Ft. – Upper Level
888 Sq. Ft. – Lower Level
29,743 Cu. Ft.

● Here is a split-level that expresses all that is warm and inviting in the traditional vein. Delightfully proportioned, the projecting wings add that desired look of distinction. The double front doors with their appealing panels open into a spacious entry hall. Straight ahead is the living room with the formal dining room but a step away. The U-shaped kitchen is strategically located with a pass-thru to the breakfast room. There is the huge family room and study, or extra bedroom, on the lower level. Note the laundry, snack bar, powder room and extra storage facilities. Four bedrooms, two full baths and plenty of closets highlight the upper level.

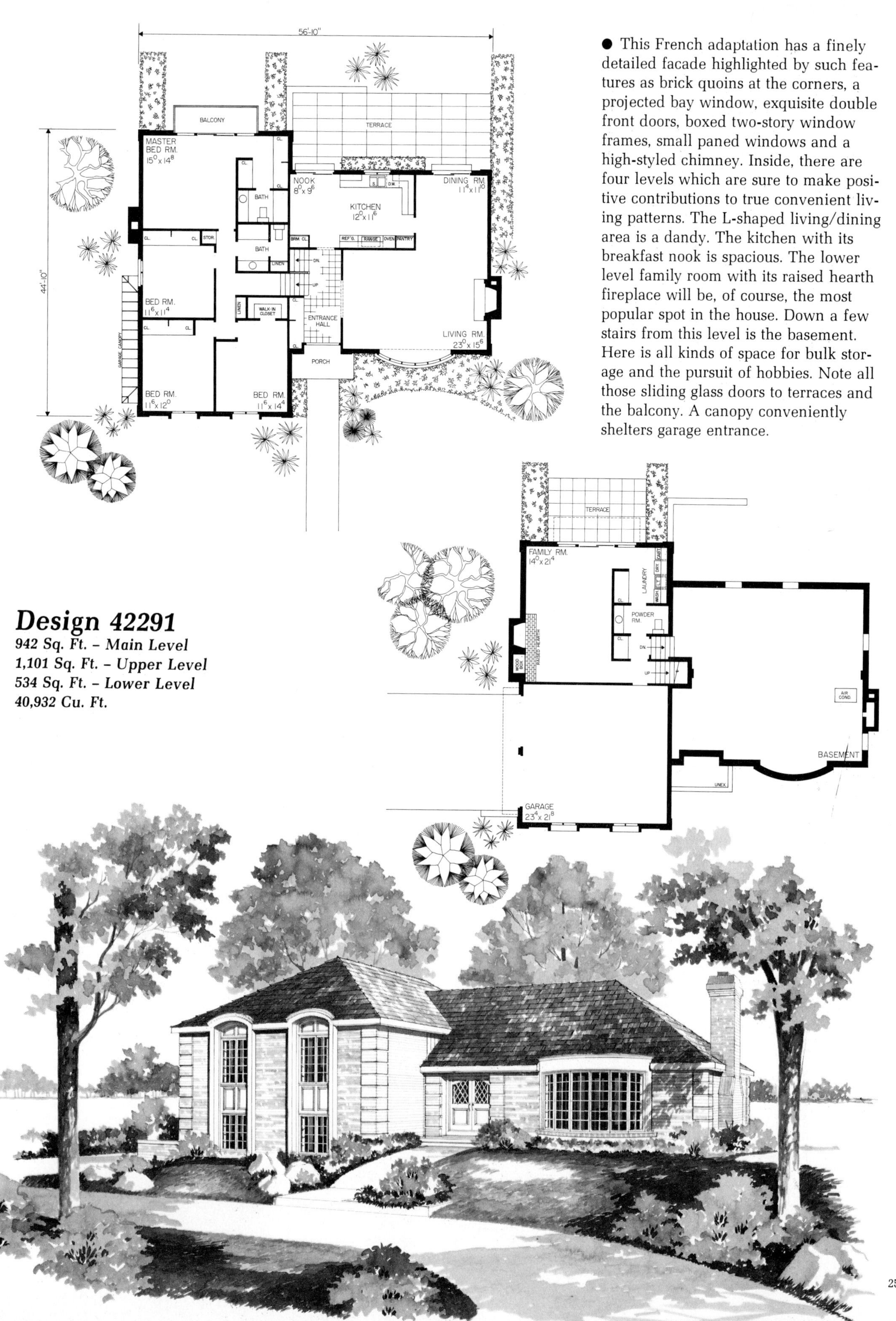

● This French adaptation has a finely detailed facade highlighted by such features as brick quoins at the corners, a projected bay window, exquisite double front doors, boxed two-story window frames, small paned windows and a high-styled chimney. Inside, there are four levels which are sure to make positive contributions to true convenient living patterns. The L-shaped living/dining area is a dandy. The kitchen with its breakfast nook is spacious. The lower level family room with its raised hearth fireplace will be, of course, the most popular spot in the house. Down a few stairs from this level is the basement. Here is all kinds of space for bulk storage and the pursuit of hobbies. Note all those sliding glass doors to terraces and the balcony. A canopy conveniently shelters garage entrance.

Design 42291

942 Sq. Ft. – Main Level
1,101 Sq. Ft. – Upper Level
534 Sq. Ft. – Lower Level
40,932 Cu. Ft.

Design 41977

896 Sq. Ft. – Main Level
884 Sq. Ft. – Upper Level
896 Sq. Ft. – Lower Level
36,718 Cu. Ft.

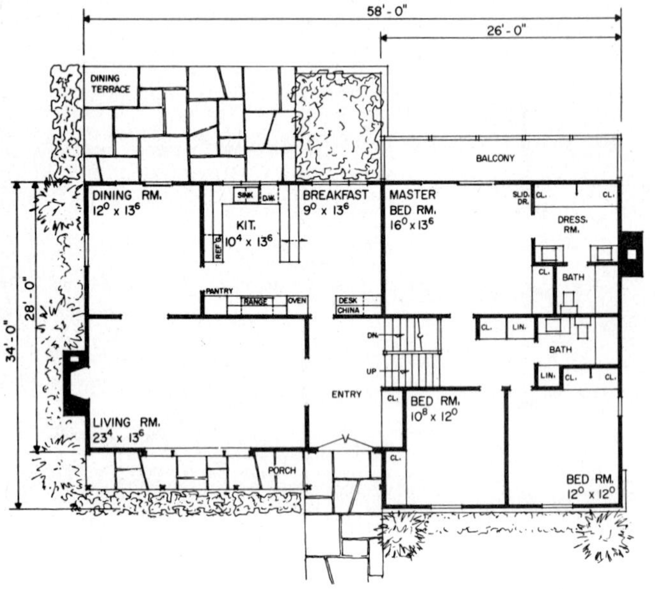

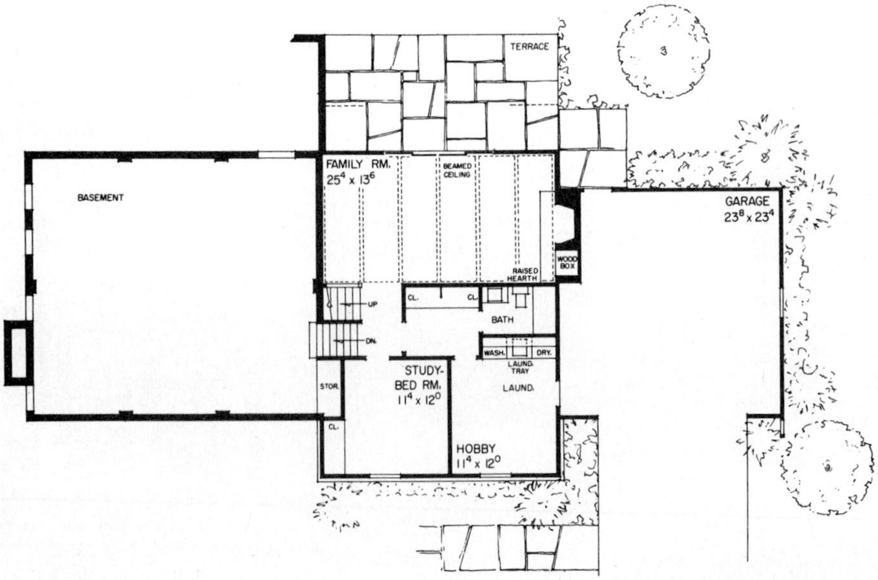

● This split-level with its impressive two-story center portion flanked by a projecting living wing on one side and a two-car garage on the other side, still maintains that very desirable ground-hugging quality. Built entirely of frame with narrow horizontal siding (brick veneer could be substituted), this home will sparkle with a New England flavor. Upon passing through the double front doors, you'll be impressed by an orderly flow of traffic. You'll go up to the sleeping zone; down to the hobby/recreation level; straight ahead to the kitchen and breakfast room; left to the quiet living room. Noteworthy is the extra baths, bedrooms and beamed ceiling family room with fireplace on lower level.

58'-0"
26'-0"
DINING TERRACE
BALCONY
DINING RM. 12^0 x 13^6
KIT. 10^4 x 13^6
SINK
D.W.
REF'G.
BREAKFAST 9^0 x 13^6
MASTER BED RM. 16^0 x 13^6
SLID. DR.
CL.
DRESS. RM.
BATH
PANTRY
RANGE
OVEN
DESK CHINA
LIN.
DN.
UP
ENTRY
34'-0"
28'-0"
LIVING RM. 23^4 x 13^6
BED RM. 10^8 x 12^0
BED RM. 12^0 x 12^0
PORCH

TERRACE
BASEMENT
FAMILY RM. 25^4 x 13^6
BEAMED CEILING
RAISED HEARTH
WOOD BOX
GARAGE 23^8 x 23^4
UP
DN.
CL.
BATH
WASH.
DRY.
LAUND. TRAY
STOR.
STUDY-BED RM. 11^4 x 12^0
LAUND.
HOBBY 11^4 x 12^0

Design 42125

728 Sq. Ft. – Main Level
672 Sq. Ft. – Upper Level
656 Sq. Ft. – Lower Level
28,315 Cu. Ft.

● This four level traditional home has a long list of features to recommend it. First of all, it is a real beauty to look at. The windows, the shutters, the doorway, the horizontal siding with corner boards and the stone work all go together with great proportion to project an image of design excellence. Inside, the livability is outstanding for such a modest home. There are three bedrooms, plus a study (make it the fourth bedroom if you wish); two full baths and a wash room; a fine kitchen with eating space; a spacious formal living and dining area; a big, all-purpose family room. Then there are such highlights as the fireplace, the laundry and the sliding glass doors.

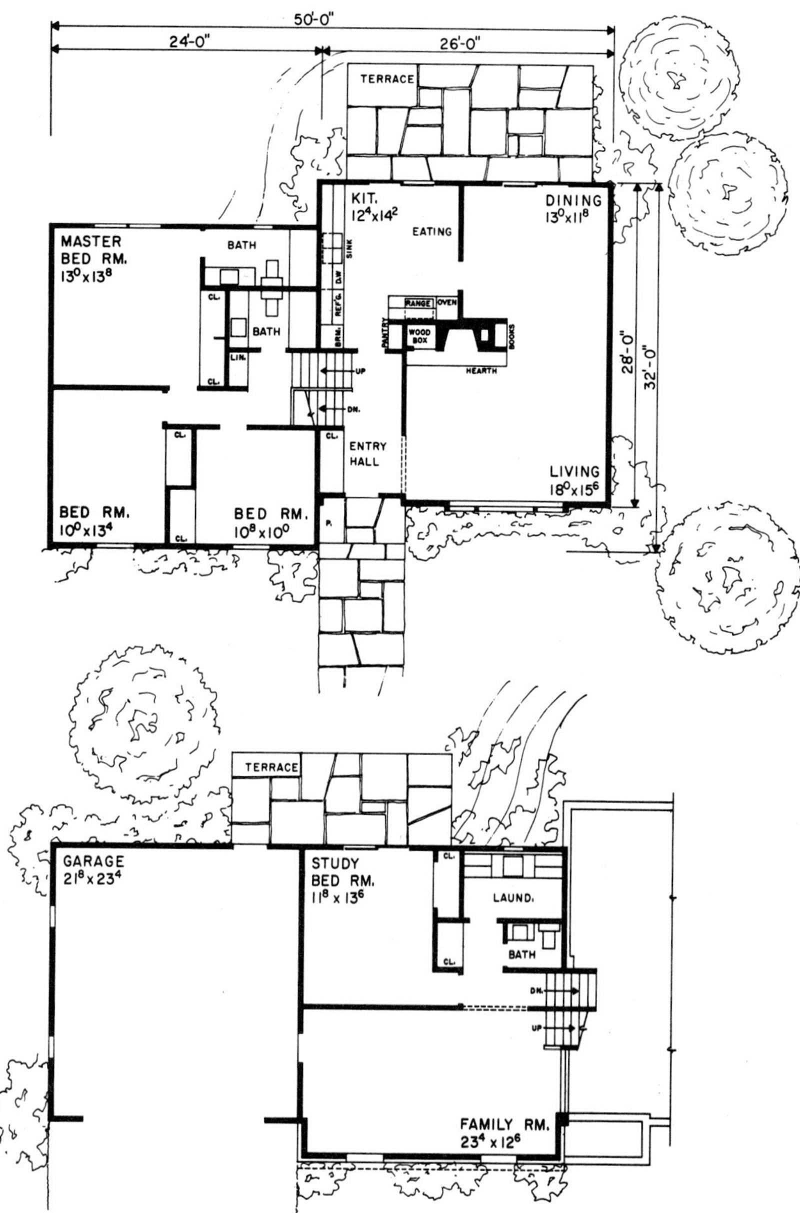

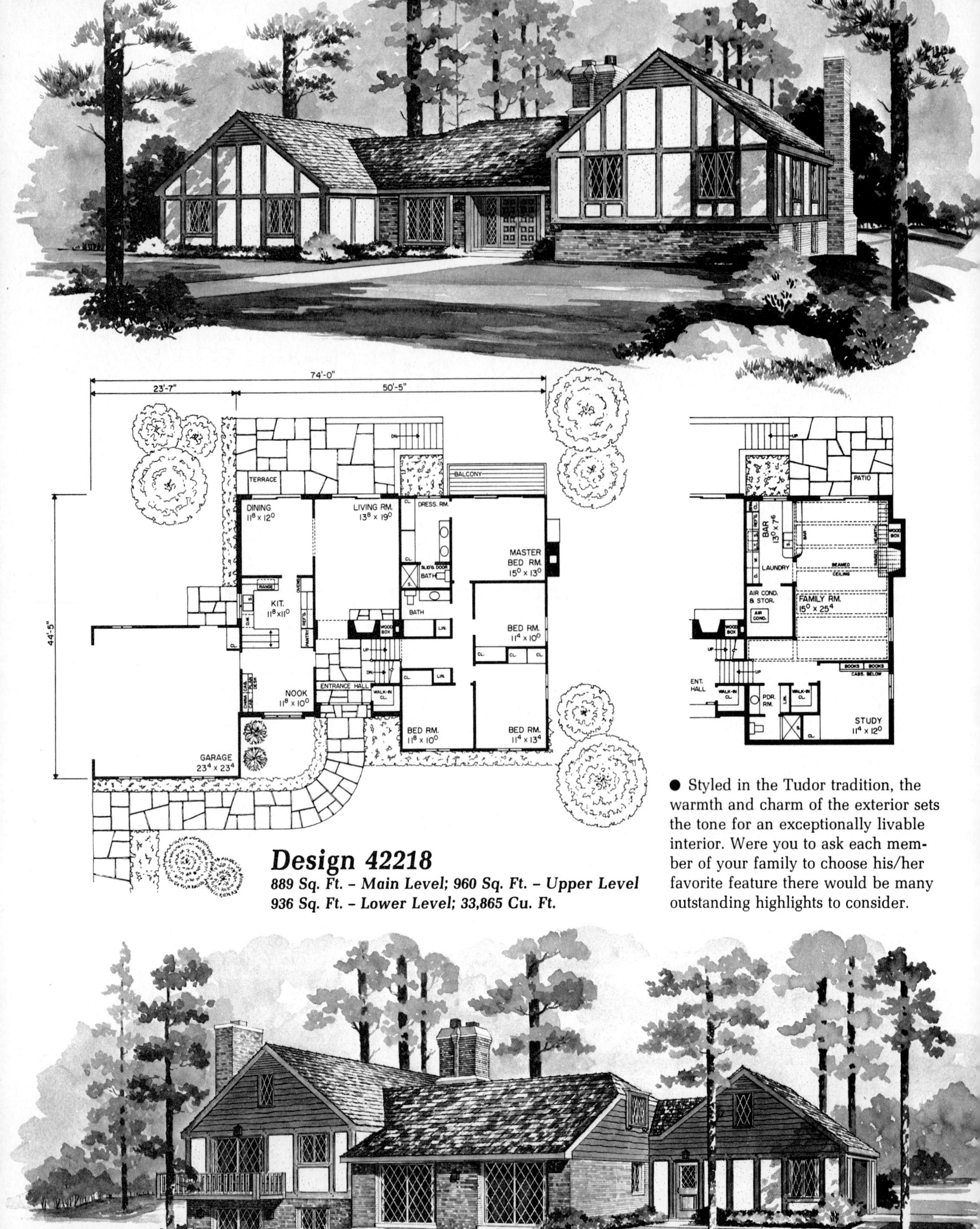

Design 42218

889 Sq. Ft. – Main Level; 960 Sq. Ft. – Upper Level
936 Sq. Ft. – Lower Level; 33,865 Cu. Ft.

● Styled in the Tudor tradition, the warmth and charm of the exterior sets the tone for an exceptionally livable interior. Were you to ask each member of your family to choose his/her favorite feature there would be many outstanding highlights to consider.

Design 42726 *1,852 Sq. Ft. - Main Level*
857 Sq. Ft. - Upper Level; 908 Sq. Ft. - Lower Level; 58,470 Cu. Ft.

● Here is a gracious Southern Colonial adaptation which functions most conveniently as a tri-level. The exterior is sure to take command in any area. The contrast of lines and material is most outstanding. The interior focal point, the circular, open staircase, leads to all of the levels. The lower front entry level features a master suite. This area includes all of the fine extras: a study, dressing room, walk-in closet, bath and private terrace. The main living level is also exceptional. Note the snack bar from the U-shaped kitchen to the sloped ceilinged gathering room. Two terraces are adjacent to the large gathering room with fireplace. Also a front living/dining room to serve the family's more formal needs. The large laundry/sewing room will be a real delight. Wash room is nearby. Numerous storage areas are available to serve your every need.

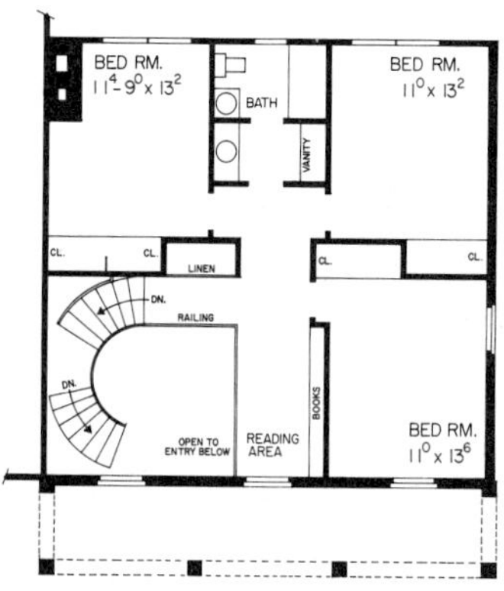

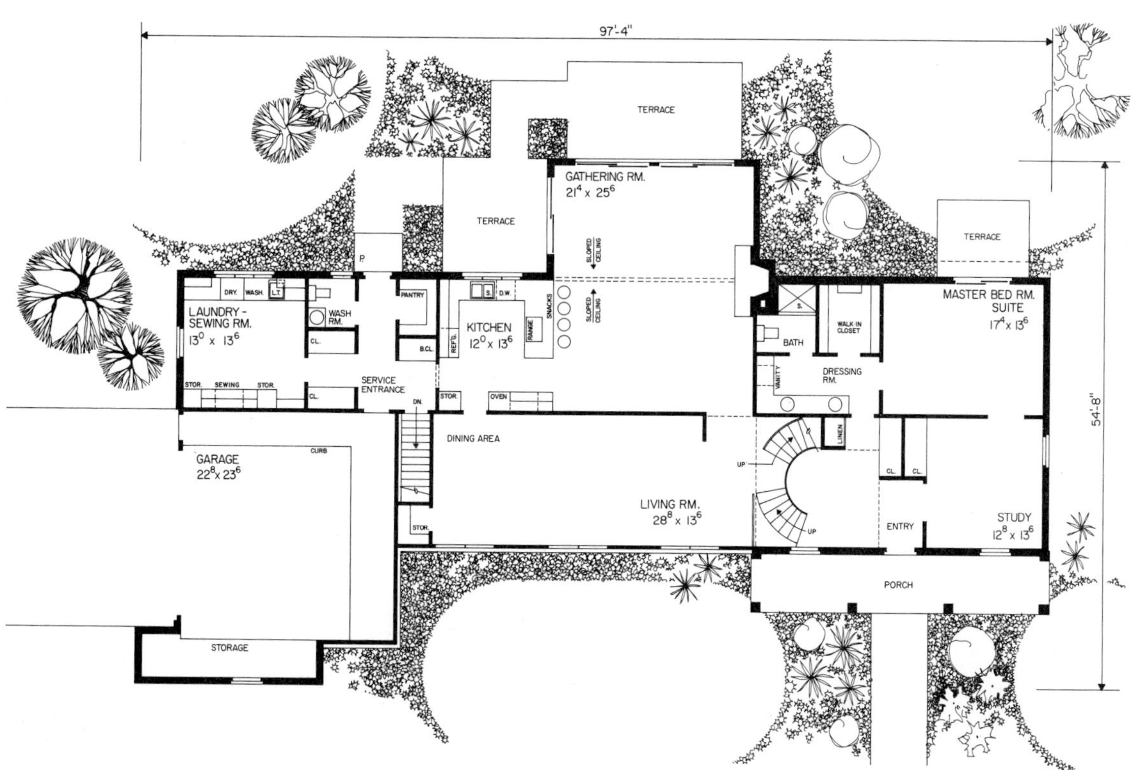

Design 42361 *257 Sq. Ft. – Entry Level; 575 Sq. Ft. – Main Level; 896 Sq. Ft. – Upper Level 304 Sq. Ft. – Lower Level; 23,500 Cu. Ft.*

Design 41324 *682 Sq. Ft. – Main Level; 672 Sq. Ft. – Upper Level; 656 Sq. Ft. – Lower Level; 24,208 Cu. Ft.*

Design 41292 *640 Sq. Ft. – Main Level; 672 Sq. Ft. – Upper Level; 646 Sq. Ft. – Lower Level; 20,537 Cu. Ft.*

● Stop and take note of this selection of three traditionally styled tri-level designs. While you study the plans be sure to list the fine qualities of each then match them to the personal needs of your family. Formal and informal living and dining areas are available in each plan.

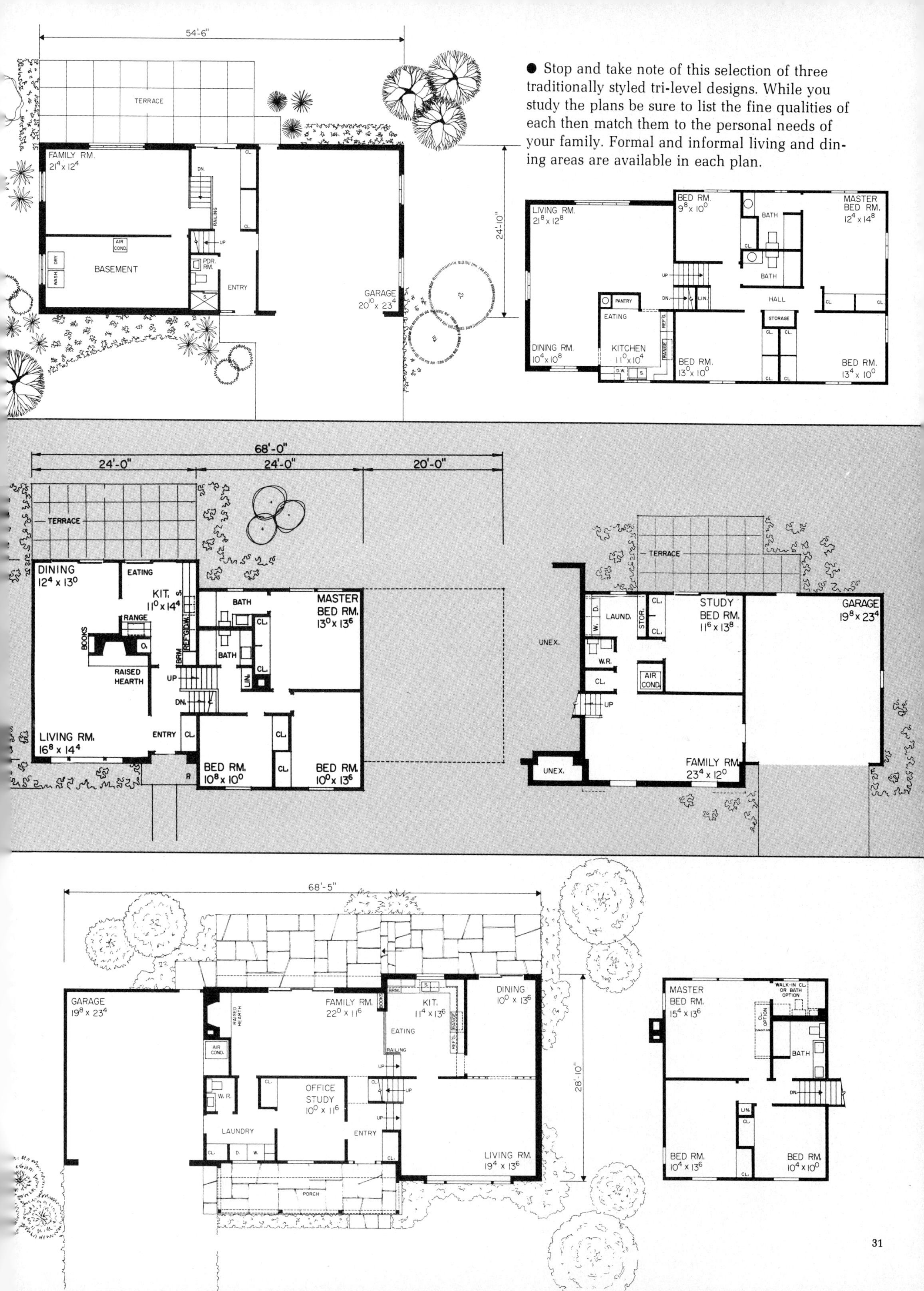

Design 42143 *832 Sq. Ft. - Main Level; 864 Sq. Ft. - Upper Level; 864 Sq. Ft. - Lower Level; 27,473 Cu. Ft.*

● Here the Spanish Southwest comes to life in the form of an enchanting multi-level home. There is much to rave about. The architectural detailing is delightful, indeed. The entrance courtyard, the twin balconies and the roof treatment are particularly noteworthy. Functioning at the rear of the house are the covered patio and the balcony with its lower patio. Well zoned, the upper level has three bedrooms and two baths; the main level has its formal living and dining rooms to the rear and kitchen area looking onto the courtyard; the lower level features the family room, study and laundry. Be sure to notice the extra wash room and the third full bath. There are two fireplaces each with a raised hearth. A dramatic house wherever built!

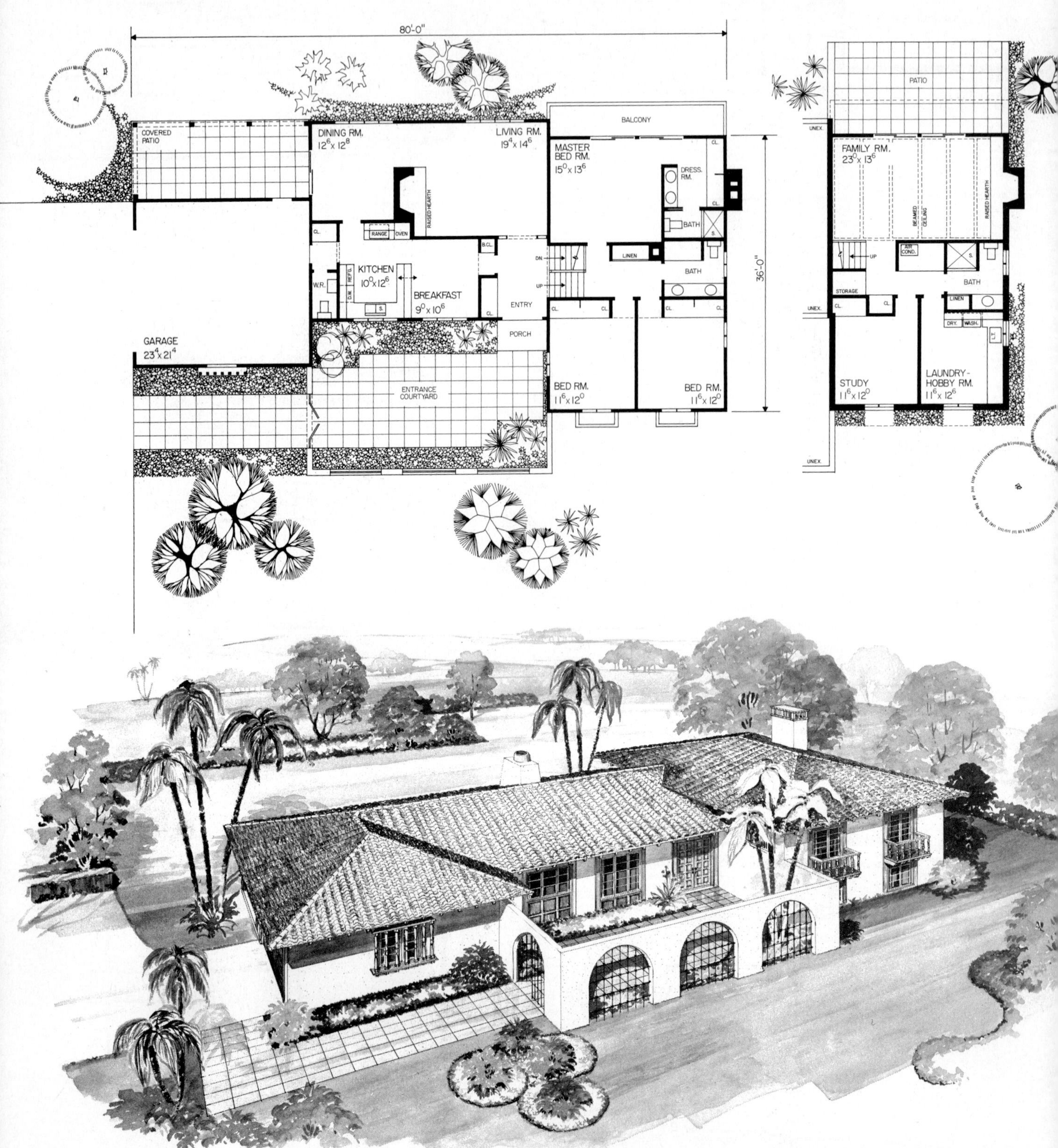

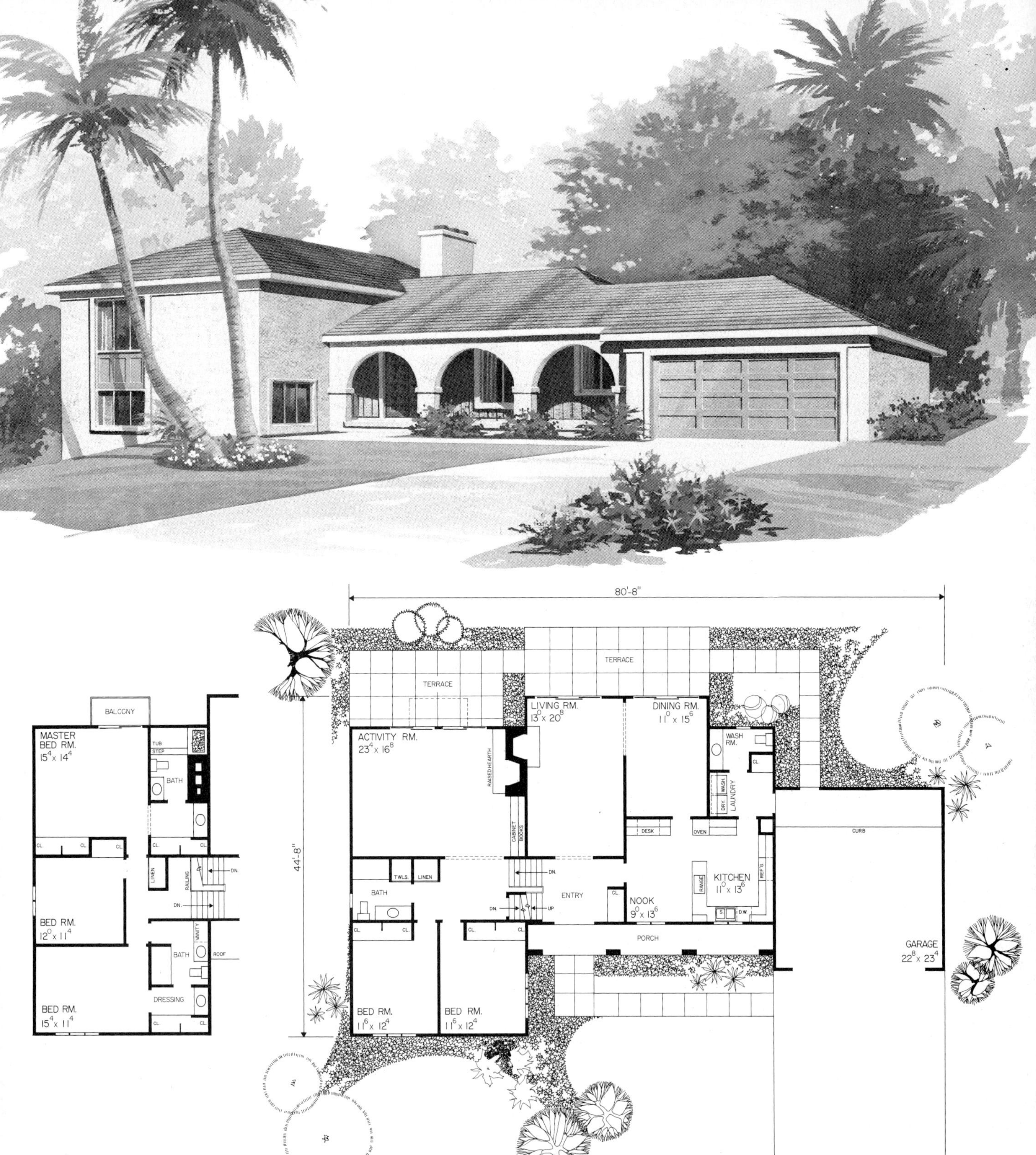

Design 42574 *984 Sq. Ft. – Main Level; 968 Sq. Ft. – Upper Level; 976 Sq. Ft. – Lower Level; 43,440 Cu. Ft.*

● Spanish flair! This home has four bedrooms plus a master suite with a private balcony, dressing room, a luxury bath with a step-up tub and four closets. Also featuring extravagant living space. The activity room is more than 23' by 16' and offers a raised hearth fireplace and a built-in bookcase. Double sliding glass doors open onto the terrace. The main level houses the formal living room with another fireplace and a formal dining room. Both with sliding glass doors that open onto a second terrace. Plus an exceptional kitchen! U-shaped for convenience, it features a built-in desk as well as an oven and range. A separate breakfast nook gaurantees to make every meal something special. Steps away, a first floor laundry to keep all your work in one area.

Design 42628

649 Sq. Ft. – Main Level; 672 Sq. Ft. – Upper Level
624 Sq. Ft. – Lower Level; 25,650 Cu. Ft.

● Traditional, yet contemporary! With lots of extras, too. Like a wet bar and game storage in the family room. A beamed ceiling, too, and a sliding glass door onto the terrace. In short, a family room designed to make your life easy and enjoyable. There's more. A living room with a traditionally styled fireplace and built-in bookshelves. And a dining room with a sliding glass door that opens to a second terrace. Here's the appropriate setting for those times when you want a touch of elegance. A sunny kitchen, too. It features a built-in oven and range, pantry, broom closet and plenty of space for a breakfast table. A convenient laundry room as well. Four large bedrooms, or three plus a study, if that arrangement suits you better. Also on the upper level is two full back-to-back baths for economical plumbing. This home has style and space to serve any family admirably.

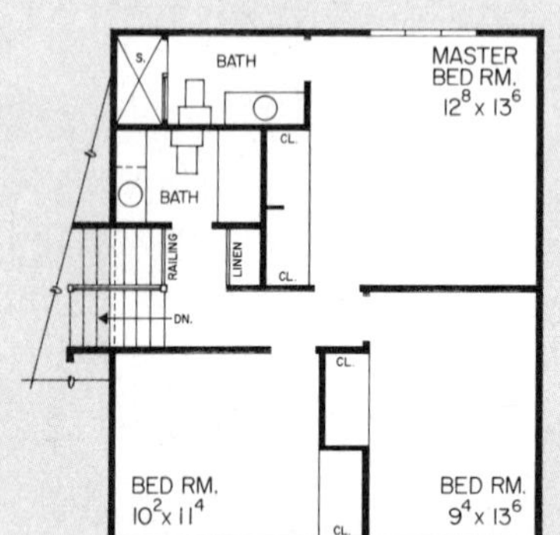

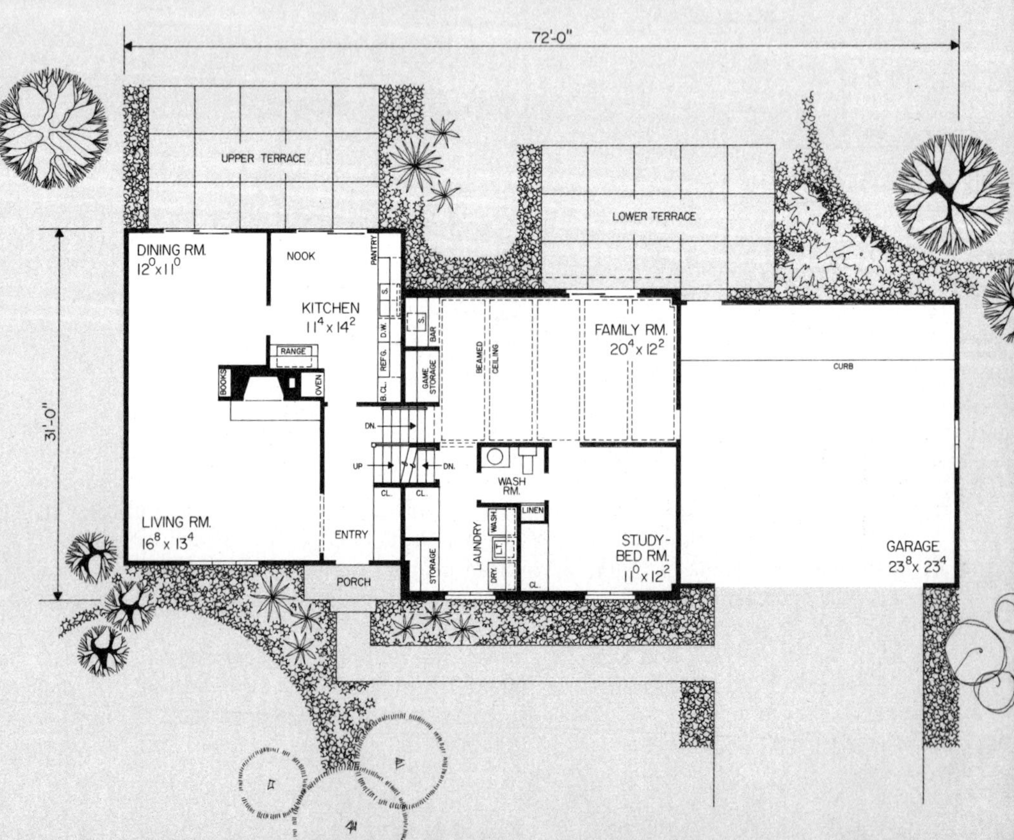

Design 42354

936 Sq. Ft. – Main Level; 971 Sq. Ft. – Upper Level 971 Sq. Ft. – Lower Level; 34,561 Cu. Ft.

● This English flavored tri-level design may be built on a flat site. Its configuration permits a flexible orientation on the site with either the garage doors or the front door facing the street. The interior offers a unique and practical floor plan layout. Flanking the spacious entrance hall is the cozy, sunken living room and the formal dining room. Looking out upon the front porch is the kitchen with its adjacent nook. A mud room is strategically located just inside the door from the garage. Opposite the front door are two flights of stairs. One leads to the upper level with its three bedrooms and two baths. The other leads to the lower level. Here is the fourth bedroom, third bathroom, a big beamed ceiling family room, a hobby room and a laundry. A real winner for family living.

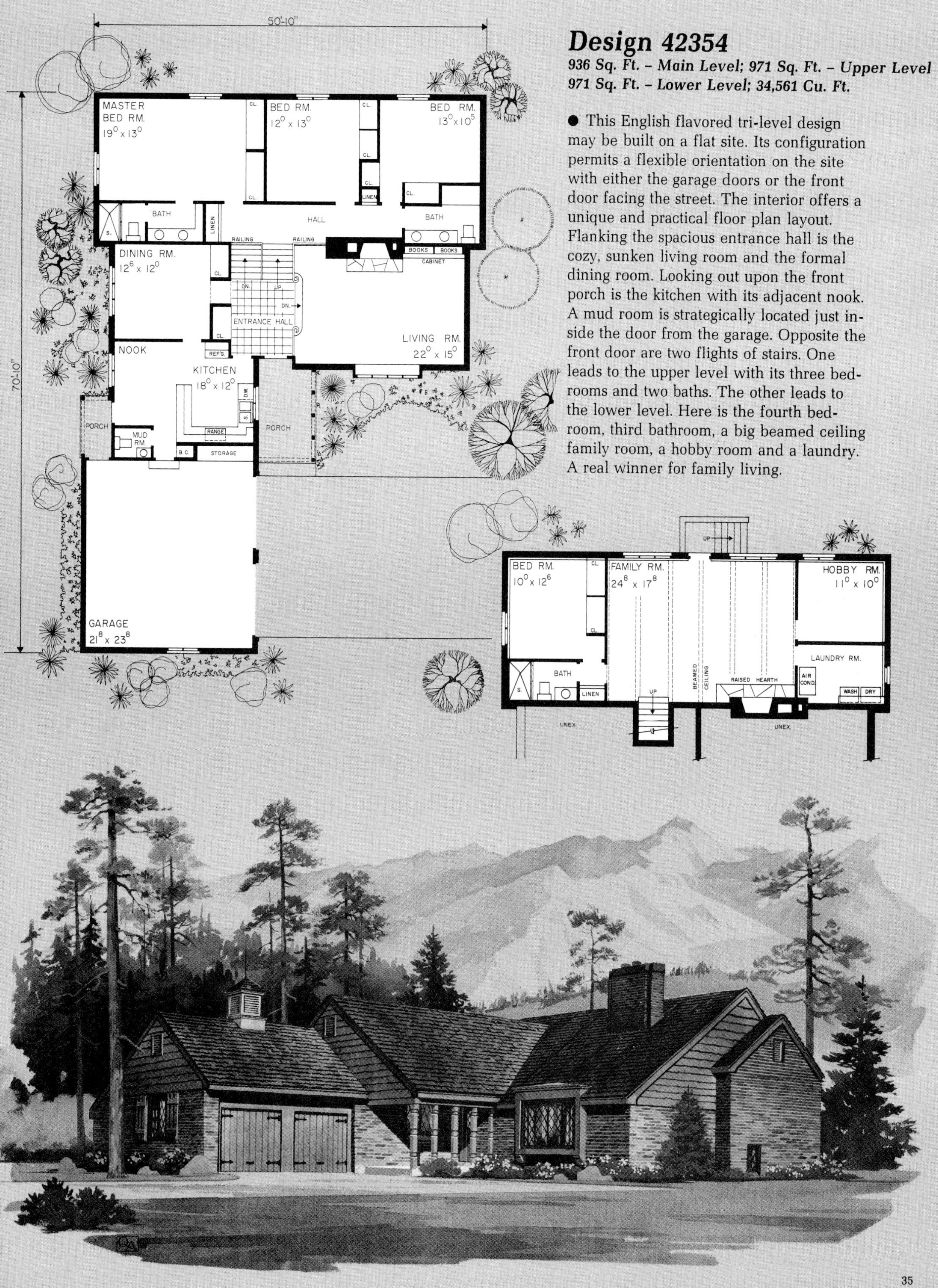

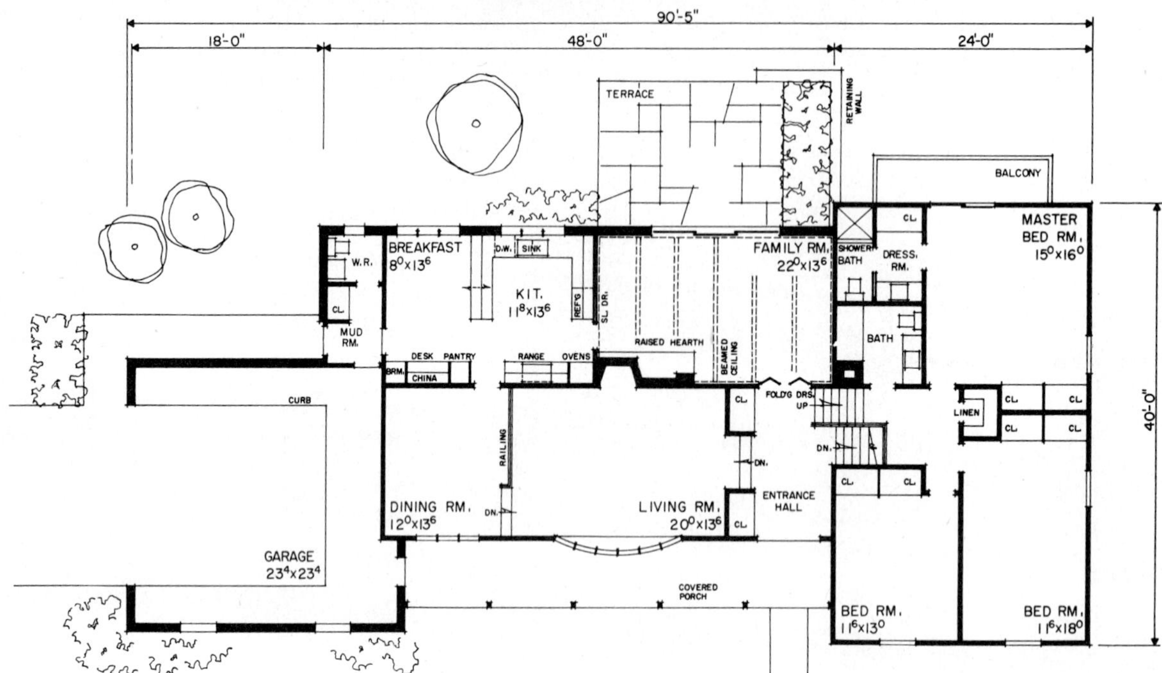

Design 41927

1,272 Sq. Ft. – Main Level; 960 Sq. Ft. – Upper Level
936 Sq. Ft. – Lower Level; 36,815 Cu. Ft.

● Living in this traditional split level home will be a great experience. For here is a design that has everything. It has good looks and an abundance of livability features. The long, low appearance is accentuated by the large covered porch which shelters the bowed window and the inviting double front doors. Whatever your preference for exterior materials they will show well on this finely proportioned home. They start with four bedrooms and three full baths and continue with: beamed ceiling family room, sunken living room, formal dining room, informal breakfast room, extra wash room, outstanding kitchen and two attractive fireplaces.

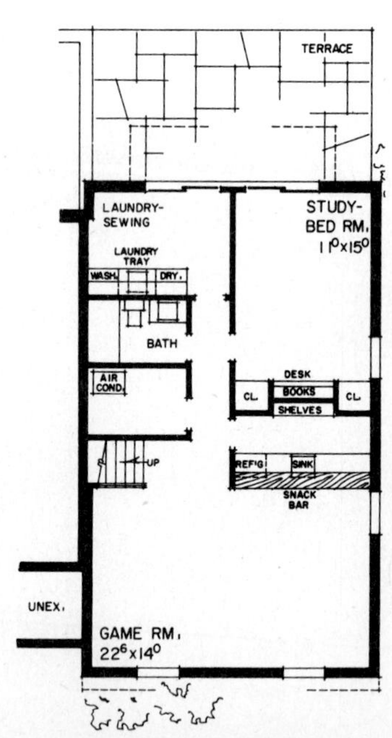

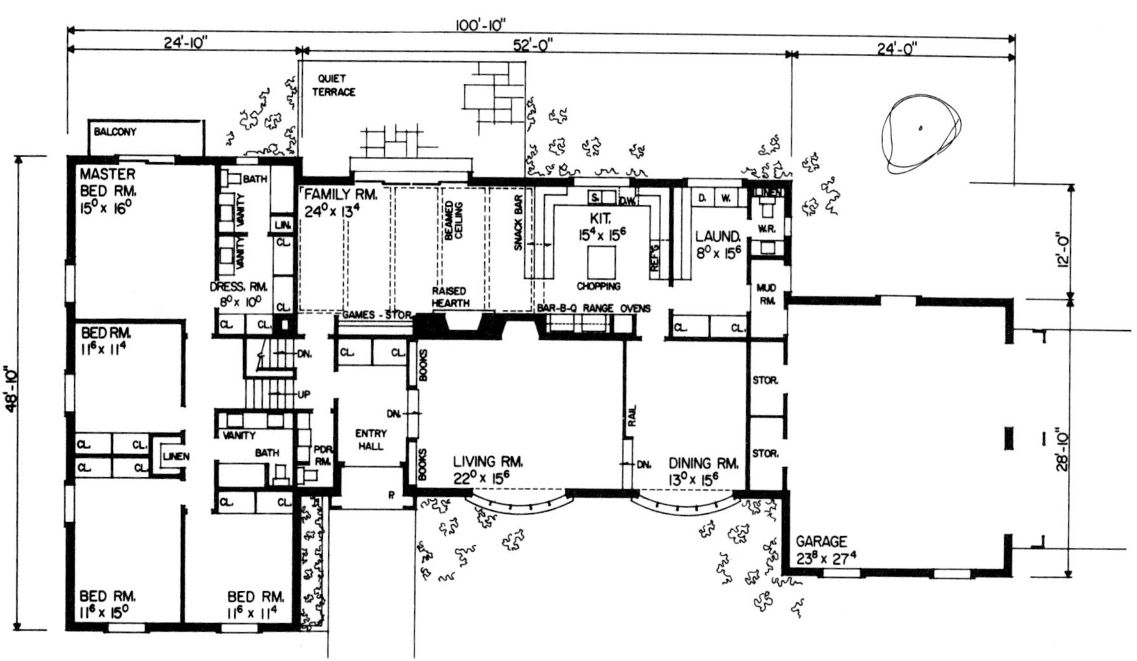

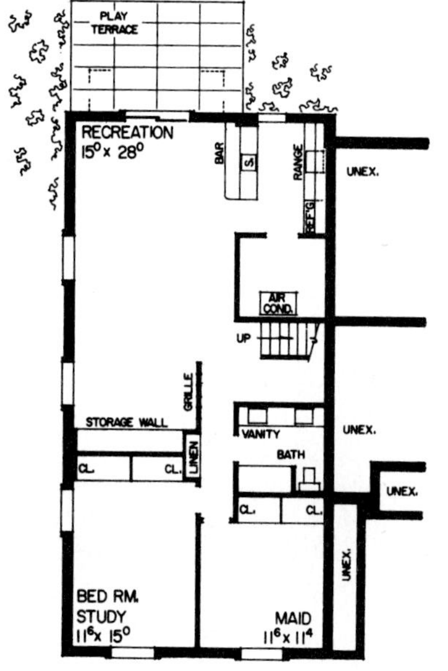

Design 41270

1,648 Sq. Ft. - Main Level; 1,200 Sq. Ft. - Upper Level
1,200 Sq. Ft. - Lower Level; 48,856 Cu. Ft.

● A French Provincial adaptation with an enormous amount of livability on three levels. Whether called upon to function as a four or six bedroom home, there will be plenty of space in which to move around. Whatever the activities of the family–formal or informal–this floor plan contains the facilities to cater to them. For instance, there is the family room of the main level and the recreation room of the lower level to more than adequately serve informal persuits. Then there is the sunken living room. The main level laundry will save many steps. There are two fireplaces and exceptional storage facilities. Four bedrooms highlights upper level.

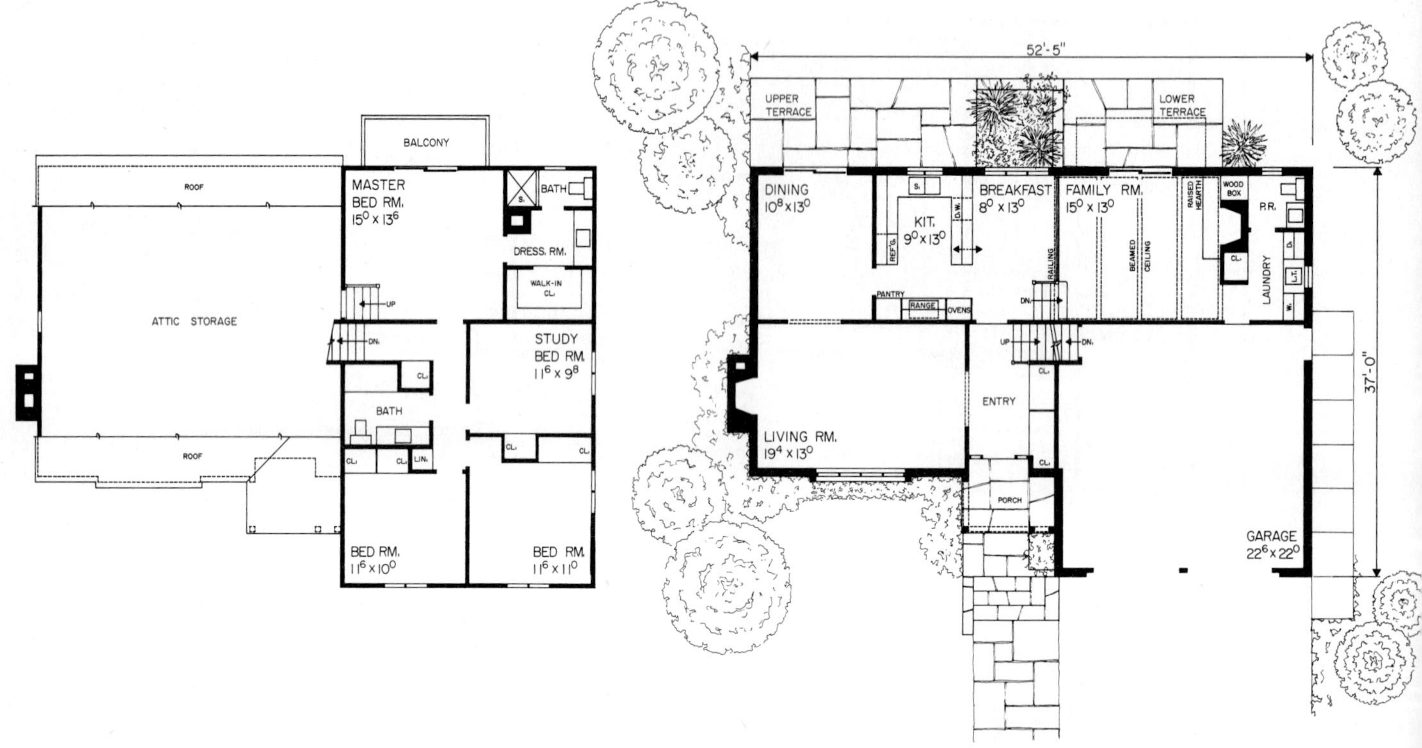

Design 42171 *795 Sq. Ft. - Main Level; 912 Sq. Ft. - Upper Level; 335 Sq. Ft. - Lower Level; 33,243 Cu. Ft.*

● This English Tudor split-level adaptation has much to recommend it. Perhaps, its most significant feature is that it can be built economically on a relatively small site. The width of the house is just over 52 feet. But its size does not inhibit it's livability features. There are many fine qualities: Observe the living room fireplace in addition to that in the family room with a wood box. Don't miss the balcony off the master bedroom. Also, worthy of note is the short flight of stairs leading to the huge attic storage area. For the development of even more space there is the basement below the main level. Access to this area is directly from the two-car garage. The breakfast room with its railing looks down into the lower level family room. Also it has a pass-thru to the kitchen.

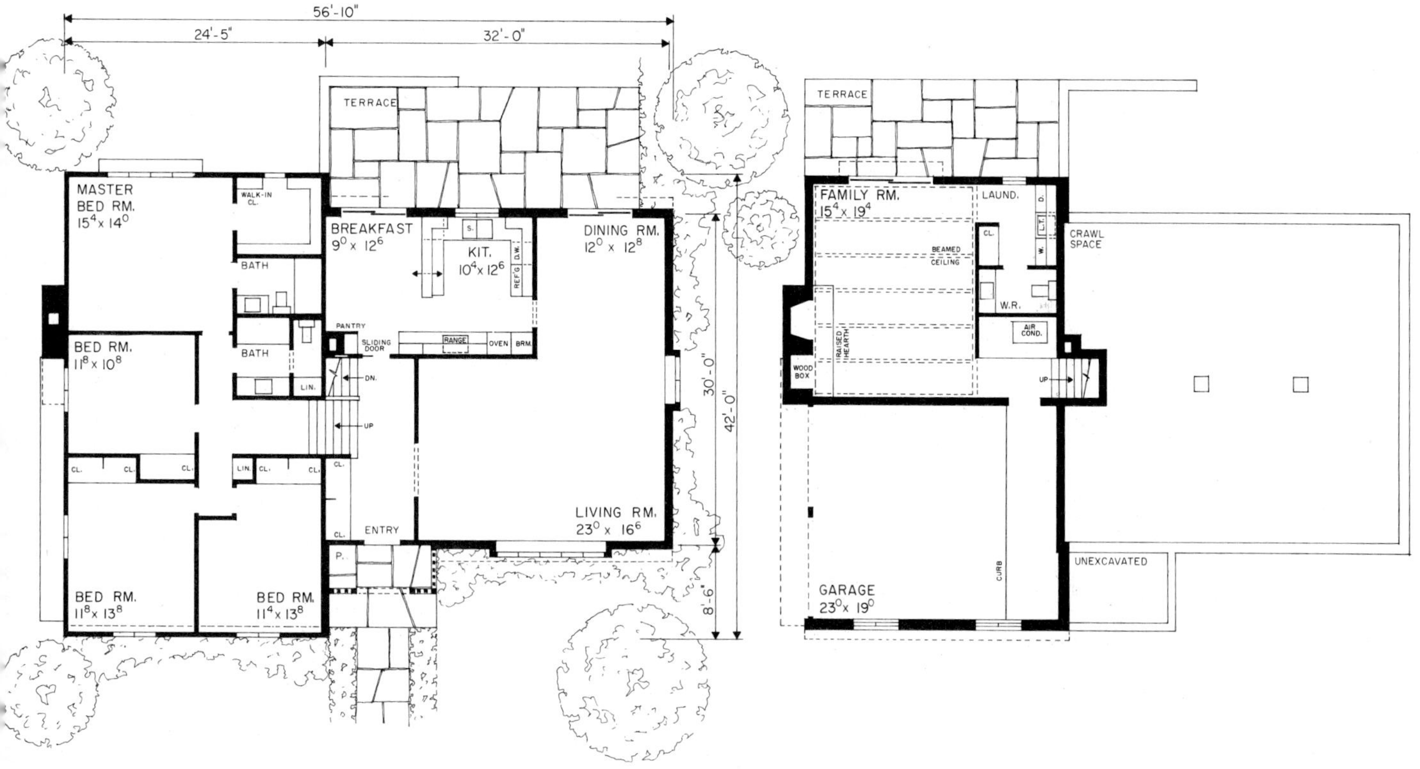

Design 42137 *987 Sq. Ft. - Main Level; 1,043 Sq. Ft. - Upper Level; 463 Sq. Ft. - Lower Level; 29,382 Cu. Ft.*

● Tudor design adapts to split level living. The result is a unique charm for all to remember. As for the livability, the happy occupants of this tri-level home will experience wonderful living patterns. A covered porch protects and adds charm to the front entry. The center hall routes traffic conveniently to the spacious formal living and dining area; the informal breakfast room and kitchen zone; the upper level bedrooms; the lower level all-purpose family room. Contributing to fine living are such highlights as 2½ baths, walk-in closet, four bedrooms, sliding glass doors, pass-thru from kitchen to breakfast room, beamed ceiling, raised hearth fireplace, separate laundry and an attached two-car garage. Note the two terraces.

Design 41200 *480 Sq. Ft. – Main Level; 560 Sq. Ft. – Upper Level; 560 Sq. Ft. – Lower Level; 17,936 Cu. Ft.*

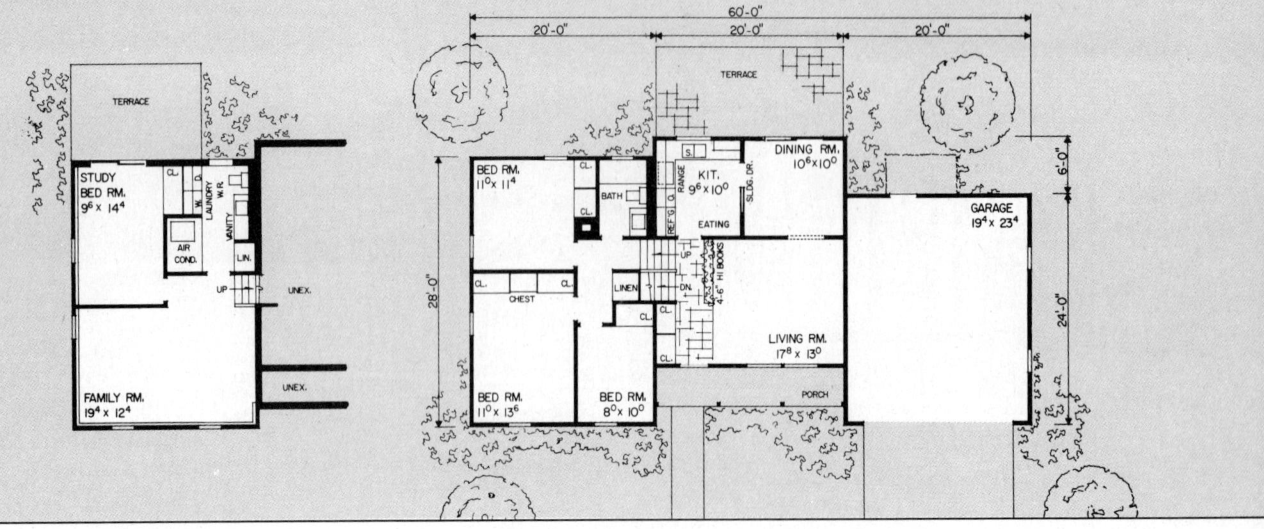

Design 41981 *784 Sq. Ft. – Main Level; 912 Sq. Ft. – Upper Level; 336 Sq. Ft. – Lower Level; 26,618 Cu. Ft.*

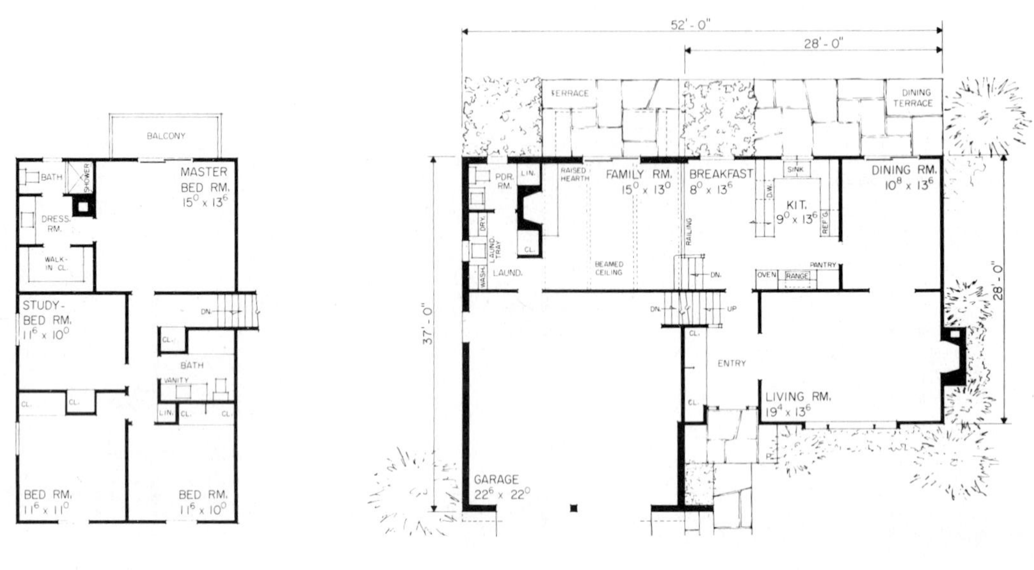

Design 41359 *814 Sq. Ft. – Main Level; 952 Sq. Ft. – Upper Level; 366 Sq. Ft. – Family Room Level*
814 Sq. Ft. – Lower Level; 29,785 Cu. Ft.

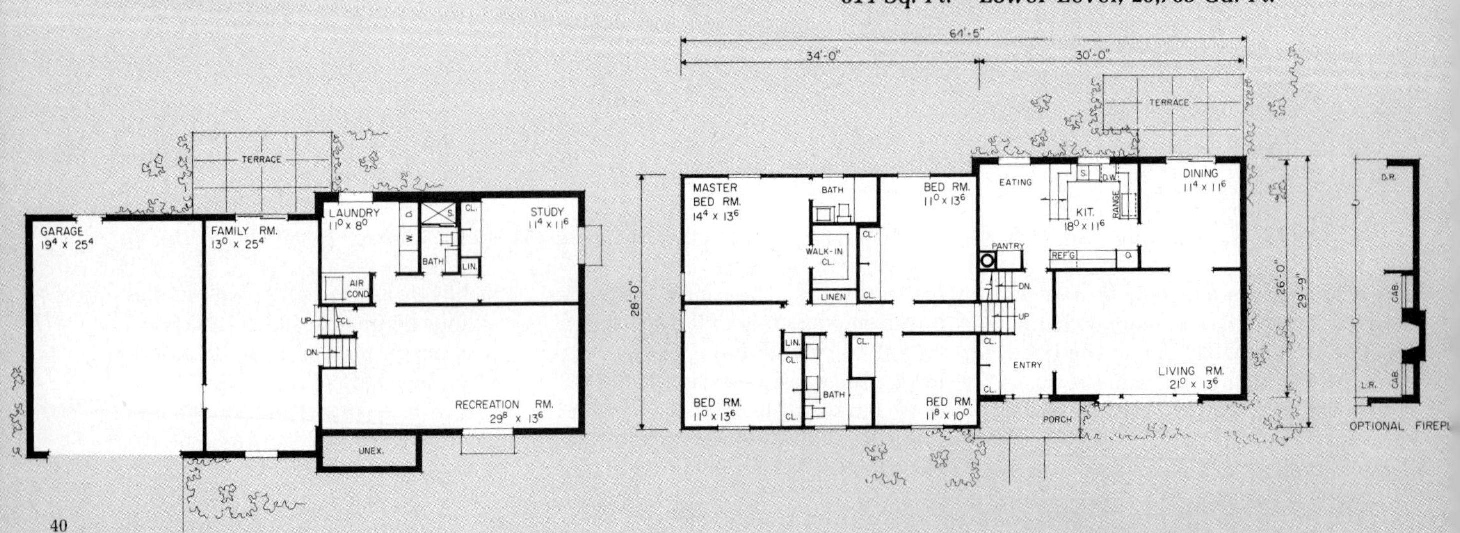

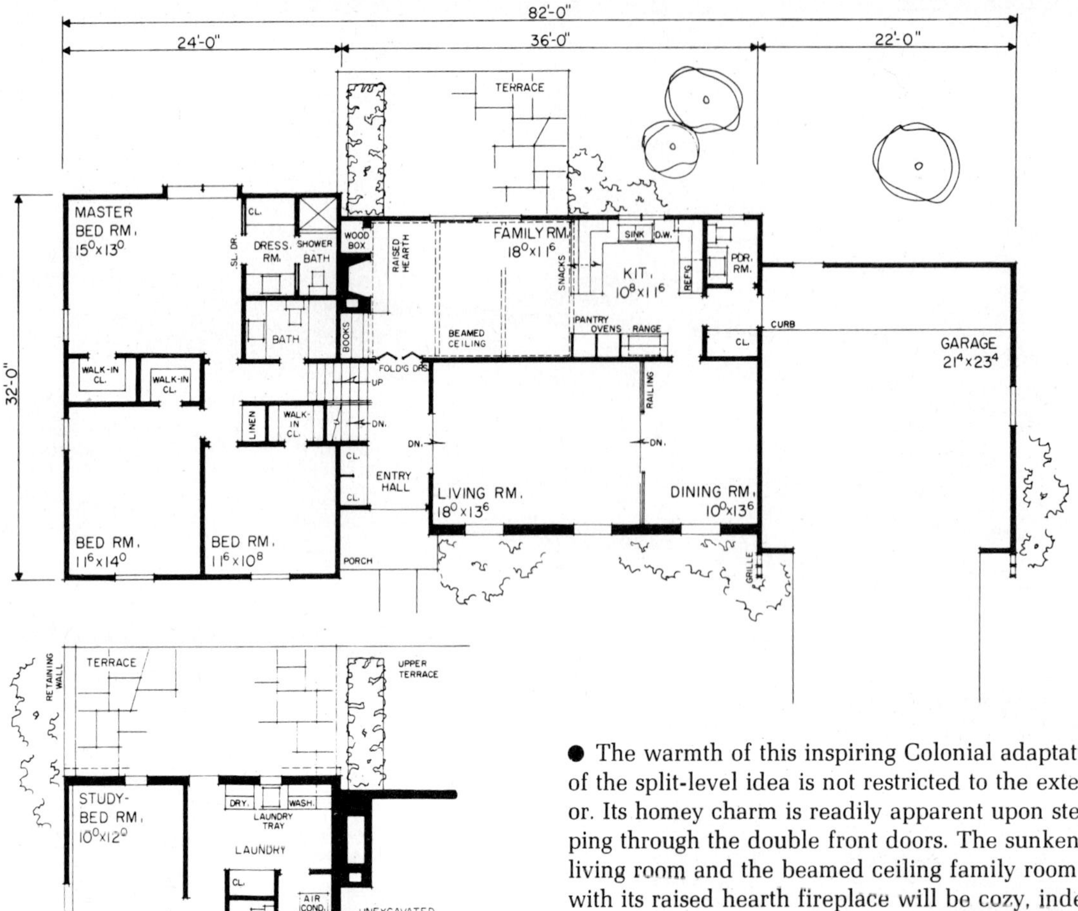

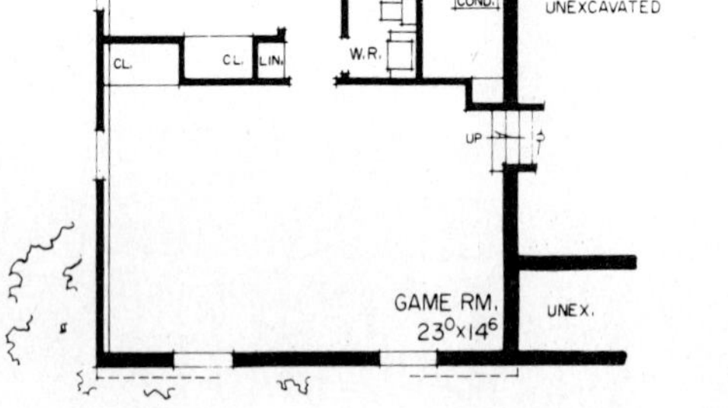

● The warmth of this inspiring Colonial adaptation of the split-level idea is not restricted to the exterior. Its homey charm is readily apparent upon stepping through the double front doors. The sunken living room and the beamed ceiling family room with its raised hearth fireplace will be cozy, indeed. The kitchen, powder room and closet just inside the door from the garage will be three added conveniences of this design that you should not overlook. There are three bedrooms on the upper level, while there is a fourth to be found on the lower level. Don't miss the big laundry and extra wash room.

Design 41930 *947 Sq. Ft. - Main Level; 768 Sq. Ft. - Upper Level; 740 Sq. Ft. - Lower Level; 25,906 Cu. Ft.*

Design 41348 *750 Sq. Ft. - Main Level; 672 Sq. Ft. - Upper Level; 664 Sq. Ft. - Lower Level; 22,143 Cu. Ft.*

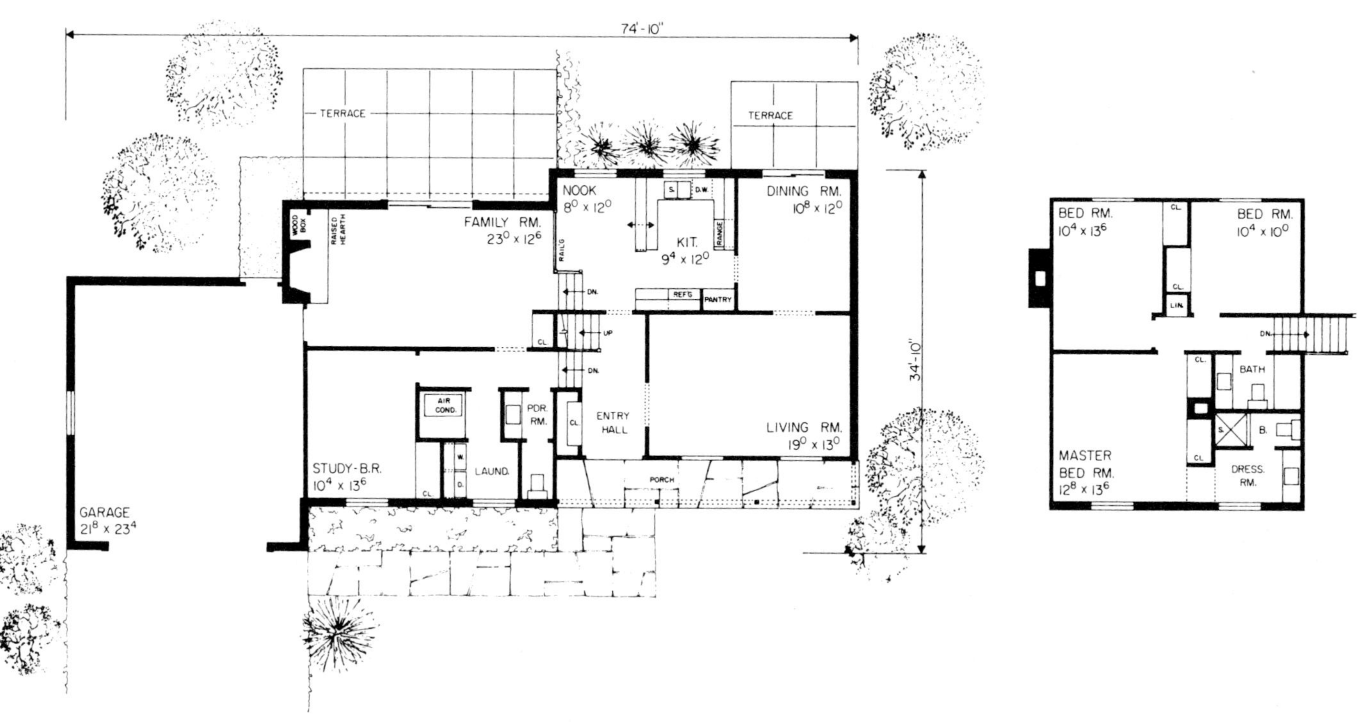

● Looking up from the street at this tri-level house there is a majestic feeling that pervades the setting. This results from the somewhat massive center section with its pediment gable and the flanking wings. The U-shaped kitchen is flanked by the separate dining room and the breakfast eating area. This informal nook looks over a railing into the lower level family room which is but four steps below the main level. A study or fourth bedroom, a laundry and an extra wash room are other important features of this lower level. The upper sleeping level will enjoy complete privacy. Here are two bedrooms and a bath for the kids, and a separate master bedroom suite for the parents. For outdoor living there are the living and dining terraces.

Design 41705

896 Sq. Ft. - Main Level
870 Sq. Ft. - Lower Level
896 Sq. Ft. - Upper Level
27,040 Cu. Ft.

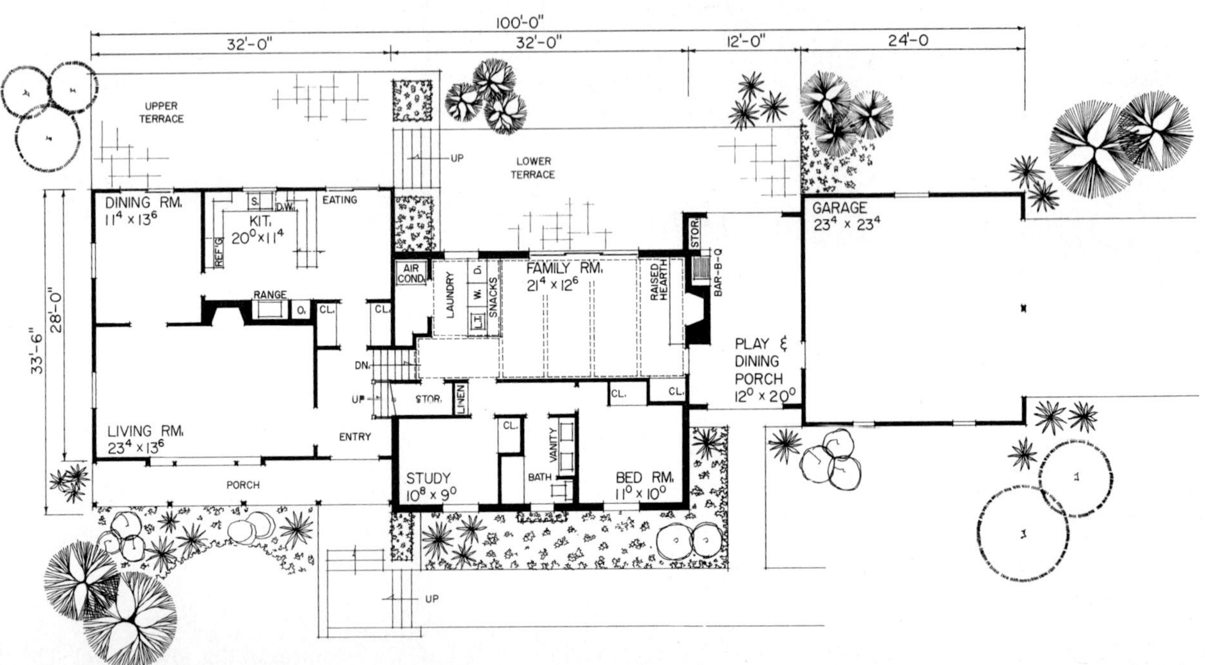

● A gently sloping, suburban site will be just the location for this superb traditional tri-level home. The main level has both a formal living room and a separate dining room. There is the informal eating area of the kitchen for additional dining. The lower level features a fine family room with snack bar, raised hearth fireplace and glass sliding doors to the lower terrace. There are also two well-lighted rooms which may function as additional bedrooms. Don't miss the full bath and the laundry area. A covered porch with a built-in barbecue unit attaches the house and two-car garage. There are four bedrooms and two baths on the upper level.

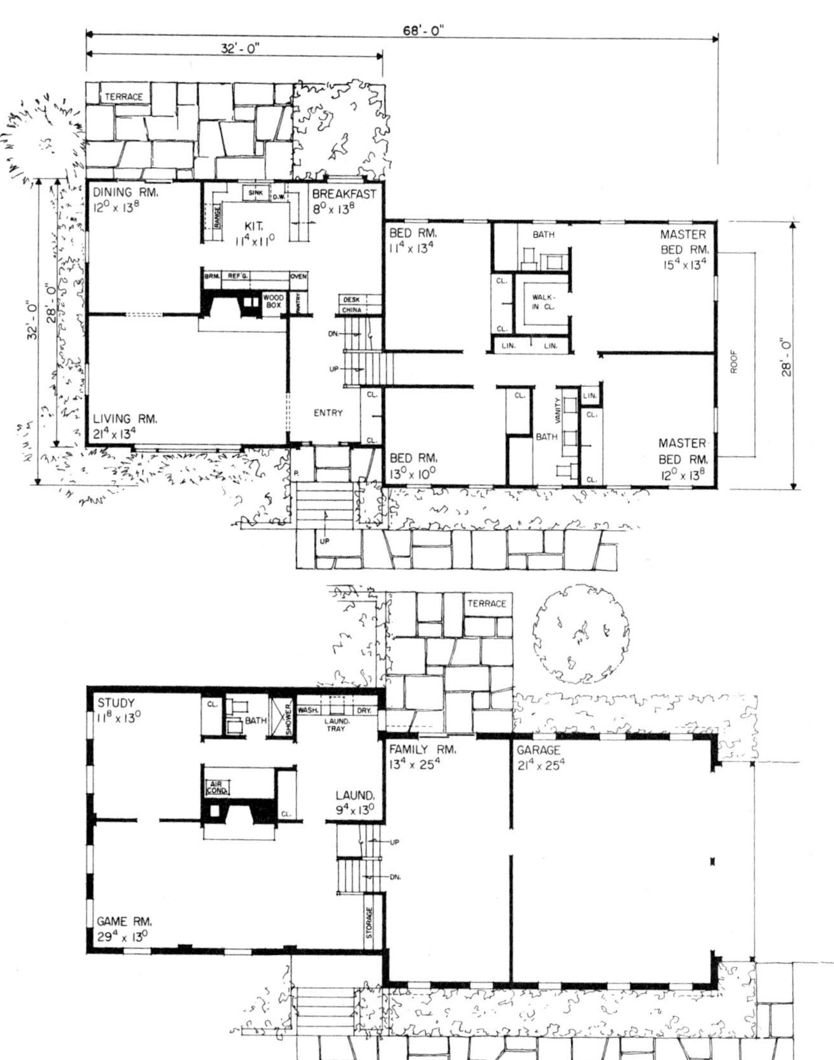

Design 41961

896 Sq. Ft. – Main Level
1,008 Sq. Ft. – Upper Level
376 Sq. Ft. – Family Room Level
896 Sq. Ft. – Lower Level
33,934 Cu. Ft.

● Here is a design that really makes a delightful impact. Its charming array of muntined, double hung windows, its contrasting frame and stone facade and its massive chimney leave a pleasing and lasting impression. The projection over the doors of the big two-car garage is a refreshing feature. And what does the inside have to offer? Livability galore. Each of the four levels makes its vital contribution to family living. If you are on one level, the chances are the children with their friends will be enjoying themselves on another level. If the adults want to converse quietly outdoors there is the upper terrace. The kids can then play on the lower terrace. Be sure to look over the outstanding lower level.

● Your building budget will return the utmost in livability when called upon to finance the construction of this tri-level design as your next home.

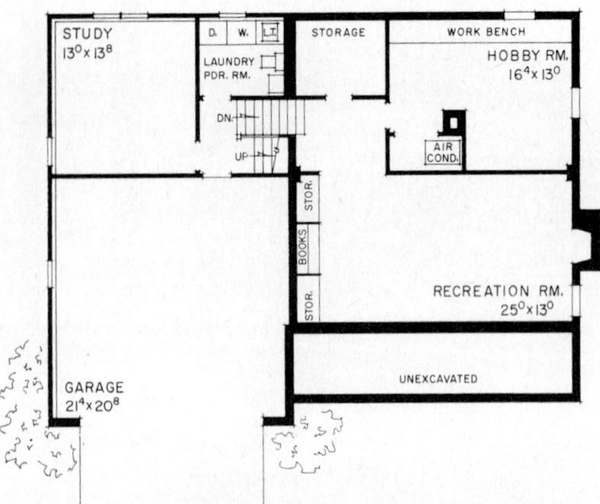

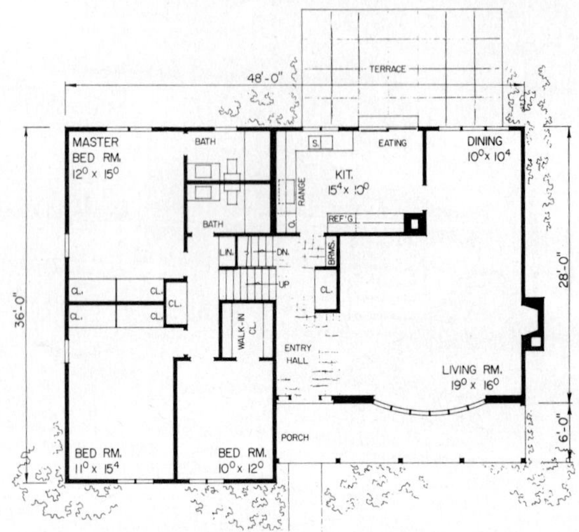

● A pleasingly traditional side-to-side three bedroom split-level which features four distinct living levels to service the activities of the active family.

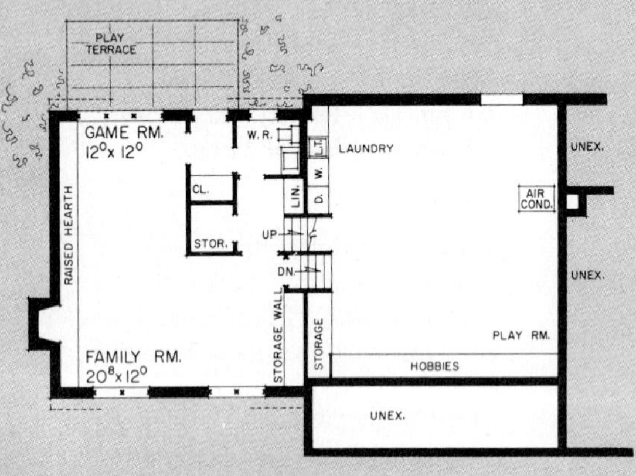

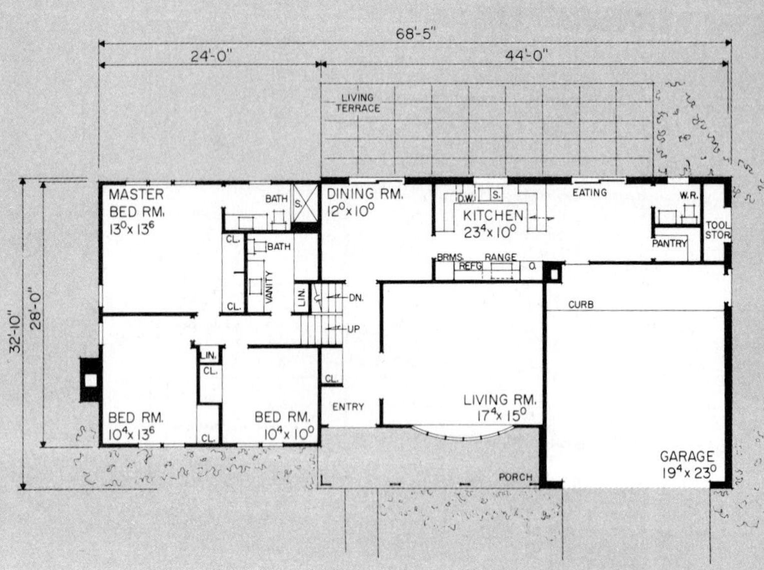

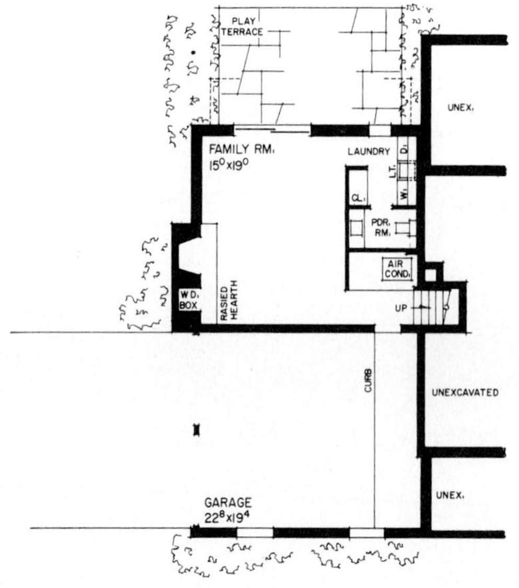

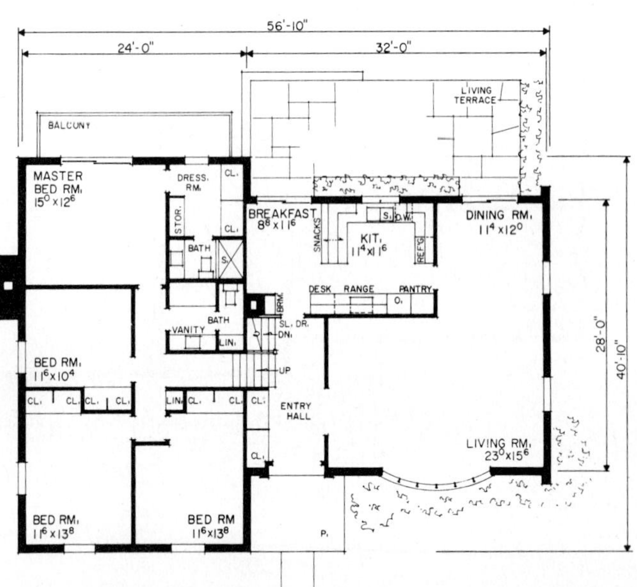

Design 41230 *728 Sq. Ft. – Main Level; 792 Sq. Ft. – Upper Level; 316 Sq. Ft. – Study Level; 728 Sq. Ft. – Recreation Level; 28,880 Cu. Ft.*

Design 41795 **804 Sq. Ft. – Main Level; 672 Sq. Ft. – Upper Level; 624 Sq. Ft. – Lower Level; 26,615 Cu. Ft.**

Design 41871 **942 Sq. Ft. – Main Level; 1,010 Sq. Ft. – Upper Level; 522 Sq. Ft. – Lower Level; 28,542 Cu. Ft.**

Design 42758

1,143 Sq. Ft. - Main Level
792 Sq. Ft. - Upper Level
770 Sq. Ft. - Lower Level
43,085 Cu. Ft.

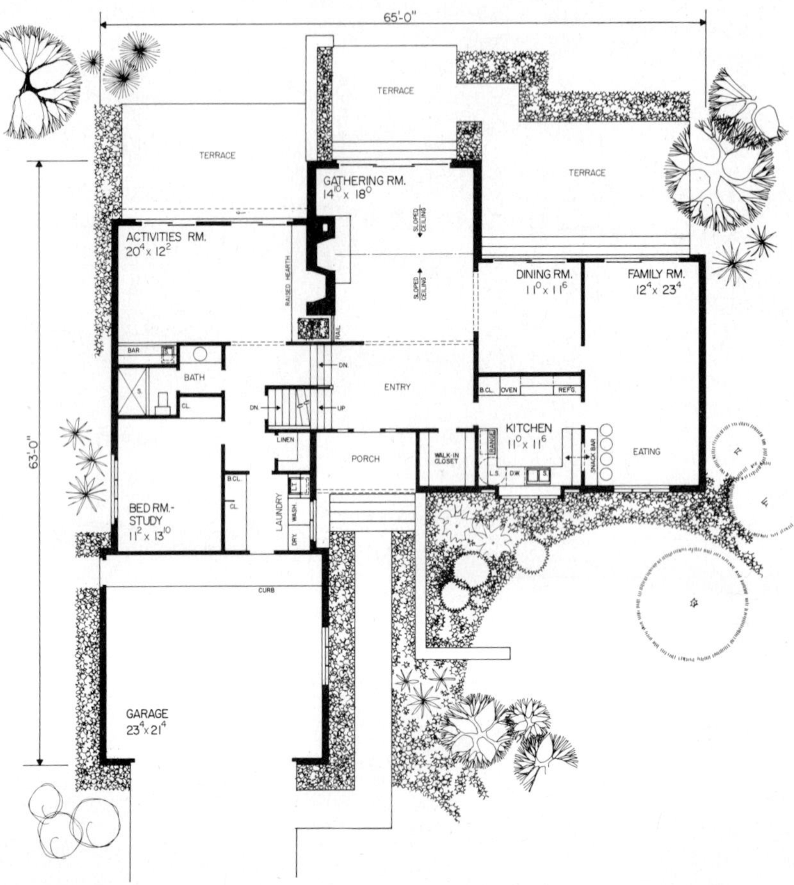

● An outstanding Tudor with three levels of exceptional livability, plus a basement. A careful study of the exterior reveals many delightful architectural details which give this home a character of its own. Notice the appealing recessed front entrance. Observe the overhanging roof with the exposed rafters. Don't miss the window treatment, the use of stucco and simulated beams, the masses of brick and the stylish chimney. Inside, the living potential is unsurpassed. Imagine, there are three living areas - the gathering, family and activities rooms. Having a snack bar, informal eating area and dining room, eating patterns can be flexible. In addition to the three bedrooms, two-bath upper level, there is a fourth bedroom with adjacent bath on the lower level.

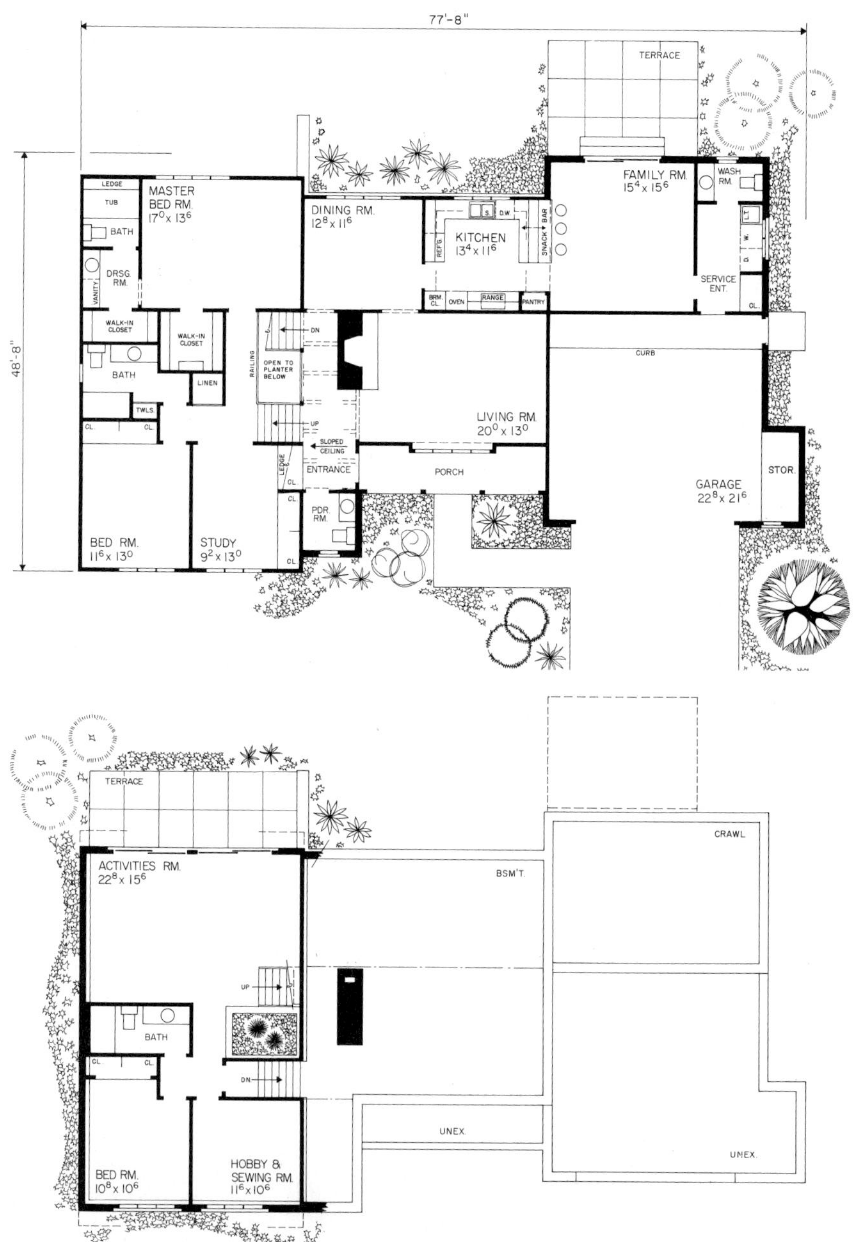

Design 42773

1,157 Sq. Ft. - Main Level
950 Sq. Ft. - Upper Level
912 Sq. Ft. - Lower Level
44,354 Cu. Ft.

● Here is another exquisitely styled Tudor tri-level designed to serve its happy occupants for many years. The contrasting use of material surely makes the exterior eye-catching. Another outstanding feature will be the covered front porch. A delightful way to enter this home. Many fine features also will be found inside this design. Formal living and dining room, U-shaped kitchen with snack bar and family room find themselves located on the main level. Two of the three bedrooms are on the upper level with two baths. Activities room, third bedroom and hobby/sewing room are on the lower level. Notice the built-in planter on the lower level which is visible from the other two levels. A powder room and a wash room both are on the main level. A study is on the upper level which is a great place for a quiet retreat. The basement will be convenient for the storage of any bulk items.

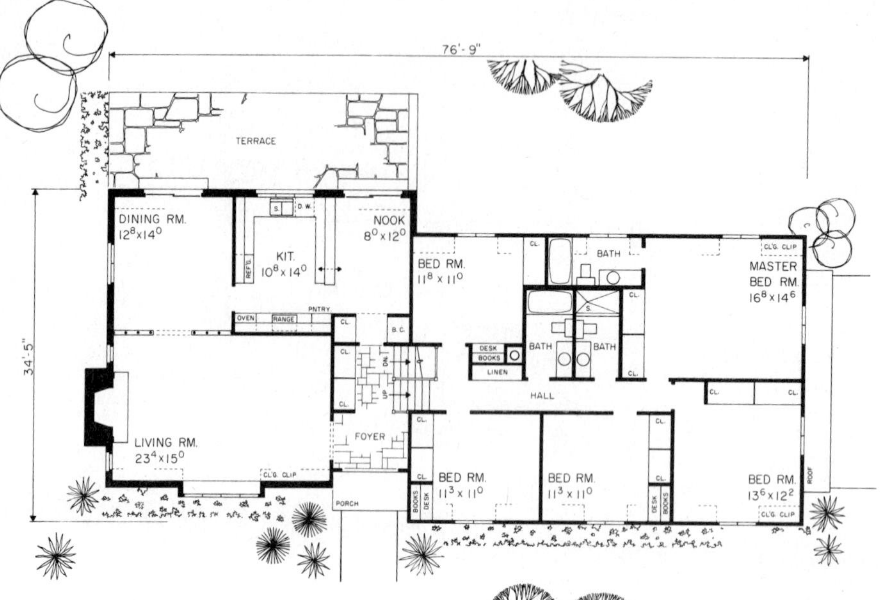

Design 42331

988 Sq. Ft. - Main Level
1,260 Sq. Ft. - Upper Level
525 Sq. Ft. - Lower Level
35,486 Cu. Ft.

● A picture to behold! The charm of this gracious Tudor adaptation will make it a sightseers favorite wherever built. The quarried stone (make it brick, if you prefer), the timber work, the stucco, the expanses of roof, the window treatment and the stolid double front doors are features which combine to achieve a truly delightful exterior. The interior of this tri-level is well-zoned to deliver excellent living patterns. There is the formal living room and the informal family room. Flanking the U-shaped work center is the separate dining room and the breakfast nook. The upper level provides the large family with five bedrooms and three full baths. Don't miss the laundry/sewing/hobby room.

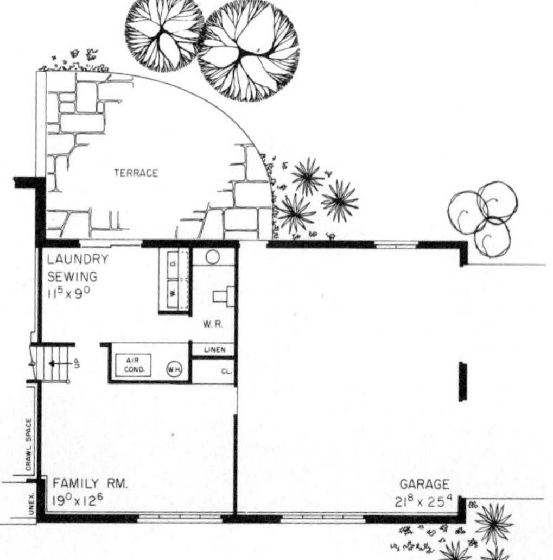

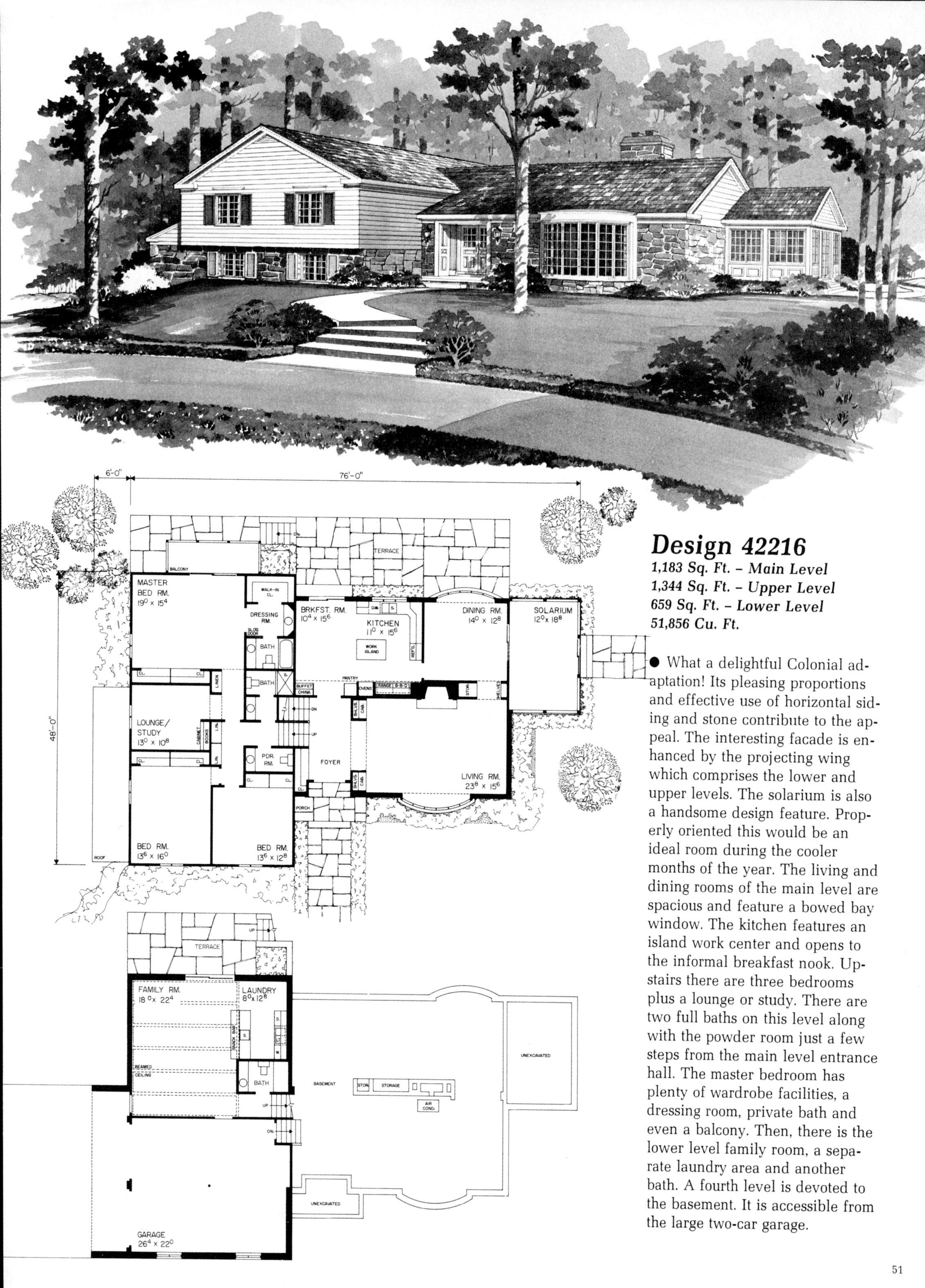

Design 42216

1,183 Sq. Ft. - Main Level
1,344 Sq. Ft. - Upper Level
659 Sq. Ft. - Lower Level
51,856 Cu. Ft.

● What a delightful Colonial adaptation! Its pleasing proportions and effective use of horizontal siding and stone contribute to the appeal. The interesting facade is enhanced by the projecting wing which comprises the lower and upper levels. The solarium is also a handsome design feature. Properly oriented this would be an ideal room during the cooler months of the year. The living and dining rooms of the main level are spacious and feature a bowed bay window. The kitchen features an island work center and opens to the informal breakfast nook. Upstairs there are three bedrooms plus a lounge or study. There are two full baths on this level along with the powder room just a few steps from the main level entrance hall. The master bedroom has plenty of wardrobe facilities, a dressing room, private bath and even a balcony. Then, there is the lower level family room, a separate laundry area and another bath. A fourth level is devoted to the basement. It is accessible from the large two-car garage.

Design 42727 *506 Sq. Ft. – Entry Level; 1,288 Sq. Ft. – Upper Level; 1,241 Sq. Ft. – Lower Level; 38,590 Cu. Ft.*

● Tri-level living at its glorious best. This Colonial facade is picturesque, indeed. The double front doors with their flanking side panels of glass are protected by the overhanging roof. The overhang of the upper level is an appealing detail and adds extra footage. Noteworthy is the size of the four bedrooms, the various storage facilities and the master bedroom balcony. Observe how the entry hall is utilized to receive traffic from both garage and front entrance. The gathering room has a dramatic planter/fireplace wall and functions through two sets of sliding glass doors with the big, L-shaped, upper terrace. The lower, main living level is wonderfully planned. Don't miss the extra bedroom, or study, located on the lower level, with a nearby powder room.

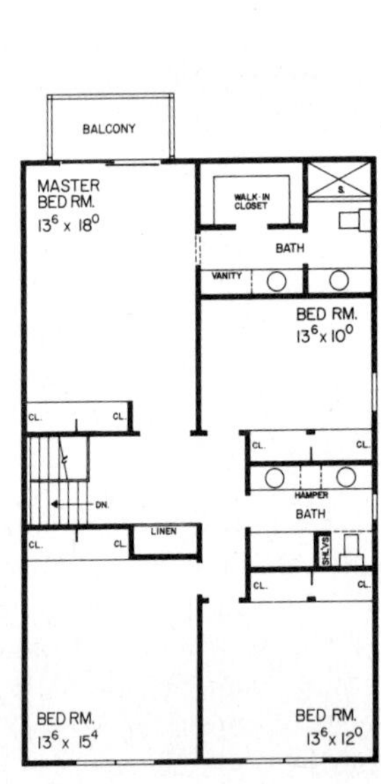

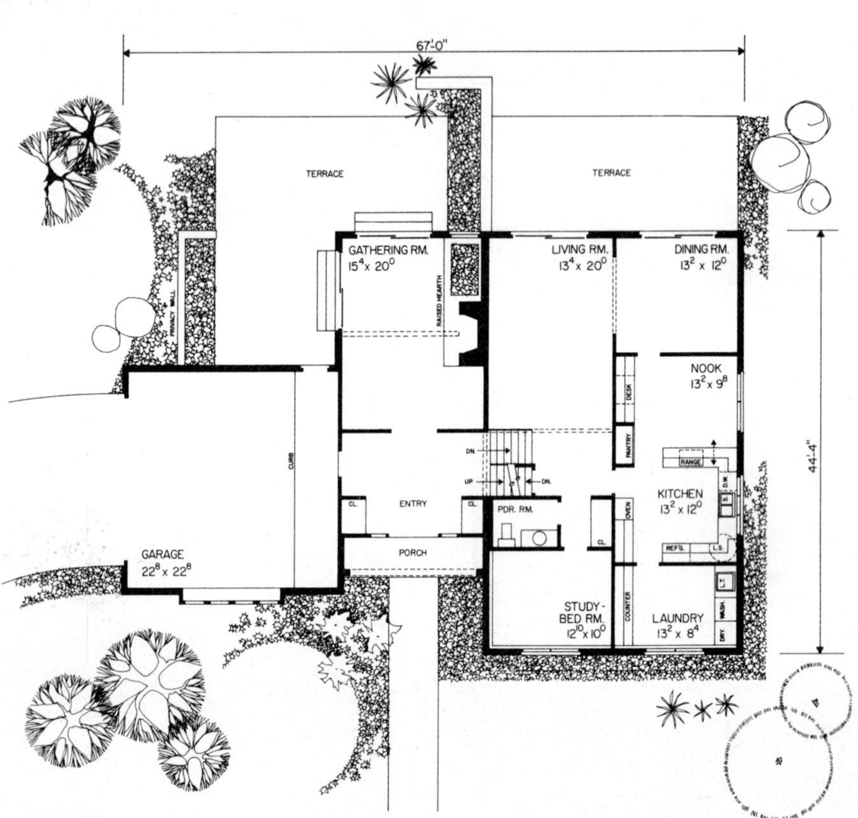

Contemporary

Split-Level Homes

Design 42712 ***1,624 Sq. Ft. – Main Level***
1,100 Sq. Ft. – Upper Level; 1,193 Sq. Ft. – Lower Level; 49,370 Cu. Ft.

● Tri-level livability plus a basement level with a contemporary facade are the characteristics of this design. This home is for the growing, active family - five bedrooms plus a study. The lounge overlooks the front entry and down on the lower level planter. The gathering room, dining room and breakfast nook function ideally with the rear terrace. There is an efficient U-shaped kitchen, planned to be free of any cross-room traffic. A pass-thru to the nook and a planning desk nearby. The service area features a powder room, handy closet and laundry. The master bedroom has a private balcony. Two of the five bedrooms are on the lower level along with the family room. Note two fireplaces. That fourth basement level will be just great for the pursuit of hobbies.

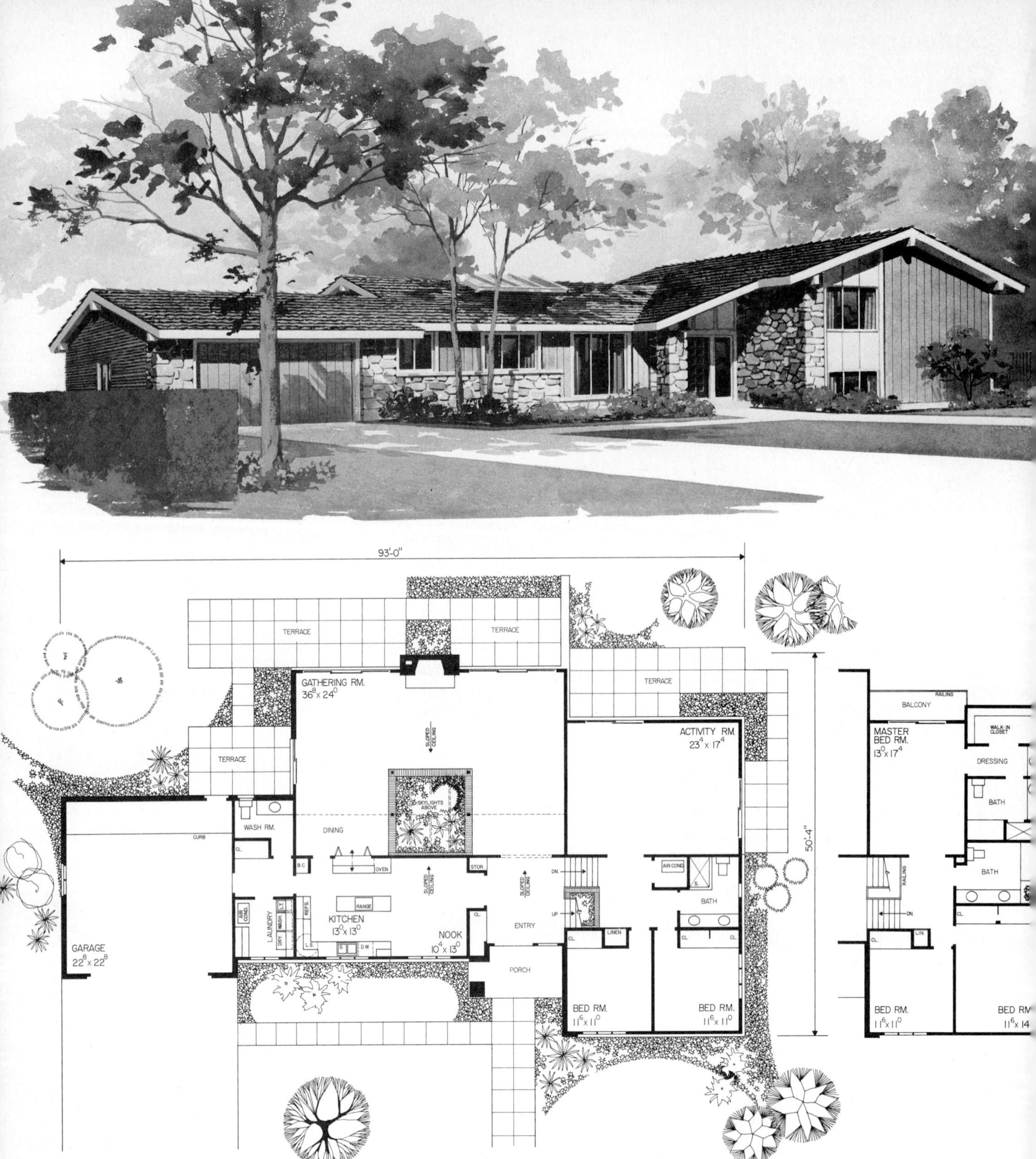

Design 42584 *1,604 Sq. Ft. – Main Level; 1,018 Sq. Ft. – Upper Level; 1,026 Sq. Ft. – Lower Level; 39,200 Cu. Ft.*

● Imagine an indoor garden with a skylight above in the huge gathering room plus a planter beside the lower level stairs. The gathering room also has a sloped ceiling, fireplace and two sets of sliding glass doors leading to the rear terrace and one set to the side terrace. That sure is luxury. But the appeal does not stop there. There are sloped ceilings in the foyer and breakfast nook. The kitchen has an island range, built-in oven and pass-thru to the dining room. Plus a large activities room. A great place for those informal activities. Five bedrooms in all to serve the large family. Including a master suite with a private balcony, dressing room, walk-in closet and bath.

Design 42588 *1,354 Sq. Ft. – Main Level; 1,112 Sq. Ft. – Upper Level; 562 Sq. Ft. – Lower Level; 46,925 Cu. Ft.*

● A thru-fireplace with an accompanying planter for the formal dining room and living room. That's old-fashioned good cheer in a contemporary home. The dining room has an adjacent screened-in porch for outdoor dining in the summertime. There are companions for these two formal areas, an informal breakfast nook and a family room. Each having sliding glass doors to separate rear terraces. Built-in desk, pantry, ample work space and island range are features of the L-shaped kitchen. The large laundry on the lower level houses the heating and cooling equipment. Three family bedrooms, bath and master bedroom suite are on the upper level.

● Such delightful living patterns could hardly have more inviting and attractive exteriors than this. The low-pitched, wide overhanging roofs, the various sized brick masses, the simple window treatment, the covered front entrance with its double doors and, of course, the always important, fine proportions are among the features that distinguish the front elevation. The highlights of the rear exterior are the covered balcony and the large glass areas. The balcony, upper terrace and the lower terrace will provide an abundance of indoor-outdoor living facilities. An additional outdoor living area is provided by the patio which functions with the study/guest room. The return of the brick wing-wall assures this little spot of its full measure of privacy. This impressive, contemporary multi-level home has features galore. Your journey through the floor plan will prove interesting.

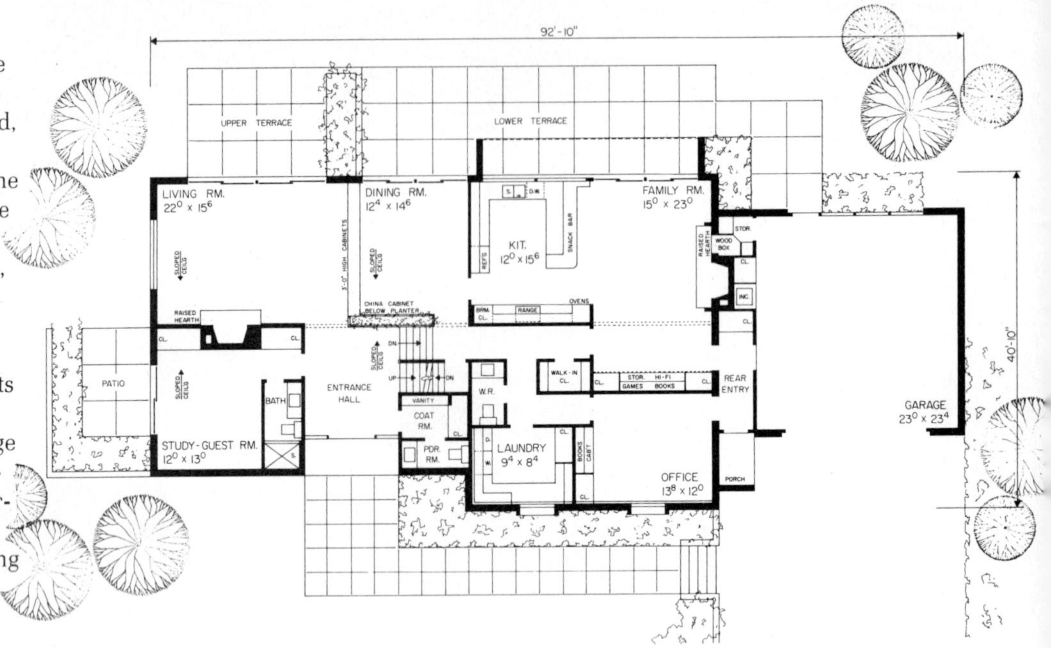

Design 41730

938 Sq. Ft. – Main Level
1,042 Sq. Ft. – Upper Level
1,340 Sq. Ft. – Lower Level
41,833 Cu. Ft.

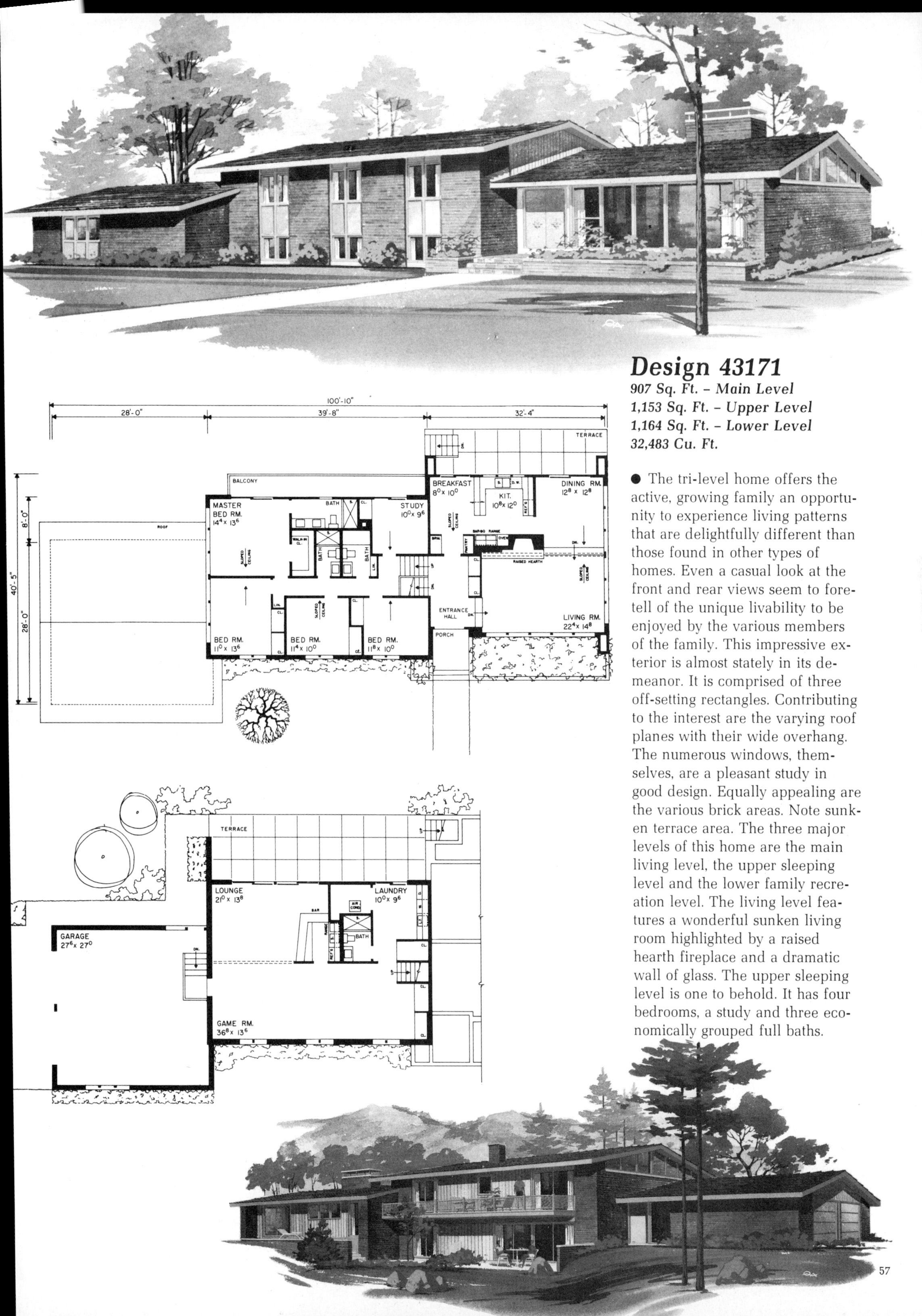

Design 43171

907 Sq. Ft. - Main Level
1,153 Sq. Ft. - Upper Level
1,164 Sq. Ft. - Lower Level
32,483 Cu. Ft.

● The tri-level home offers the active, growing family an opportunity to experience living patterns that are delightfully different than those found in other types of homes. Even a casual look at the front and rear views seem to foretell of the unique livability to be enjoyed by the various members of the family. This impressive exterior is almost stately in its demeanor. It is comprised of three off-setting rectangles. Contributing to the interest are the varying roof planes with their wide overhang. The numerous windows, themselves, are a pleasant study in good design. Equally appealing are the various brick areas. Note sunken terrace area. The three major levels of this home are the main living level, the upper sleeping level and the lower family recreation level. The living level features a wonderful sunken living room highlighted by a raised hearth fireplace and a dramatic wall of glass. The upper sleeping level is one to behold. It has four bedrooms, a study and three economically grouped full baths.

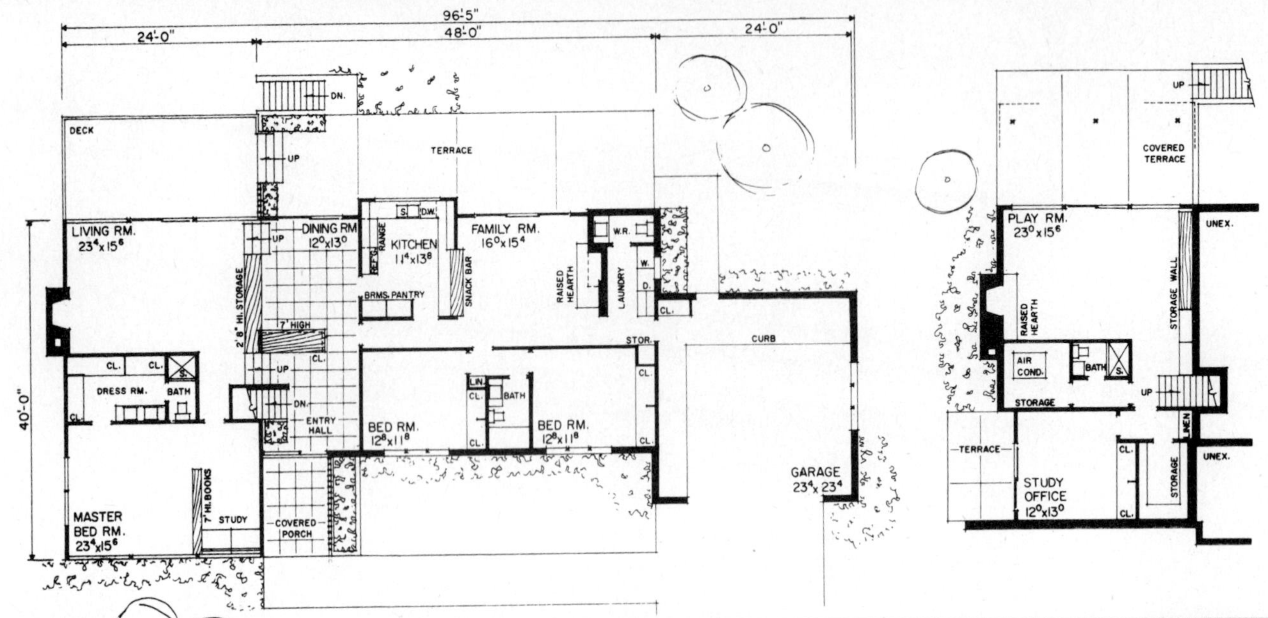

Design 41789 *1,486 Sq. Ft. – Main Level; 1,200 Sq. Ft. – Upper Level; 1,200 Sq. Ft. – Lower Level; 39,516 Cu. Ft.*

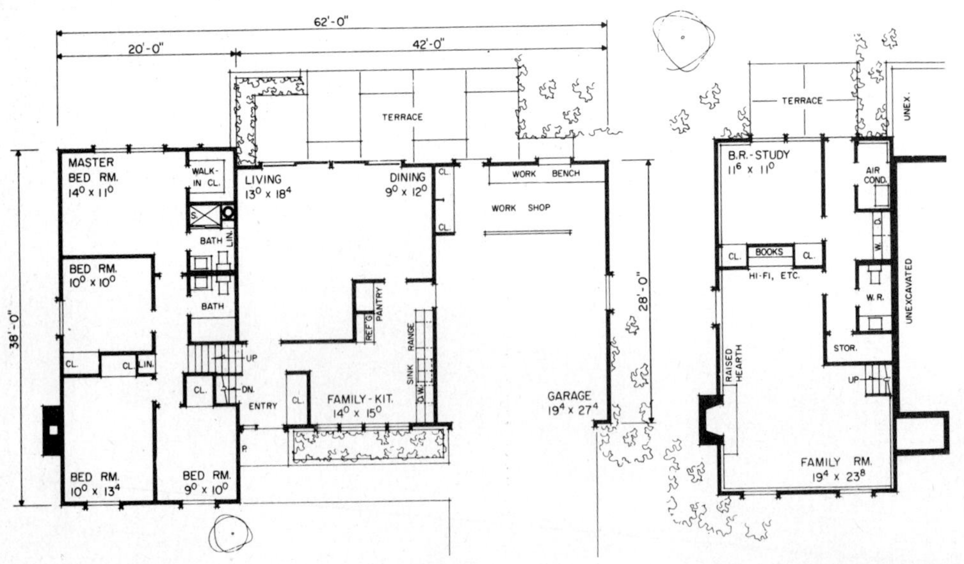

● Here are three sparkling contemporary split-levels each with all the livability any active family would require in a house. Design 41769 right, is the smallest of the three designs. Yet, it features four bedrooms, 2-1/2 baths, a study, a family room, a huge formal living and dining area, a big family kitchen and an oversized garage. Design 41823, top, offers unique living patterns. This house will really be fun in which to live. Above, is Design 41789 - an ideal home for the large family. Imagine, six bedrooms and three full baths. There is also an extra powder room and wash room. Note family room and recreation room of main and upper levels. Don't miss fireplaces and sunken living room.

Design 41823 *1,344 Sq. Ft. – Main Level; 960 Sq. Ft. – Upper Level; 884 Sq. Ft. – Lower Level; 31,090 Cu. Ft.*

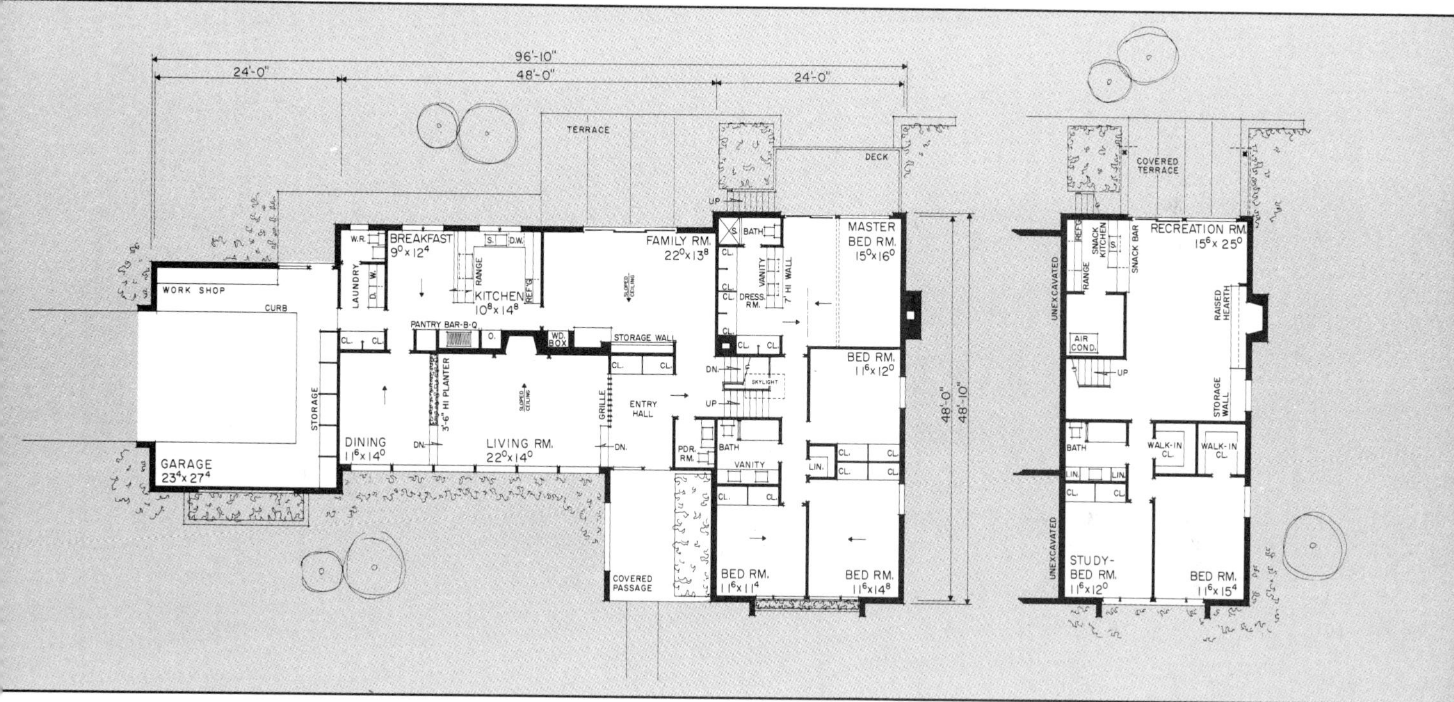

Design 41769 *572 Sq. Ft – Main Level; 760 Sq. Ft. – Upper Level; 760 Sq. Ft. – Lower Level; 20,352 Cu. Ft.*

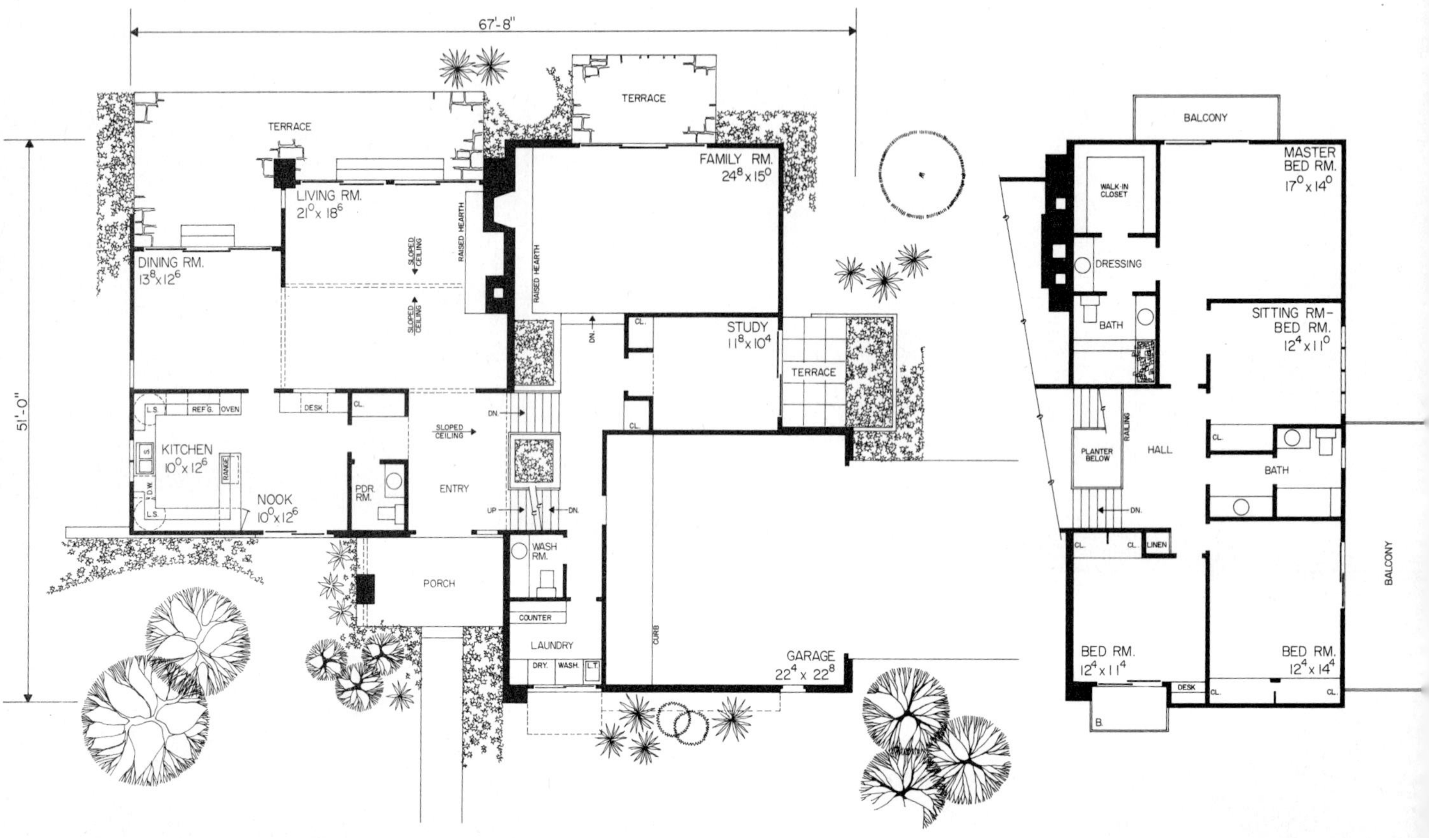

Design 42536 *1,077 Sq. Ft. - Main Level; 1,319 Sq. Ft. - Upper Level; 914 Sq. Ft. - Lower Level; 31,266 Cu. Ft.*

● Here are three levels of outstanding livability all packed in a delightfully contemporary exterior. The low pitched roof has a wide overhang with exposed rafter tails. The stone masses contrast effectively with the vertical siding and the glass areas. The extension of the sloping roof provides the recessed feature of the front entrance with the patterned double doors. The homemaker's favorite highlight will be the layout of the kitchen. No crossroom traffic here. Only a few steps from the formal and informal eating areas, it is the epitome of efficiency. A sloping beamed ceiling, sliding glass doors and a raised hearth fireplace enhance the appeal of the living room. The upper level offers the option of a fourth bedroom or a sitting room functioning with the master bedroom. Note the three balconies. On the lower level, the big family room, quiet study, laundry and extra washroom are present.

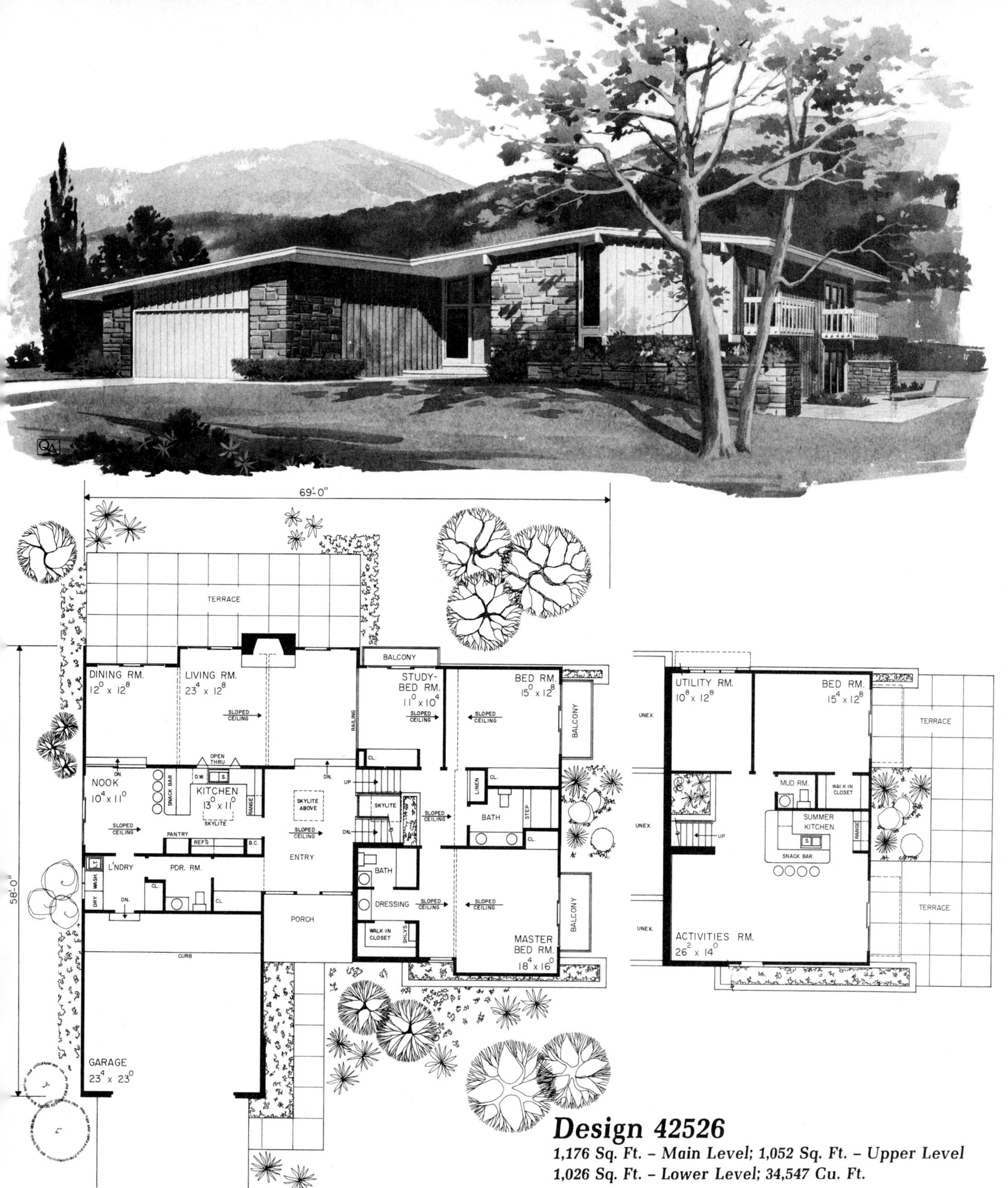

Design 42526

1,176 Sq. Ft. – Main Level; 1,052 Sq. Ft. – Upper Level
1,026 Sq. Ft. – Lower Level; 34,547 Cu. Ft.

● Here is an eye-catching multi-level which offers living patterns that will surely please the whole family. Flanking the front door, and above as well, are large glass panels which provide with the help of the skylite, plenty of natural light to the spacious entry. From this generous area traffic flows conveniently. To the left is the well-planned kitchen area. The breakfast nook, laundry, powder room and plenty of storage are nearby. Straight ahead is the sunken living room with the adjacent dining room. Having a sloping beamed ceiling and an abundance of glass, this will be another cheerful area. To the right, stairs lead to the upper level. Here two bedrooms, study and two baths are housed. Note the exceptional features including a planter, balcony and sunken tub. Also from the hall, there is access to the featured-packed lower level. How do you like that huge activities room?

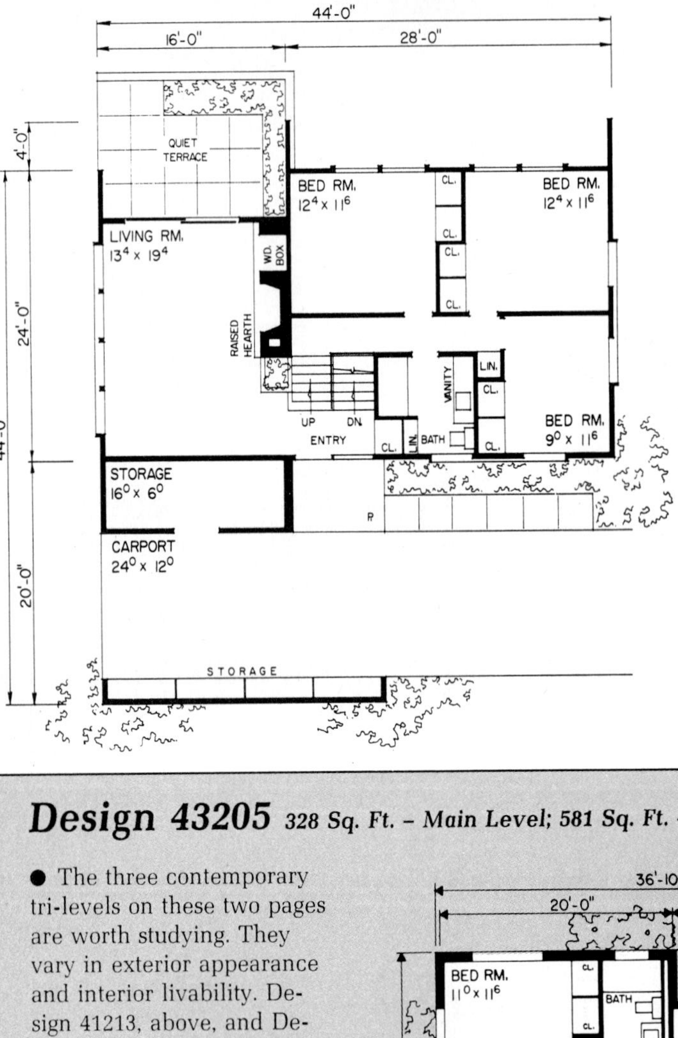

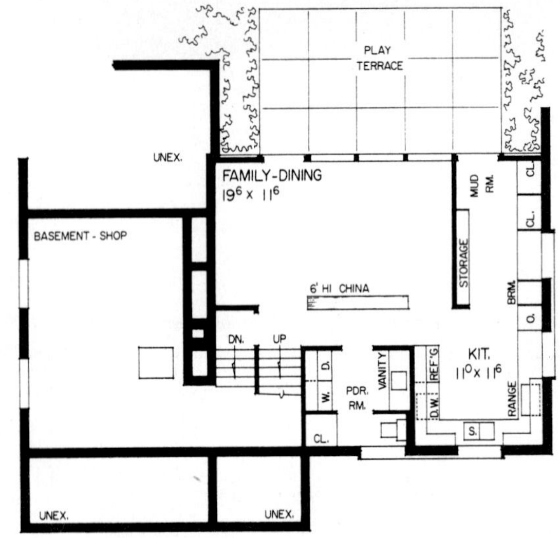

Design 41213

320 Sq. Ft. - Living Level
672 Sq. Ft. - Upper Level
672 Sq. Ft. - Lower Level
18,650 Cu. Ft.

Design 43205 *328 Sq. Ft. - Main Level; 581 Sq. Ft. - Upper Level; 600 Sq. Ft. - Lower Level; 14,769 Cu. Ft.*

● The three contemporary tri-levels on these two pages are worth studying. They vary in exterior appearance and interior livability. Design 41213, above, and Design 43205, right, both have a three bedroom upper sleeping level while Design 41229, above right, has a more spacious four bedroom, two bath upper sleeping level with a fifth bedroom and a study on the lower level. Each of the three designs has formal and informal living and rear terraces for outdoor enjoyment.

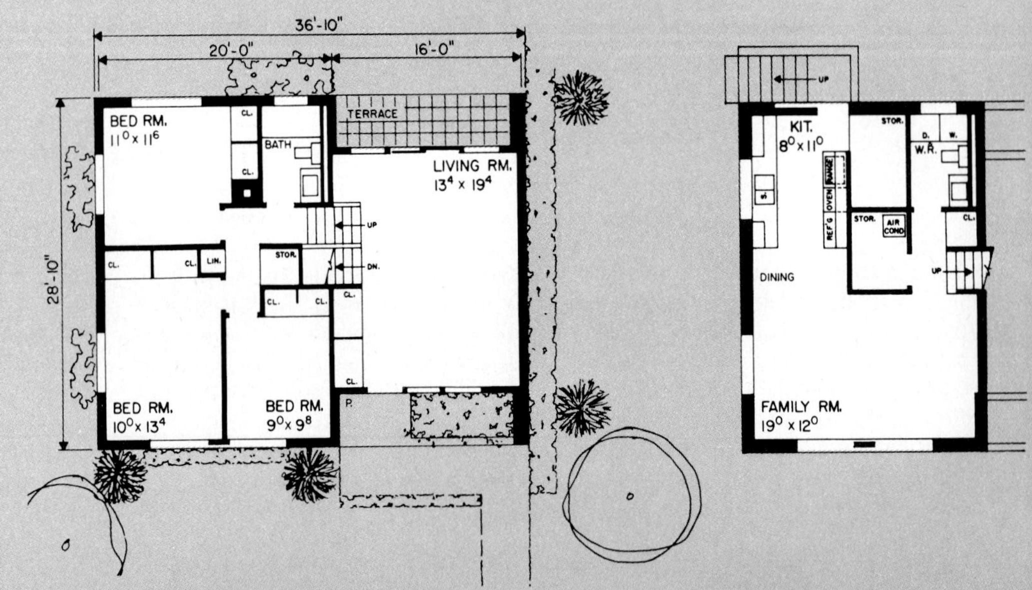

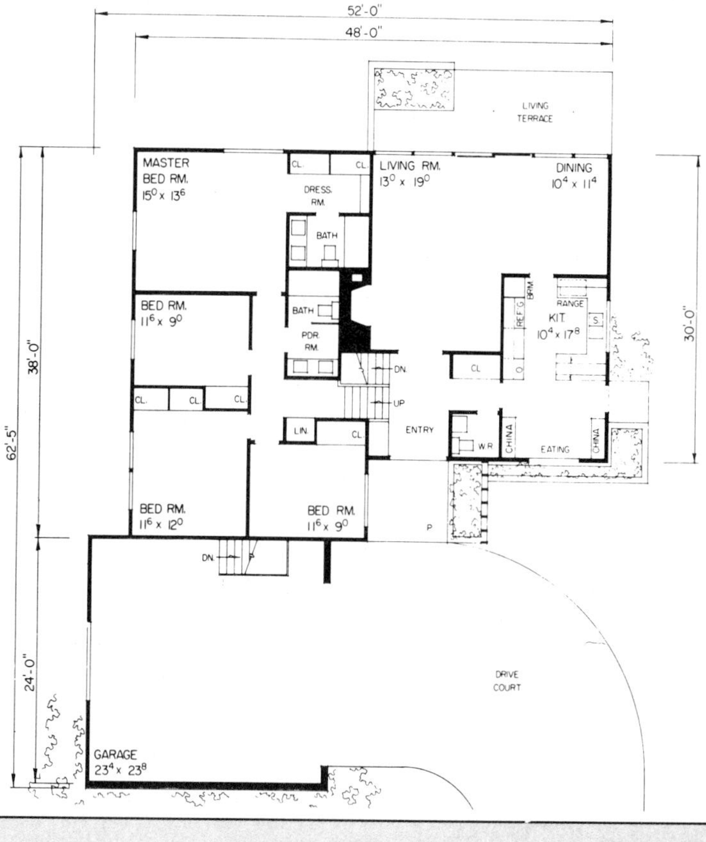

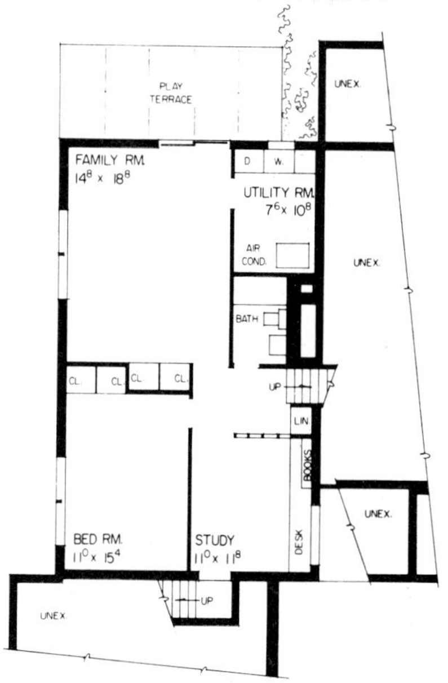

Design 41229

720 Sq. Ft. – Main Level
912 Sq. Ft. – Upper Level
912 Sq. Ft. – Lower Level
26,506 Cu. Ft.

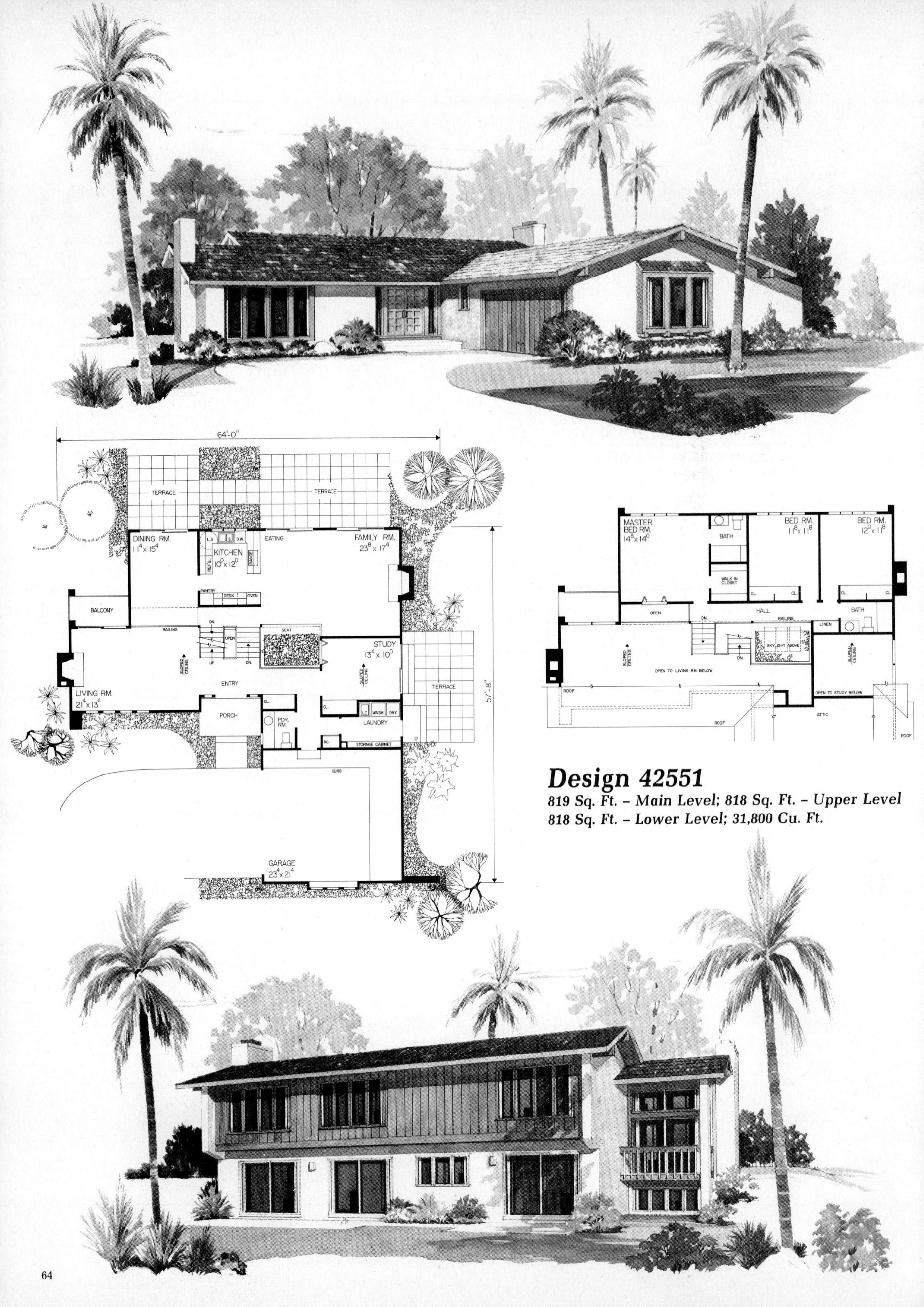

Design 42551

819 Sq. Ft. – Main Level; 818 Sq. Ft. – Upper Level
818 Sq. Ft. – Lower Level; 31,800 Cu. Ft.

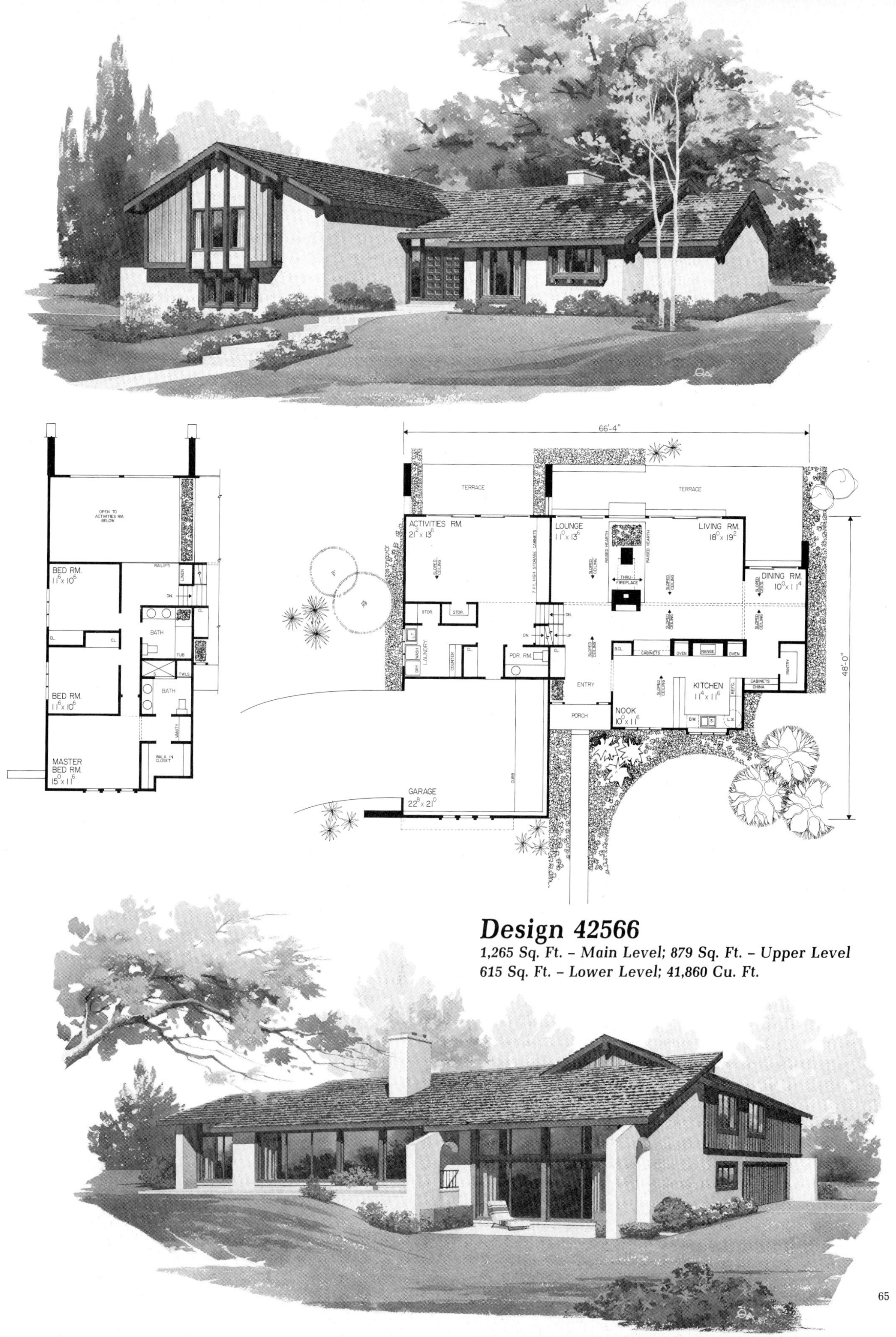

Design 42566

1,265 Sq. Ft. – Main Level; 879 Sq. Ft. – Upper Level
615 Sq. Ft. – Lower Level; 41,860 Cu. Ft.

Design 41093

654 Sq. Ft. – Main Level; 768 Sq. Ft. – Upper Level
492 Sq. Ft. – Lower Level; 18,762 Cu. Ft.

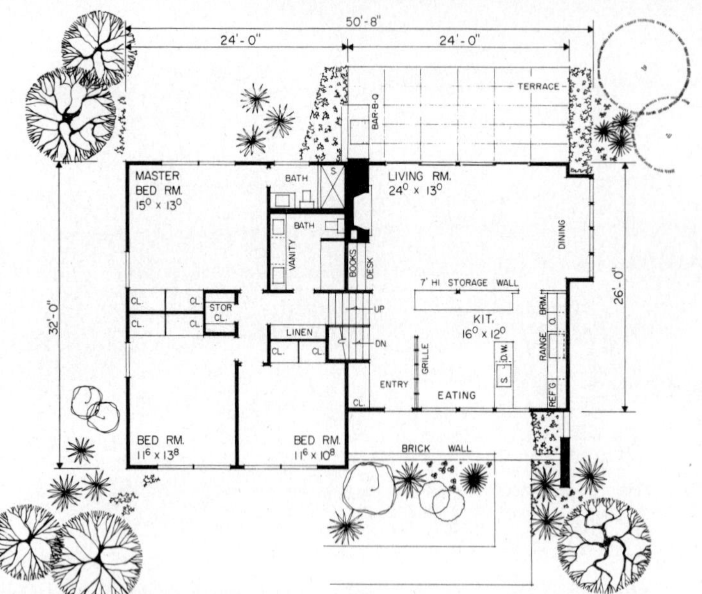

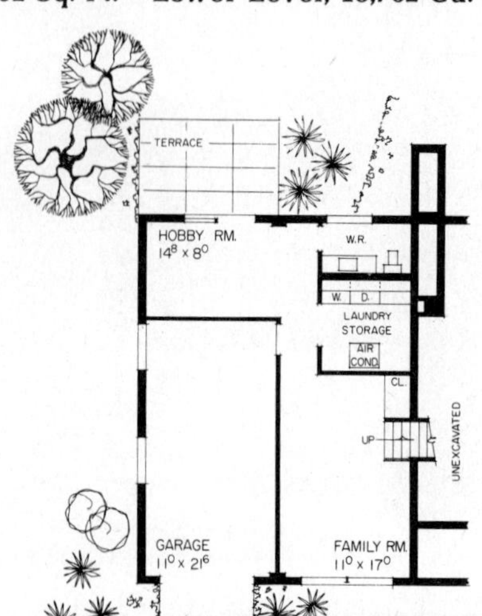

Design 41978

1,280 Sq. Ft. – Main Level
960 Sq. Ft. – Upper Level; 24,031 Cu. Ft.

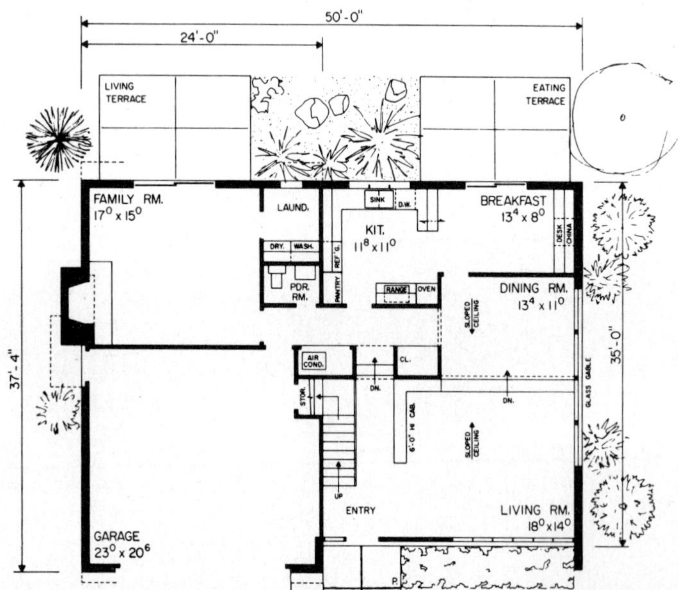

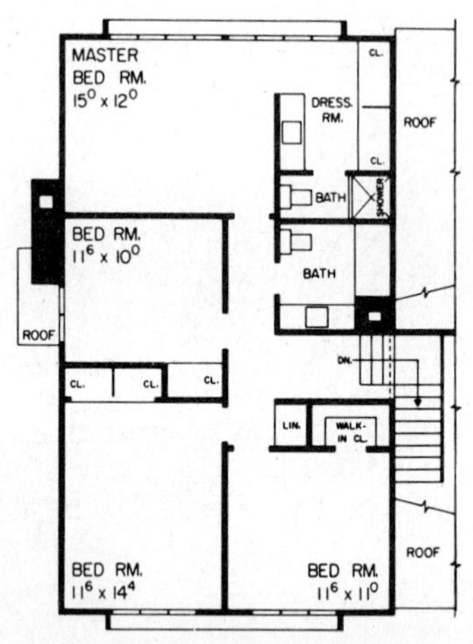

Design 41217 *712 Sq. Ft. – Main Level*

720 Sq. Ft. – Upper Level; 720 Sq. Ft. – Lower Level
26,673 Cu. Ft.

● The large family will experience years of fun living in this multi-level house. Occupying five bedrooms, sleeping accommodations are excellent. There are two sizable living areas – the formal living room of the main level and the all-purpose family room of the lower level. Housing a convenient breakfast area, the formal dining room and the outdoor terrace, eating patterns will be anything but routine. There is both an indoor and an outdoor barbecue unit to further your flexibility.

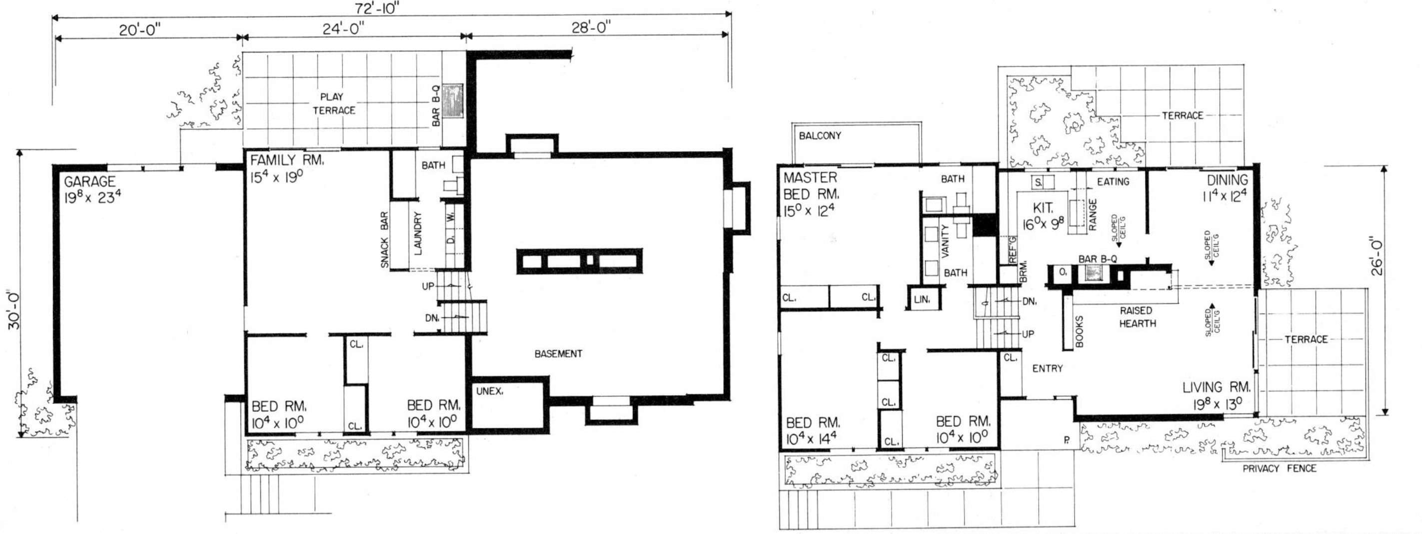

Design 41353 *484 Sq. Ft. – Main Level*

624 Sq. Ft. – Upper Level; 300 Sq. Ft. – Lower Level
13,909 Cu. Ft.

● This three bedroom contemporary is zoned for efficiency without a bit of wasted space. Note the two back-to-back baths, the separate dining room and the kitchen eating area. The lower level has a family room and the utility room.

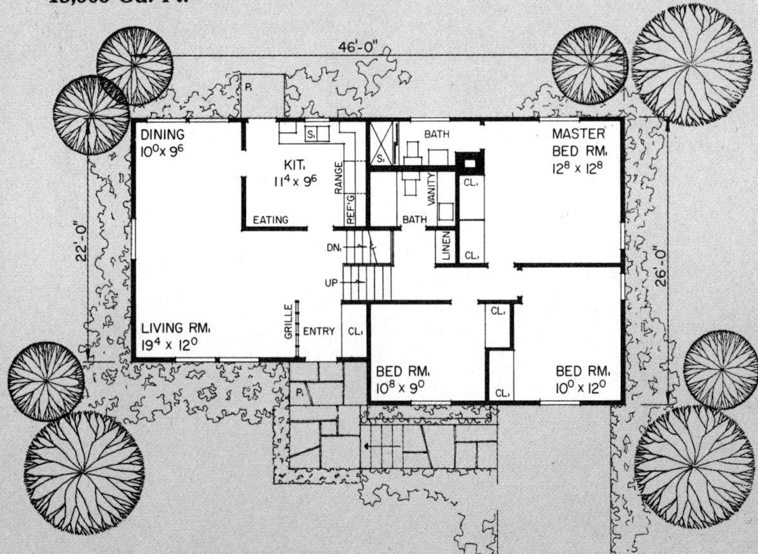

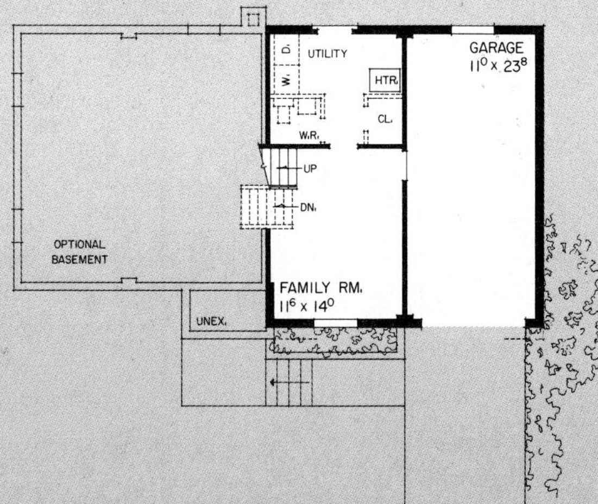

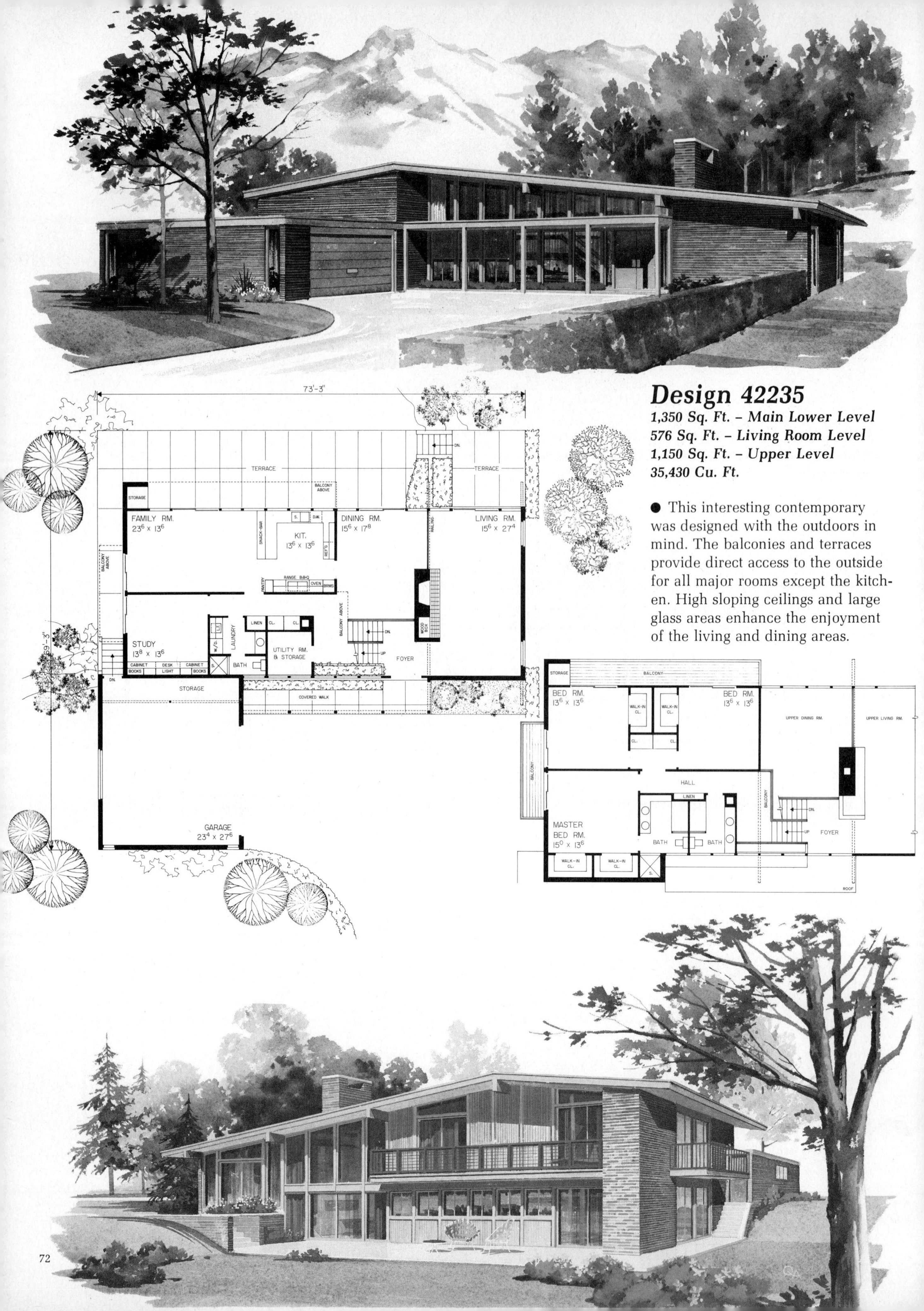

Design 42235

1,350 Sq. Ft. – Main Lower Level
576 Sq. Ft. – Living Room Level
1,150 Sq. Ft. – Upper Level
35,430 Cu. Ft.

● This interesting contemporary was designed with the outdoors in mind. The balconies and terraces provide direct access to the outside for all major rooms except the kitchen. High sloping ceilings and large glass areas enhance the enjoyment of the living and dining areas.

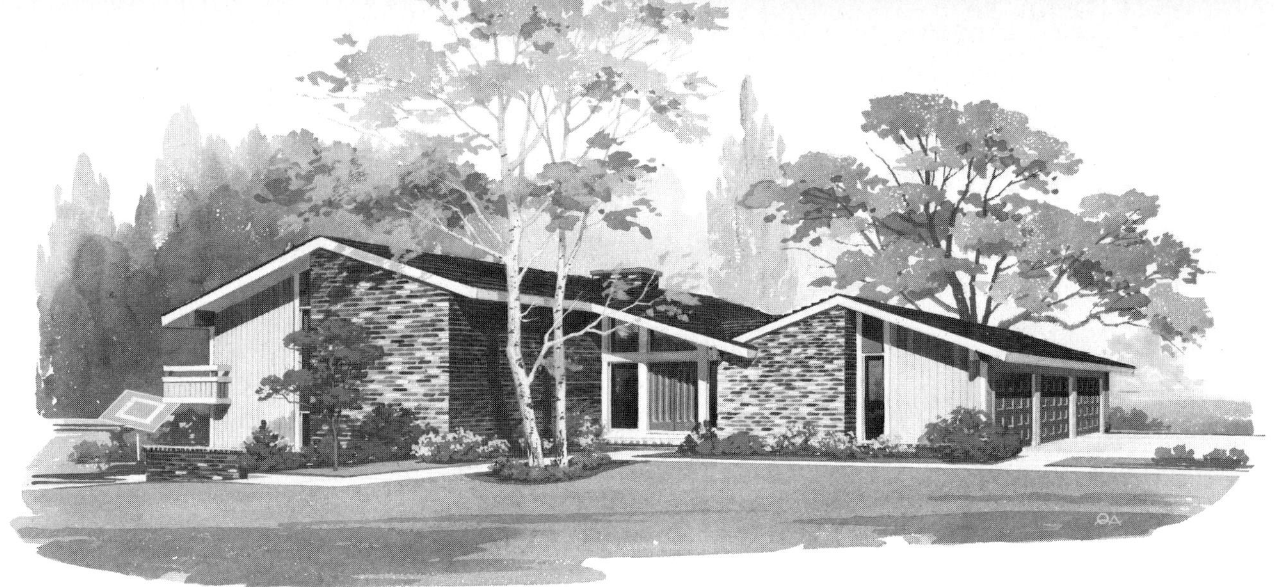

Design 42736

212 Sq. Ft. – Entry Level; 1,263 Sq. Ft. – Upper Level 1,848 Sq. Ft. – Lower Level; 43,090 Cu. Ft.

● Contemporary exterior styling with an overhanging roof are definitely appealing characteristics for this home. Indoor-outdoor living will be excellent. Two balconies and four terraces to be enjoyed by all members of the active family. Features including built-in planters, thru-fireplace and open lounge are located throughout the plan.

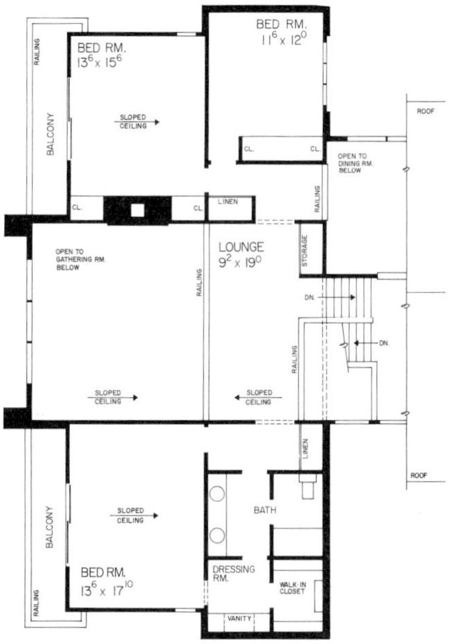

BED RM. 13^6 x 15^6
BED RM. 11^6 x 12^0
BALCONY
RAILING
SLOPED CEILING
CL.
LINEN
OPEN TO DINING RM. BELOW
ROOF
OPEN TO GATHERING RM. BELOW
LOUNGE 9^2 x 19^0
STORAGE
DN.
BATH
BED RM. 13^6 x 17^{10}
DRESSING RM.
WALK-IN CLOSET
VANITY

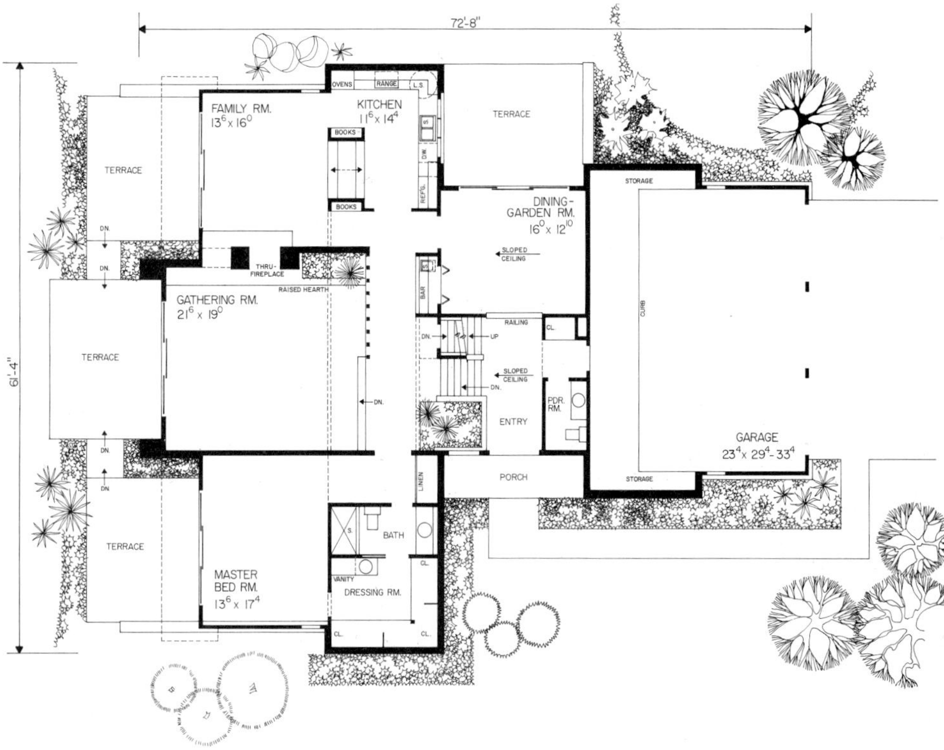

72'-8"
61'-4"
TERRACE
FAMILY RM. 13^6 x 16^0
OVENS
RANGE
KITCHEN 11^6 x 14^4
BOOKS
TERRACE
STORAGE
DINING-GARDEN RM. 16^0 x 12^{10}
SLOPED CEILING
THRU-FIREPLACE
RAISED HEARTH
GATHERING RM. 21^6 x 19^0
BAR
RAILING
DN.
UP
PDR. RM.
ENTRY
GARAGE 23^4 x 29^4-33^4
CURB
PORCH
LINEN
BATH
MASTER BED RM. 13^6 x 17^4
VANITY
DRESSING RM.
CL.

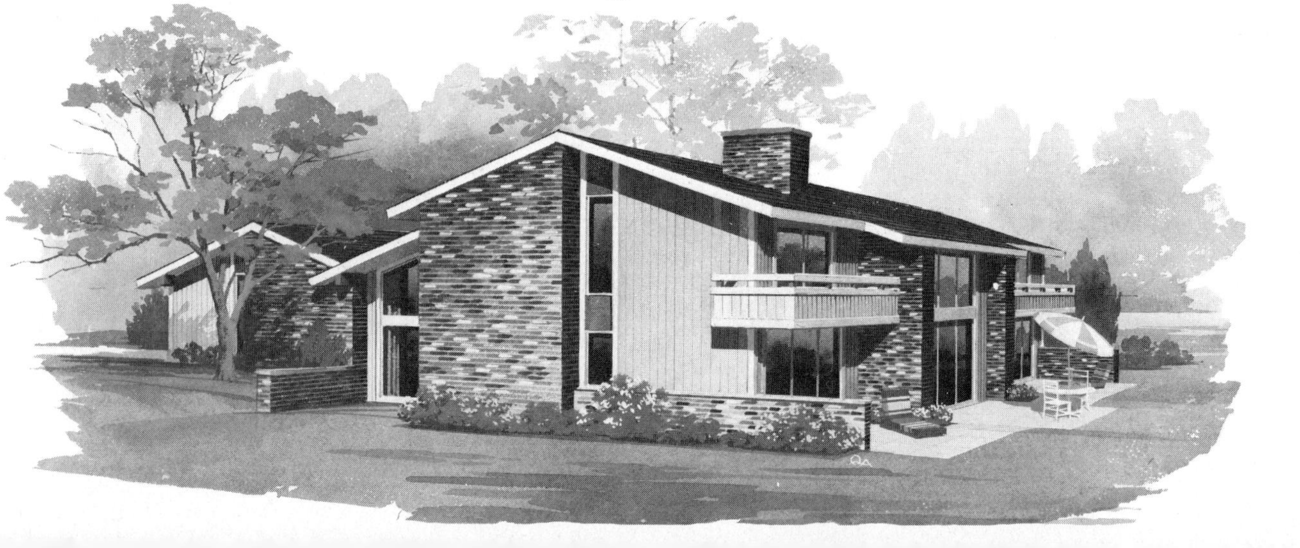

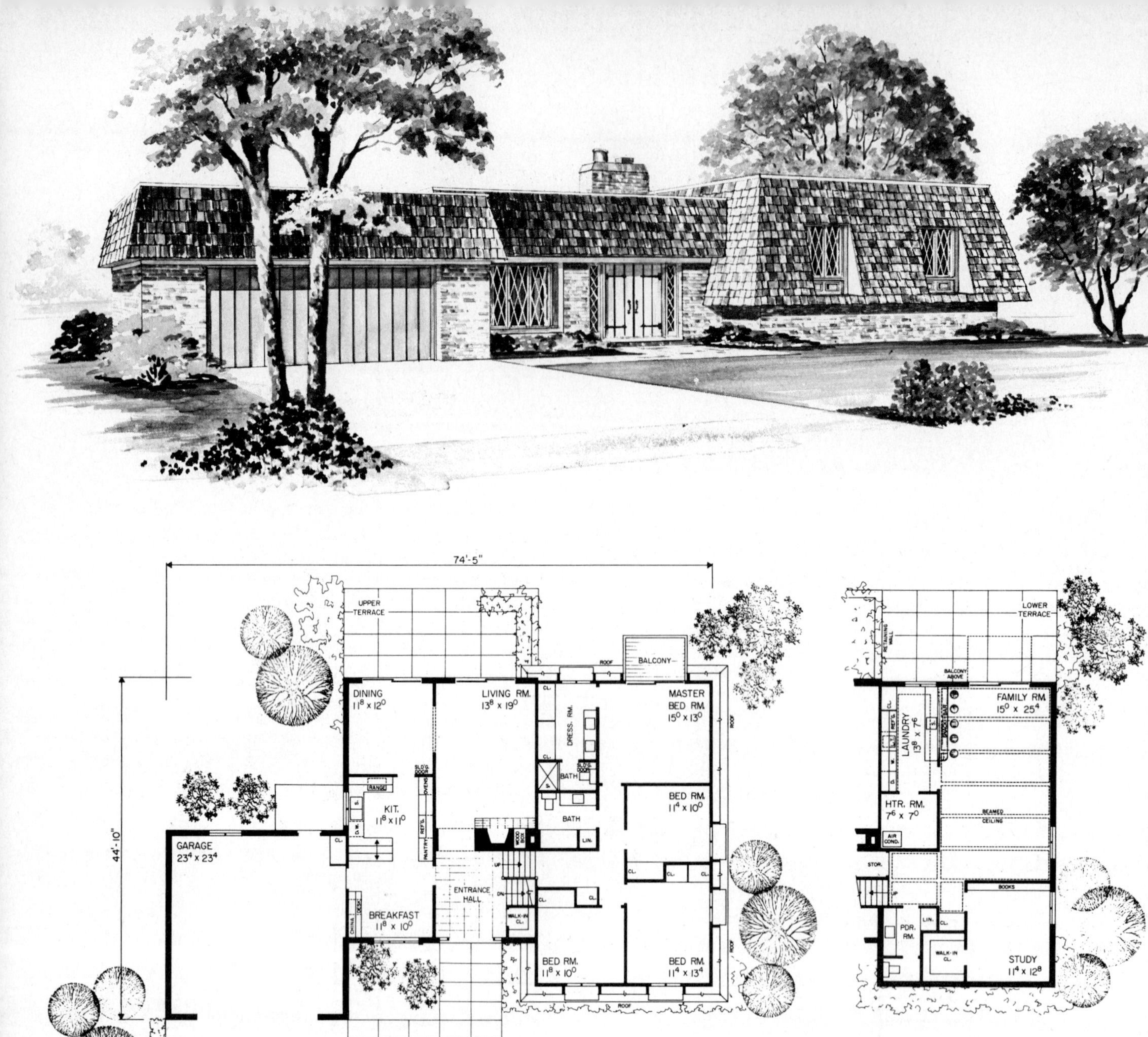

Design 42186

915 Sq. Ft. – Main Level; 960 Sq. Ft. – Upper Level
960 Sq. Ft. – Lower Level; 33,130 Cu. Ft.

● Here is a French Mansard type of roof adapted to a split-level design. The result is a most distinctive one. Notice the wide overhang, the double front doors and the diamond lite windows. Inside, there is all the livability a growing family would require. The upper level features four good sized bedrooms and two baths. The master bedroom has a dressing room and access to an outdoor rear balcony. The lower level highlights a big family room with beamed ceiling, snack bar and sliding glass doors to the lower terrace. Also, there is the complete laundry, an extra powder room and a quiet study. The main level is wonderful, also. The efficient work center is flanked by the breakfast room and the dining room. The entrance hall is spacious and leads to the living room with its dramatic fireplace and built-in wood box. Count the many storage areas on each level.

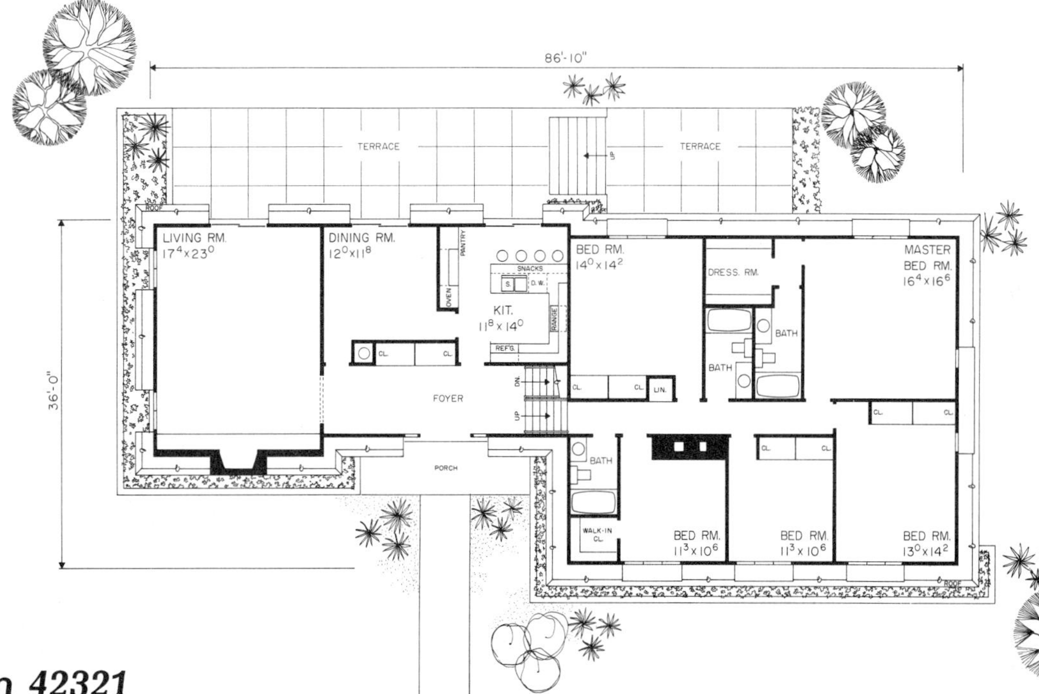

Design 42321

1,073 Sq. Ft. - Main Level
1,428 Sq. Ft. - Upper Level
646 Sq. Ft. - Lower Level
35,057 Cu. Ft.

● This contemporary adaptation of the mansard roof strikes a dramatic note, indeed. The brick mass of the chimney, the double doors flanked by the glass panels, the design of the recessed windows and the overhanging side walls are distinguishing features of the exterior. Inside the air of distinction is equally apparent. Notice how each of the main level rooms functions with the adjacent terrace. On the upper level there are five bedrooms serviced by three baths and plenty of closets. Don't overlook the family room which features built-in cabinets, beamed ceiling and fireplace and the fourth bath on the lower level. The kitchen is efficient and has a snack bar for those very informal meals. Observe the size of the garage.

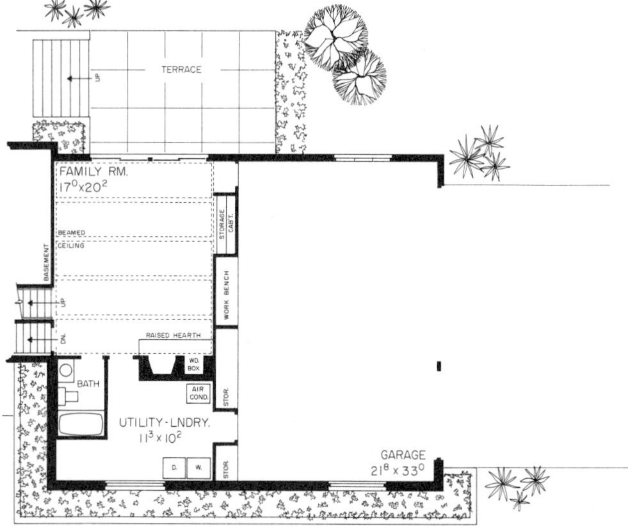

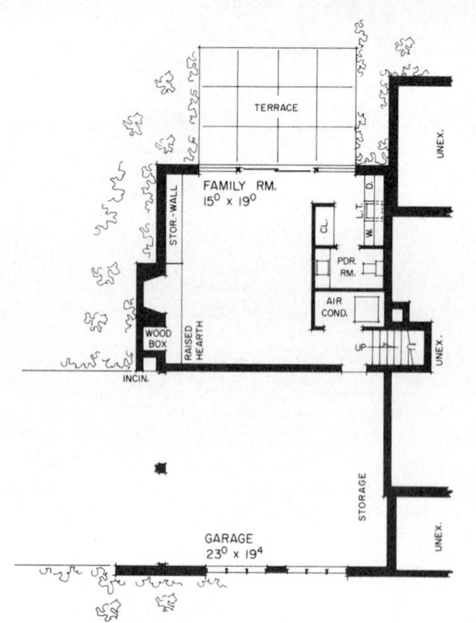

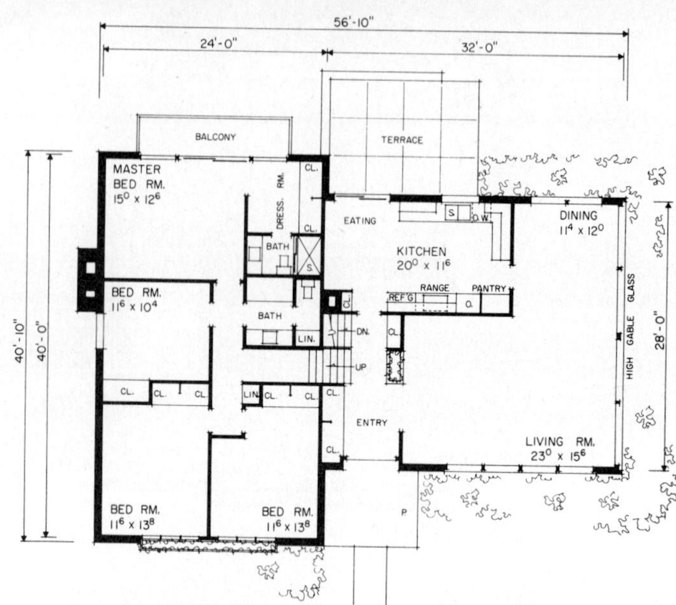

Design 41893 *814 Sq. Ft. – Main Level; 952 Sq. Ft. – Upper Level 366 Sq. Ft. – Family Room Level; 814 Sq. Ft. – Lower Level; 29,080 Cu. Ft.*

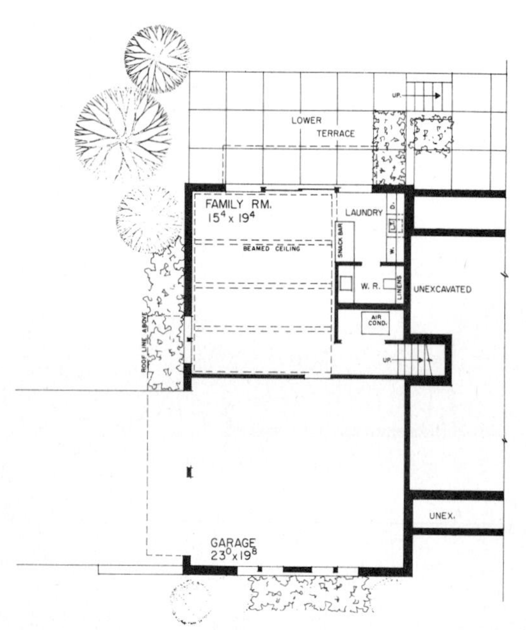

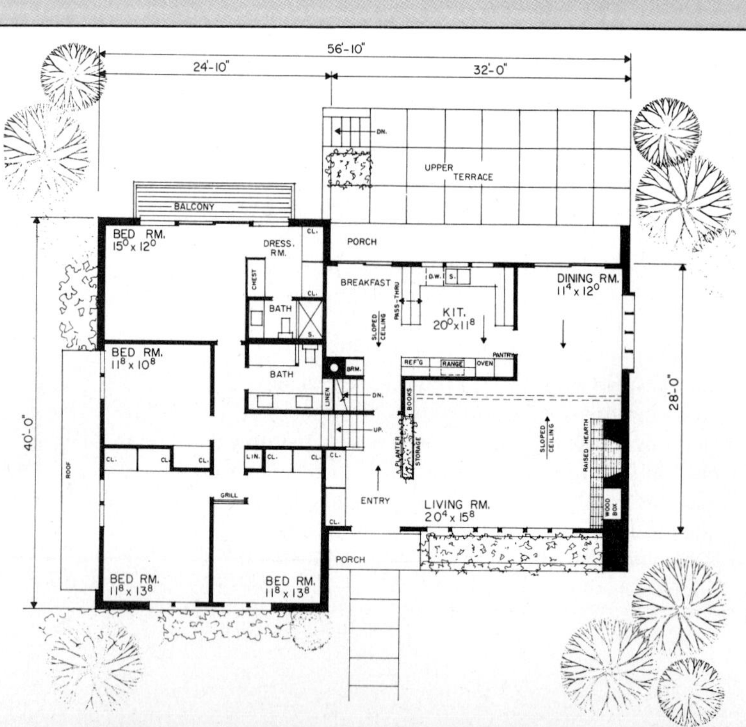

Design 41771 *942 Sq. Ft. – Main Level; 1,010 Sq. Ft. – Upper Level; 522 Sq. Ft. – Lower Level; 26,024 Cu. Ft.*

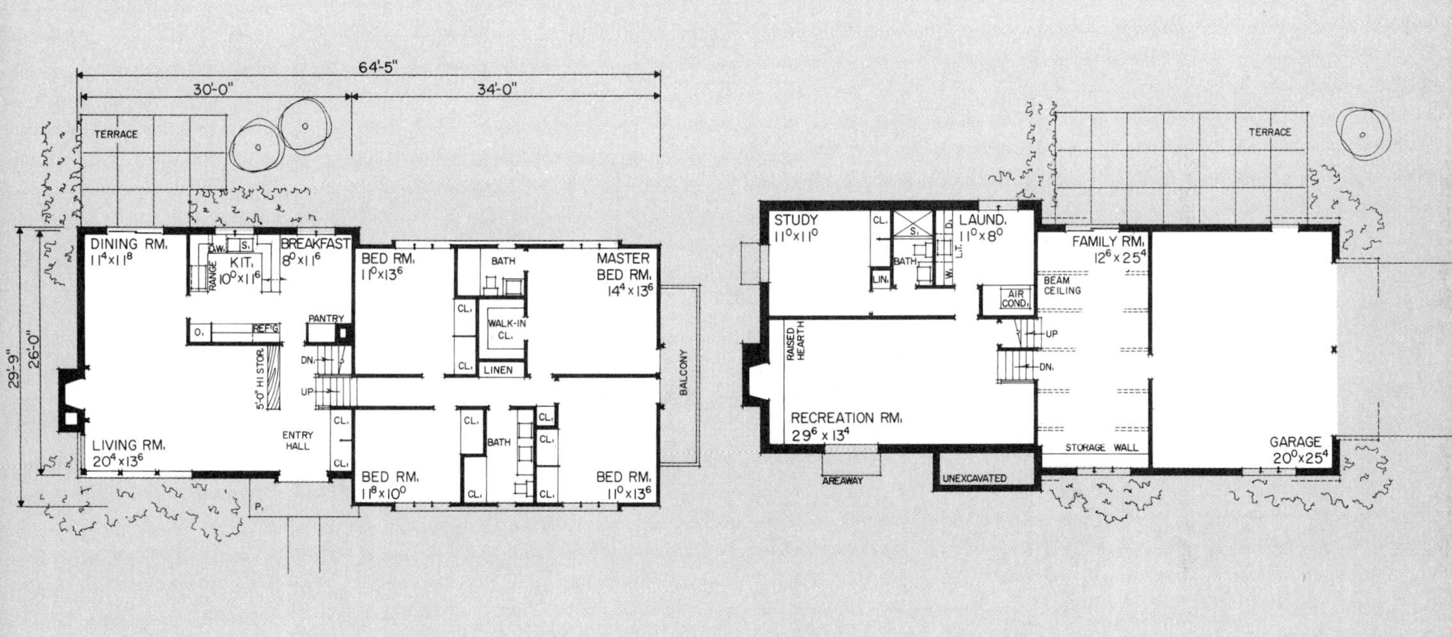

Design 43154 *908 Sq. Ft. – Main Level; 980 Sq. Ft. – Upper Level; 496 Sq. Ft. – Lower Level; 26,399 Cu. Ft.*

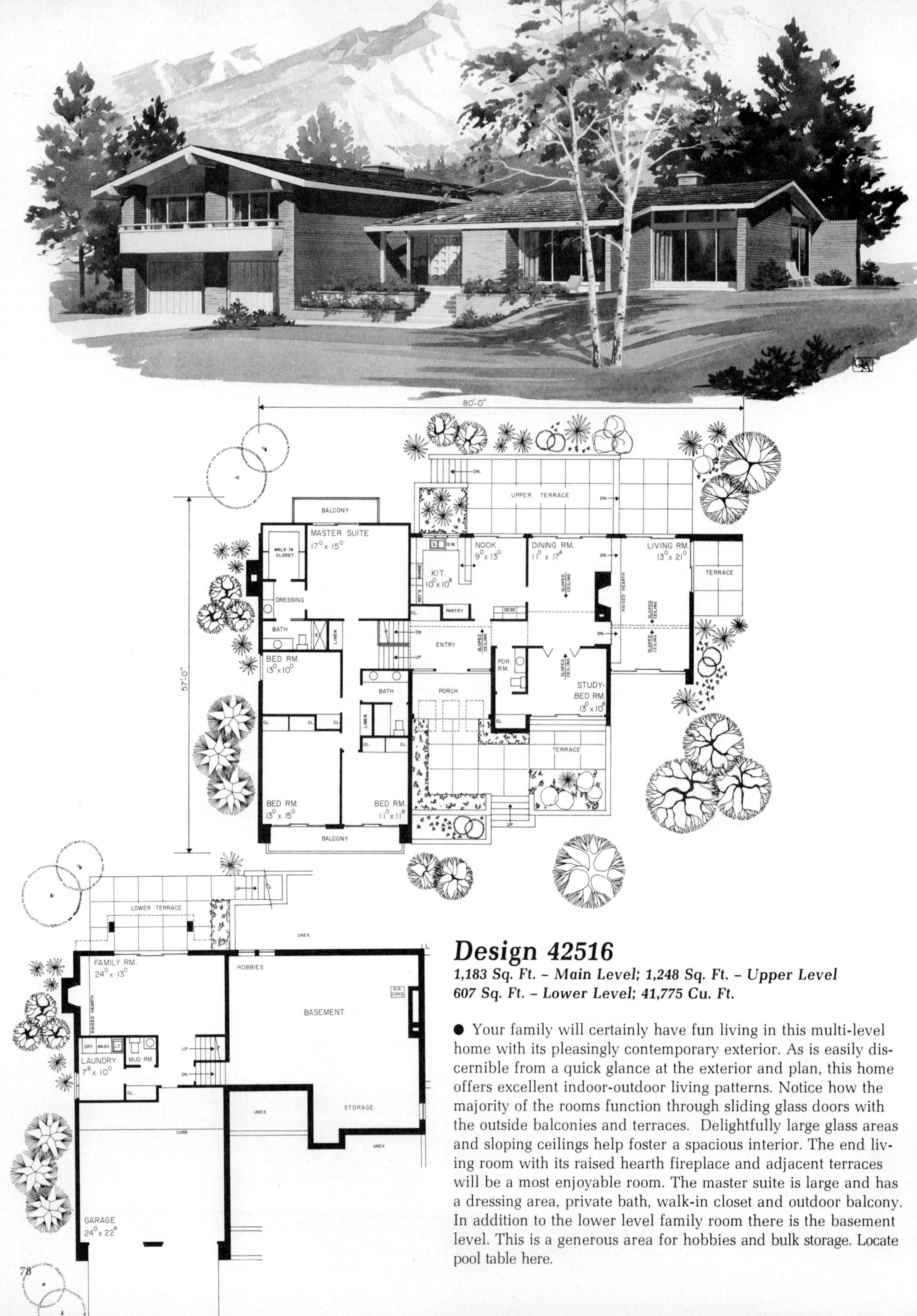

Design 42516

1,183 Sq. Ft. – Main Level; 1,248 Sq. Ft. – Upper Level
607 Sq. Ft. – Lower Level; 41,775 Cu. Ft.

● Your family will certainly have fun living in this multi-level home with its pleasingly contemporary exterior. As is easily discernible from a quick glance at the exterior and plan, this home offers excellent indoor-outdoor living patterns. Notice how the majority of the rooms function through sliding glass doors with the outside balconies and terraces. Delightfully large glass areas and sloping ceilings help foster a spacious interior. The end living room with its raised hearth fireplace and adjacent terraces will be a most enjoyable room. The master suite is large and has a dressing area, private bath, walk-in closet and outdoor balcony. In addition to the lower level family room there is the basement level. This is a generous area for hobbies and bulk storage. Locate pool table here.

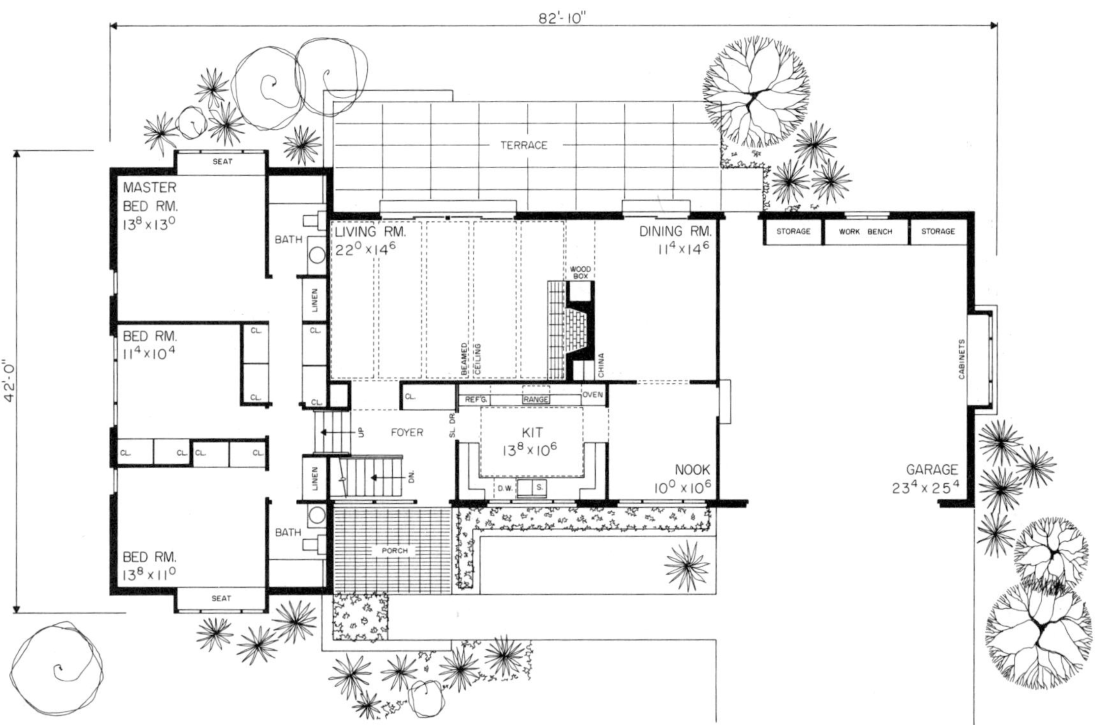

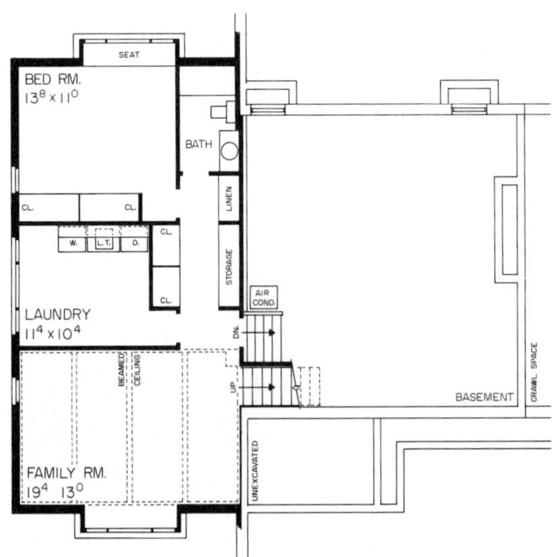

Design 42314

974 Sq. Ft. – Main Level; 826 Sq. Ft. – Upper Level
832 Sq. Ft. – Lower Level; 32,063 Cu. Ft.

● Similar in character to its split-level companion on the opposing page, this contemporary design's living patterns are, interestingly enough, quite different. Living patterns flow to the rear with both the living and dining rooms functioning through sliding glass doors with the big terrace. Separating these two rooms is the attractive fireplace. A handy wood box is at one end; a convenient china closet at the other. The upper level has three bedrooms, two baths and lots of closets. The lower level highlights a fourth bedroom, a large laundry room and beamed ceilinged family room. A basement level permits the development of additional recreational and hobby space for your family's enjoyment.

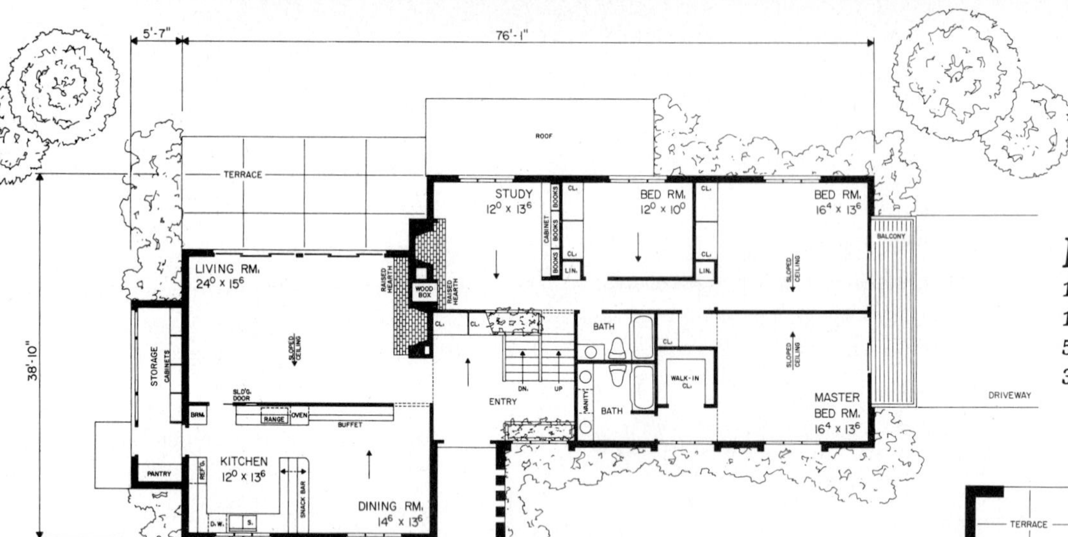

Design 42267

1,063 Sq. Ft. – Main Level
1,197 Sq. Ft. – Upper Level
513 Sq. Ft. – Lower Level
33,557 Cu. Ft.

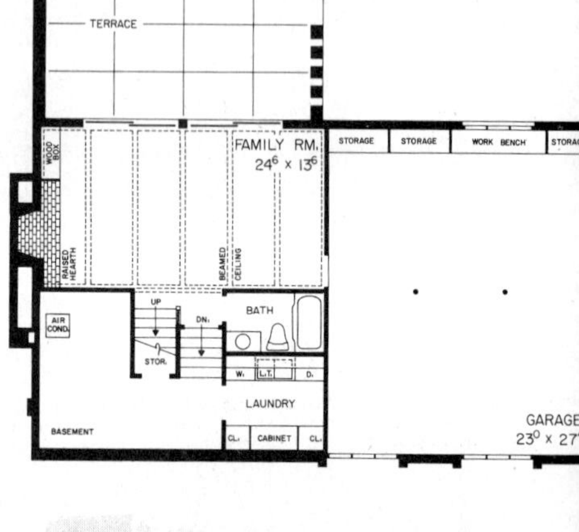

● Your investment in this distinctive four level is sure to be one of the most astute you will ever make. It will serve you and your family for years on end. And it will return dividends in pleasurable daily family living. Of course, as a hedge against inflation the purchase of a new home is an excellent choice. Your money will buy an impressive looking home with a simple, straight-forward appeal. The interesting roof lines, the effective window treatment, the masses of brick and the recessed front entrances are features that strike the note of distinction. Your money will also buy 2,773 square feet of tremendous livability.

Design 42207

960 Sq. Ft. - Main Level
720 Sq. Ft. - Upper Level
640 Sq. Ft. - Lower Level
25,224 Cu. Ft.

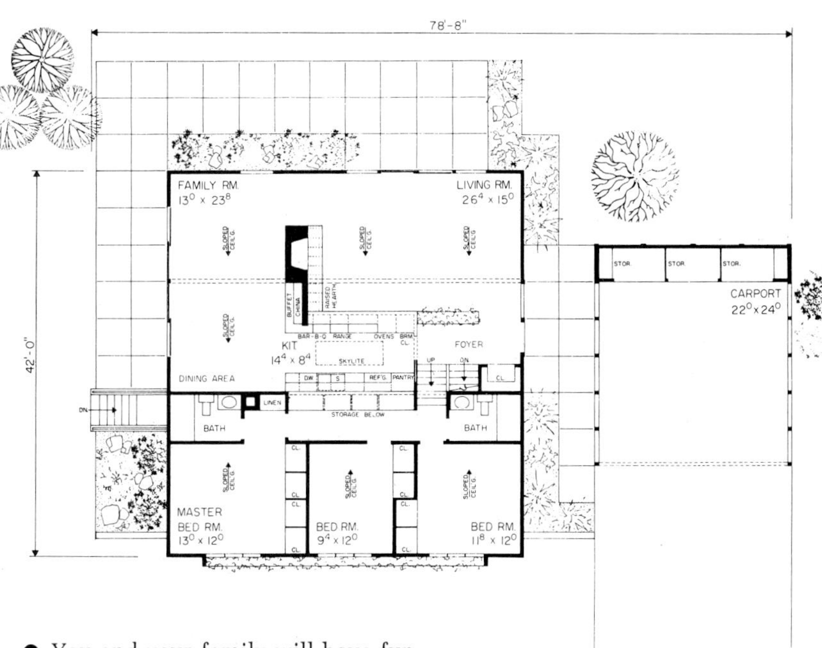

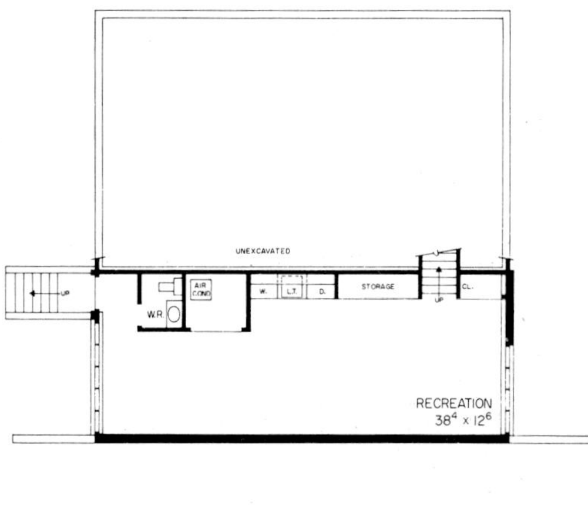

● You and your family will have fun adapting to the living patterns offered by this exciting tri-level. Note the sloping ceilings, kitchen and recreation room to name a few of the many features highlighted in this design.

● Here is an interesting home that has much to offer - both in the manner of exterior design and interior livability. Impressive areas of brick and vertical siding contrast effectively with appealing applications of glass. This interior is, indeed, spacious and forever conscious of the surrounding countryside. Notice how each of the major rooms function through sliding glass doors with the outside. The living room and dining-kitchen area share one big terrace. The upper level bedrooms have their long balcony. And the lower level study (or fourth bedroom, if you wish) and family room have an equally long terrace. Observe the basement level, the sloping ceilings, the sunken kitchen area, the raised hearth fireplace and all the various storage facilities. Don't miss the 3½ baths.

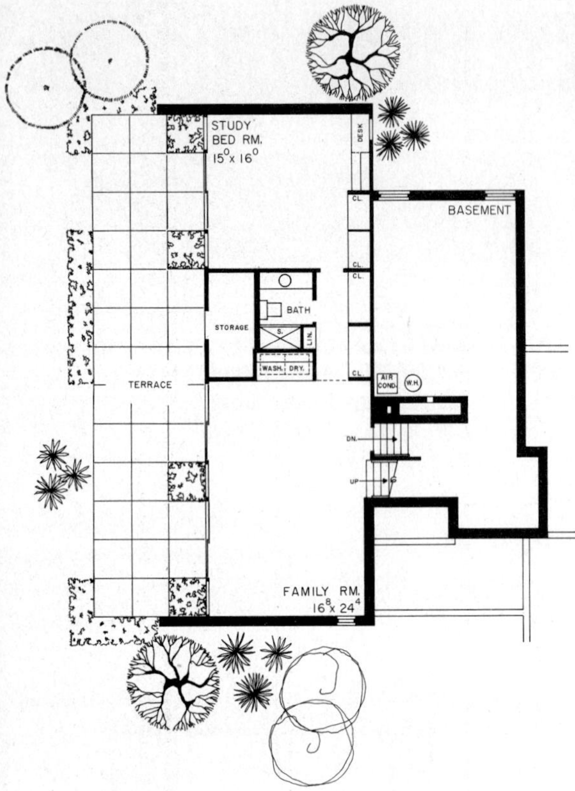

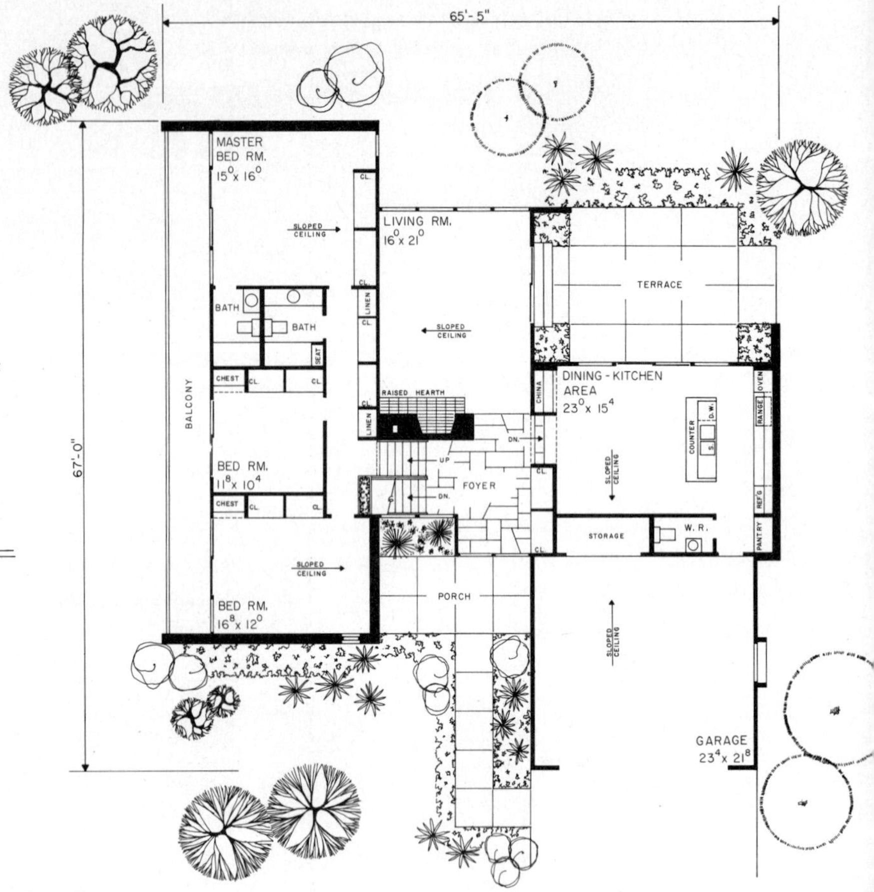

Design 42328

1,036 Sq. Ft. - Main Level; 972 Sq. Ft. - Upper Level
972 Sq. Ft. - Lower Level; 36,877 Cu. Ft.

● If you have been looking for a refreshing contemporary home with unique zoning which caters to the younger generation's activities, give this design much thought. The upper level, reached by passing through the bright and cheerful gallery, is for the teenage set. The playroom area is surrounded by four bedrooms and a vanity - bath - powder room area. Far removed is the parents' master bedroom zone on the main level. The children may pass from the family room directly to their own sleeping quarters, while the parents' quarters are directly accessible from the living room. The projecting living room enjoys the use of both outdoor terraces. The U-shaped kitchen is but a step from the dining room and within reach of the snack bar. An outstanding kitchen in which to function.

Design 42345

1,840 Sq. Ft. - Main Level
1,008 Sq. Ft. - Upper Level
38,260 Cu. Ft.

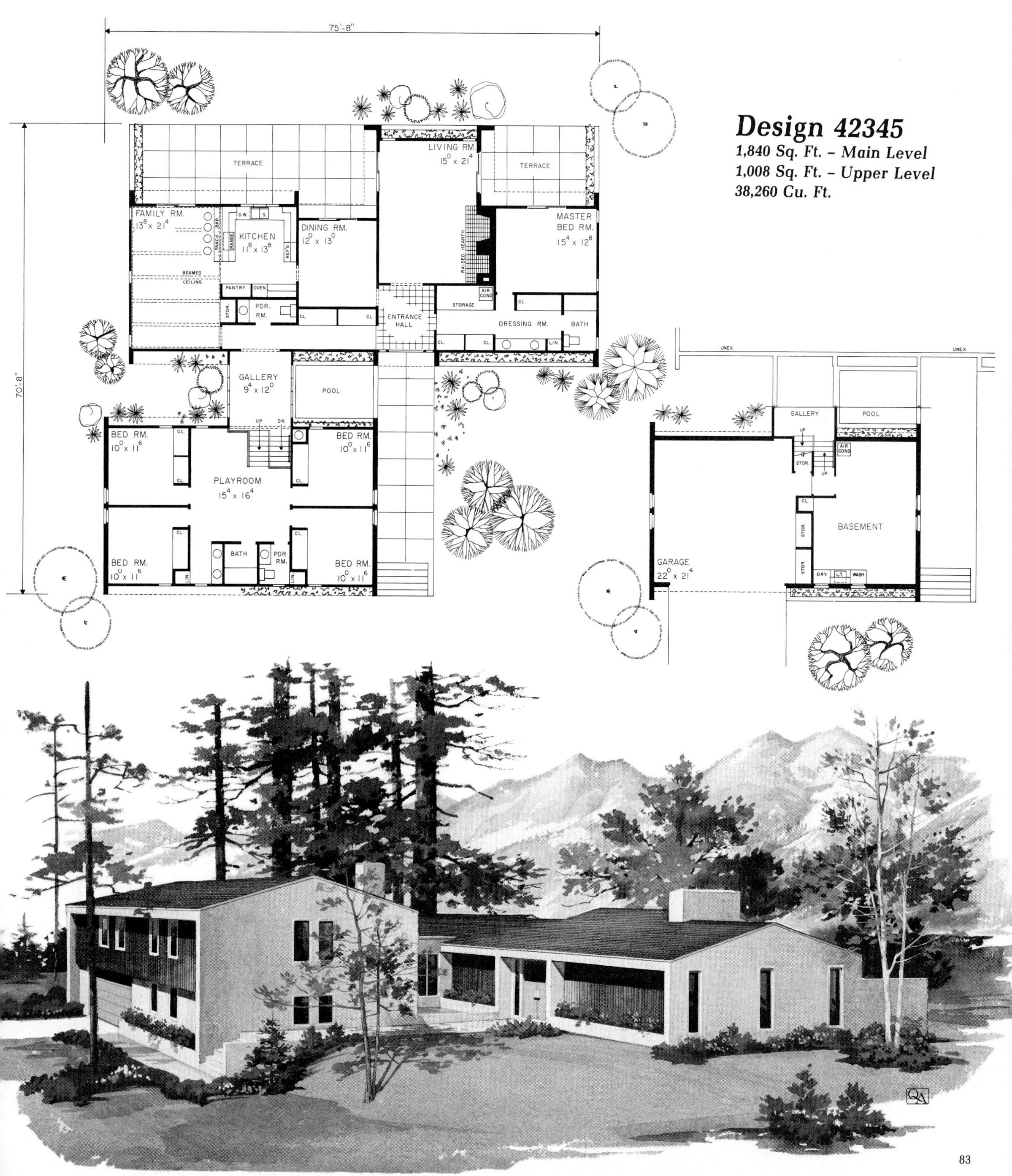

Design 43150

624 Sq. Ft. – Main Level; 768 Sq. Ft. – Upper Level
480 Sq. Ft. – Lower level; 19,157 Cu. Ft.

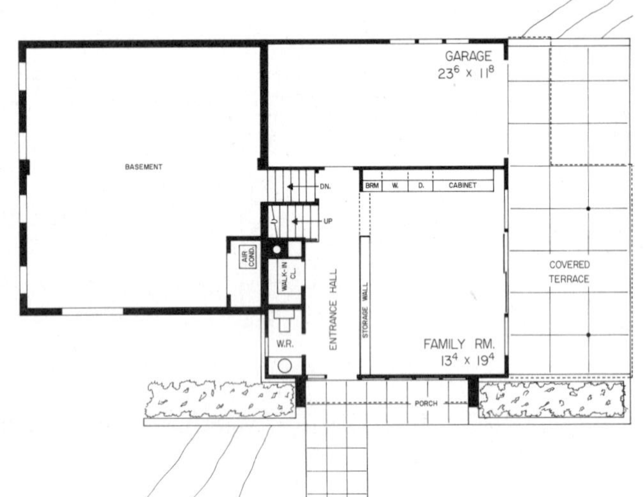

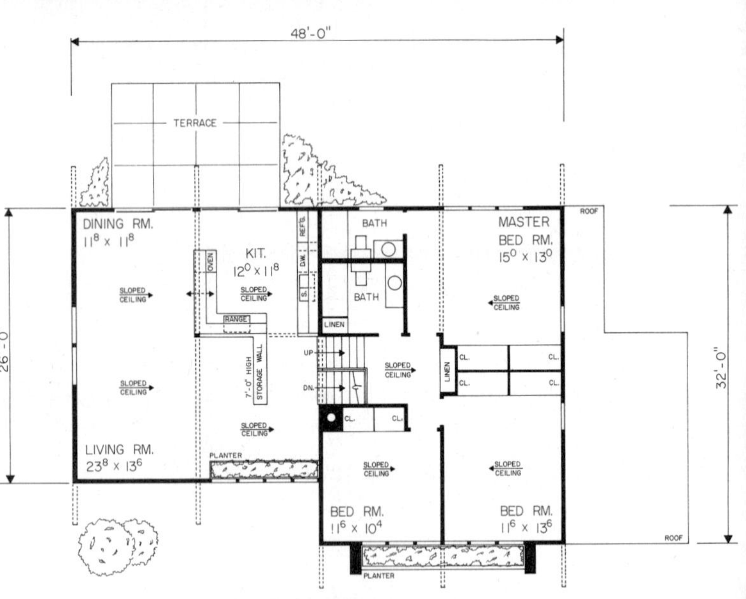

Design 41059

656 Sq. Ft. - Main Level; 952 Sq. Ft. - Upper Level
576 Sq. Ft. - Lower level; 27,232 Cu. Ft.

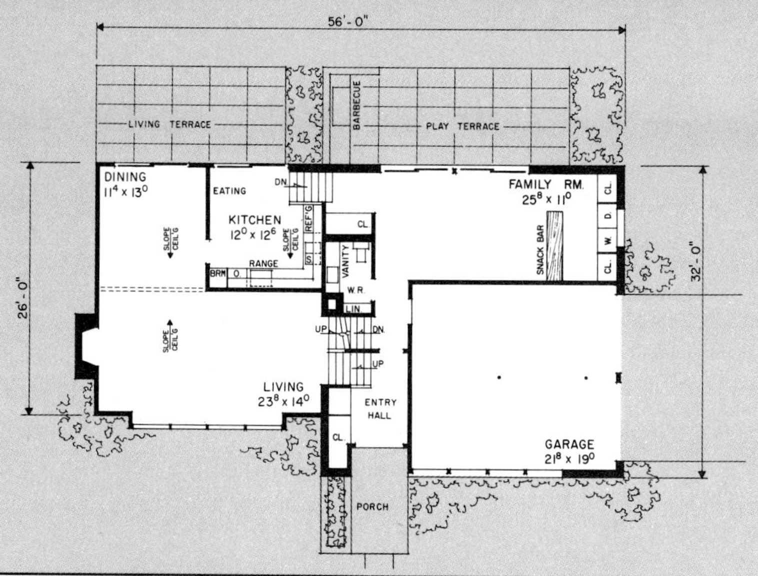

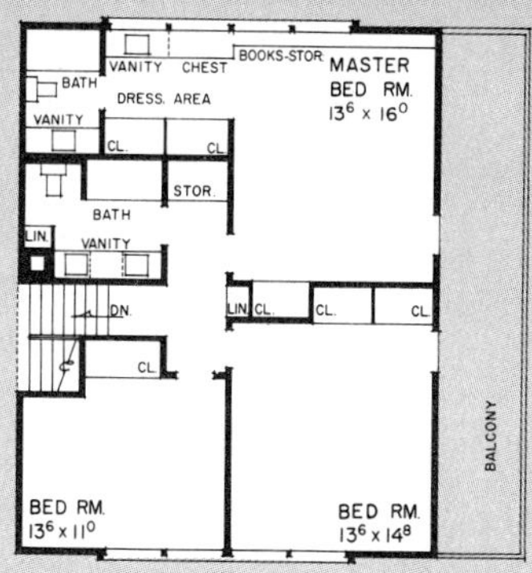

Design 43182

525 Sq. Ft. - Main Level; 639 Sq. Ft. - Upper Level
586 Sq. Ft. - Lower Level; 17,000 Cu. Ft.

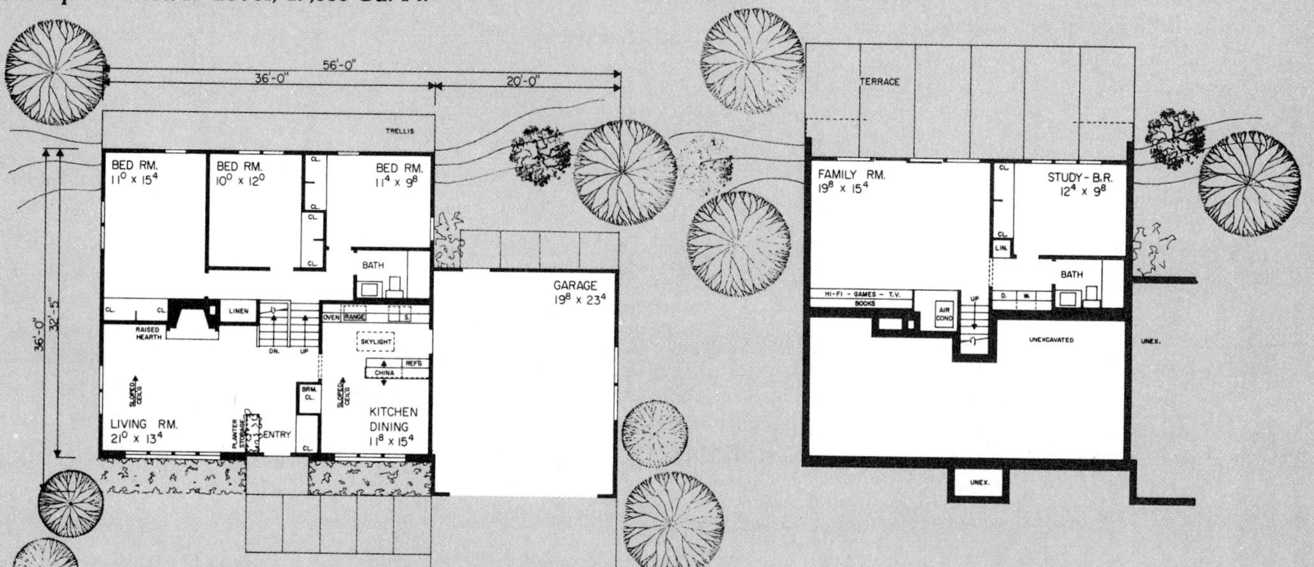

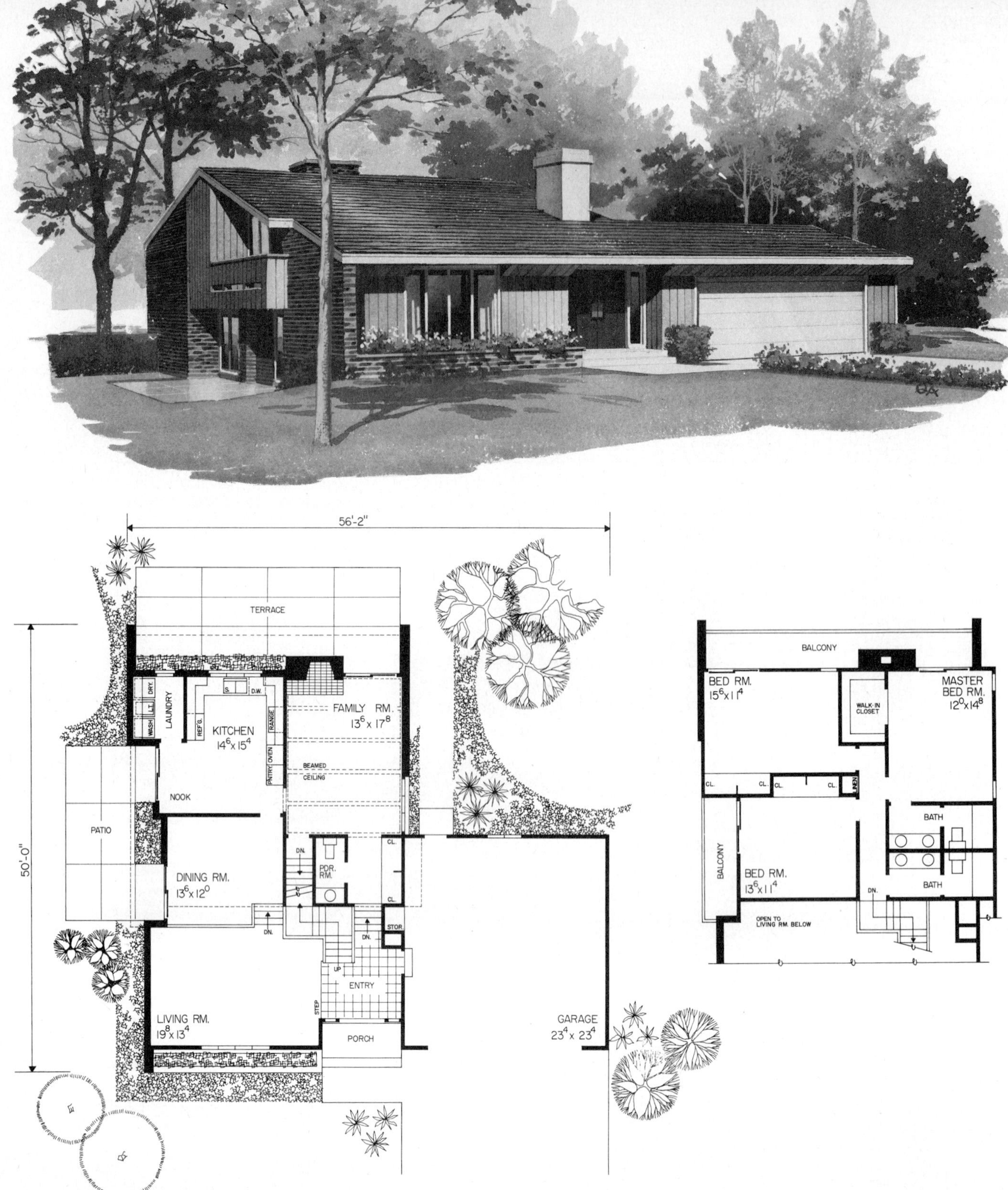

Design 42393 *392 Sq. Ft. – Entry Level; 841 Sq. Ft. – Upper Level; 848 Sq. Ft. – Lower Level; 24,980 Cu. Ft.*

● For those with a flair for something refreshingly contemporary both inside and out. This modest sized multi-level has a unique exterior and an equally interesting interior. The low-pitched, wide-overhanging roof protects the inviting double front doors and the large picture window. The raised planter and the side balcony add an extra measure of appeal. Inside, the living patterns will be delightful! The formal living room will look down into the dining room. Like the front entry, the living room has direct access to the lower level. The kitchen is efficient and spacious enough to accommodate an informal breakfast eating area. The laundry room is nearby. The all-purpose family room has beamed ceiling, fireplace and sliding glass doors to rear terrace. The angular, open stairwell to the upper level is dramatic, indeed. Notice how each bedroom has direct access to an outdoor balcony.

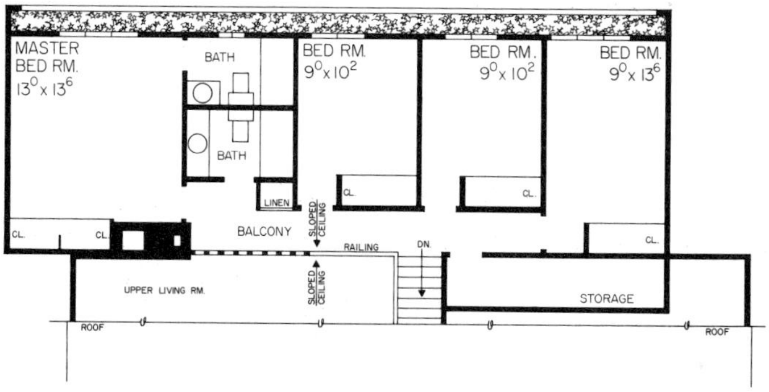

MASTER BED RM. 13^0 x 13^6
BATH
BATH
LINEN
BED RM. 9^0 x 10^2
BED RM. 9^0 x 10^2
BED RM. 9^0 x 13^6
CL.
BALCONY
SLOPED CEILING
RAILING
DN.
UPPER LIVING RM.
STORAGE
ROOF

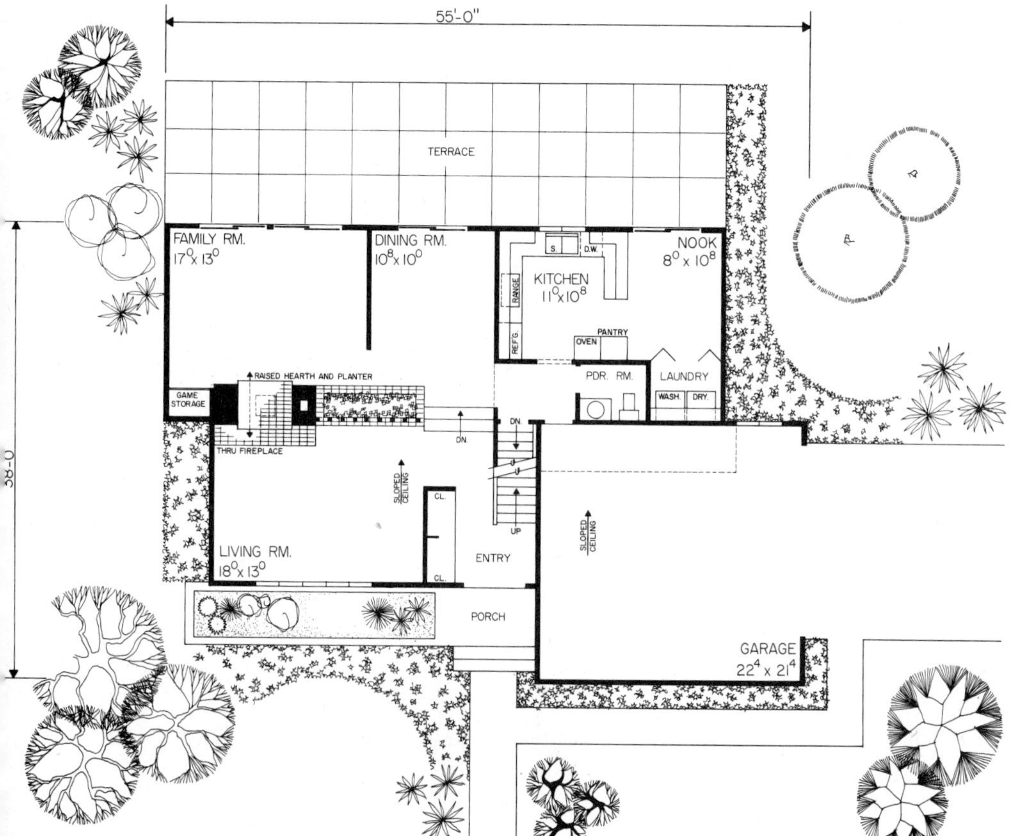

55'-0"

TERRACE
FAMILY RM. 17^0 x 13^0
DINING RM. 10^8 x 10^0
KITCHEN 11^0 x 10^8
NOOK 8^0 x 10^8
RANGE
REFG.
S.
D.W.
PANTRY
OVEN
PDR. RM.
LAUNDRY
WASH.
DRY.
GAME STORAGE
RAISED HEARTH AND PLANTER
THRU FIREPLACE
DN.
UP
LIVING RM. 18^0 x 13^0
ENTRY
PORCH
GARAGE 22^4 x 21^4

Design 42377

388 Sq. Ft. – Living Room Level
782 Sq. Ft. – Main Level
815 Sq. Ft. – Upper Level
22,477 Cu. Ft.

● What an impressive up-to-date multi-level home this is. Its refreshing configuration will command a full measure of attention. Separating the living and slightly lower levels is a thru-fireplace which has a raised hearth in the family room. An adjacent planter with vertical members provides additional interest and beauty. The rear terrace is accessible from nook, family and dining rooms. Notice the powder room, the convenient laundry area and the basement stairs. Four bedrooms serviced by two full baths comprise the upper level which looks down into the living room. A large walk-in storage closet will be ideal for those seasonal items. An attractive outdoor planter extends across the rear just outside the bedroom windows. This will surely be a house that will be fun in which to live.

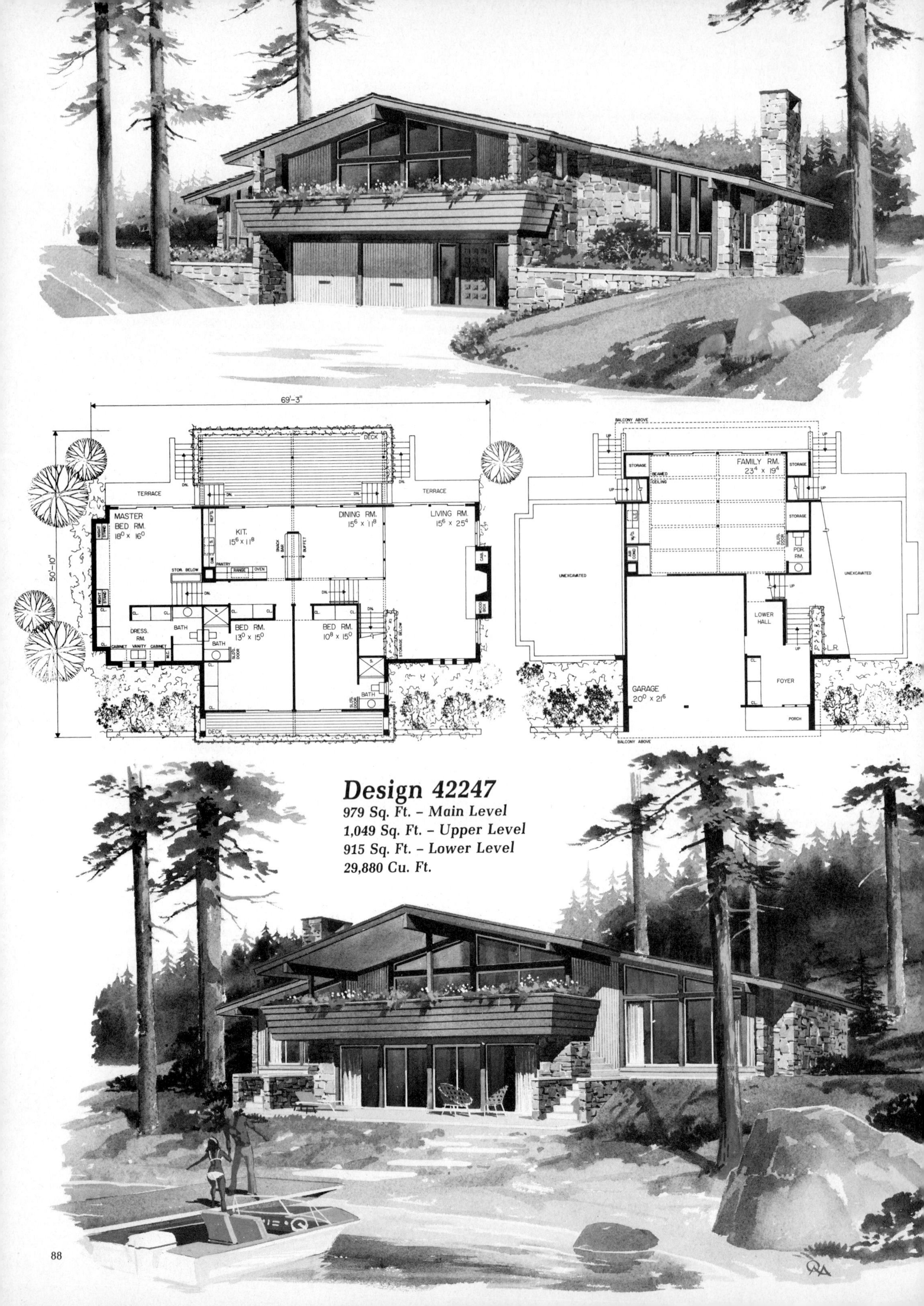

Design 42247

979 Sq. Ft. – Main Level
1,049 Sq. Ft. – Upper Level
915 Sq. Ft. – Lower Level
29,880 Cu. Ft.

Design 42105

419 Sq. Ft. – Main Level; 864 Sq. Ft. – Upper Level
886 Sq. Ft. – Lower Level; 24,168 Cu. Ft.

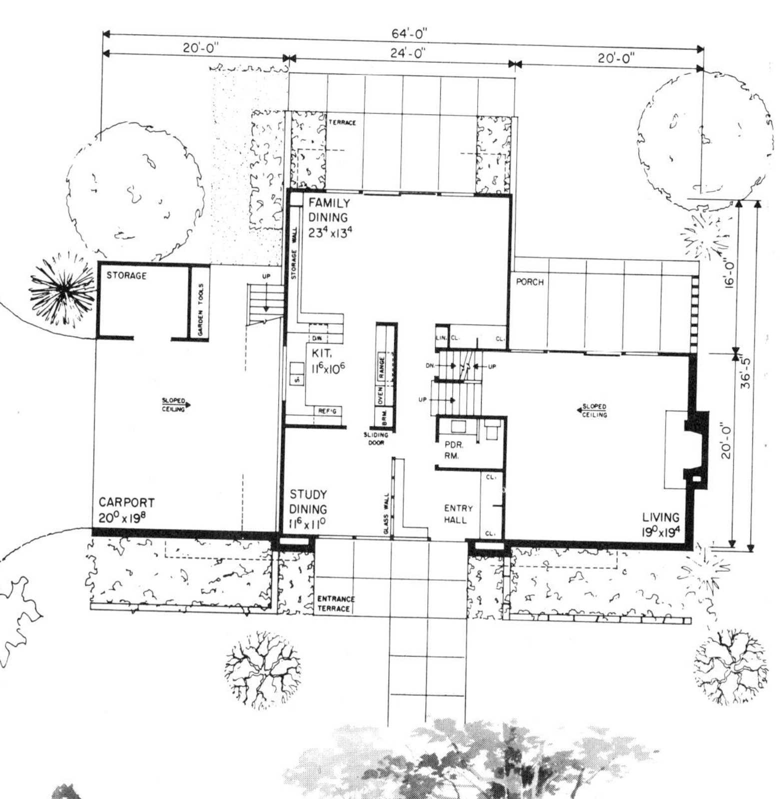

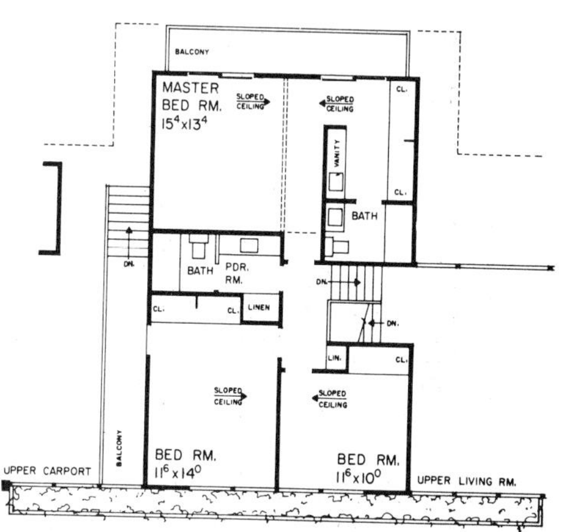

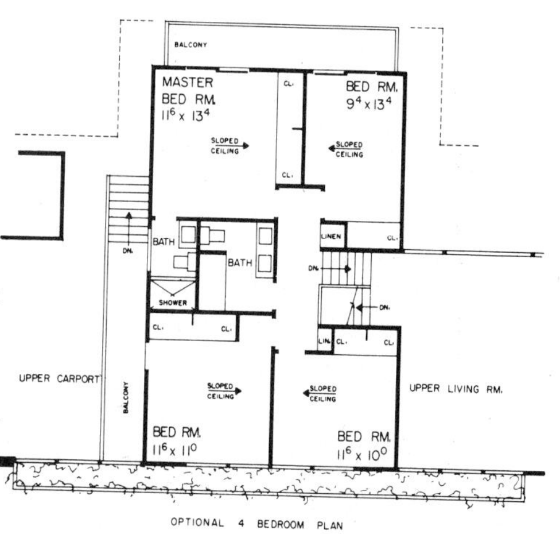

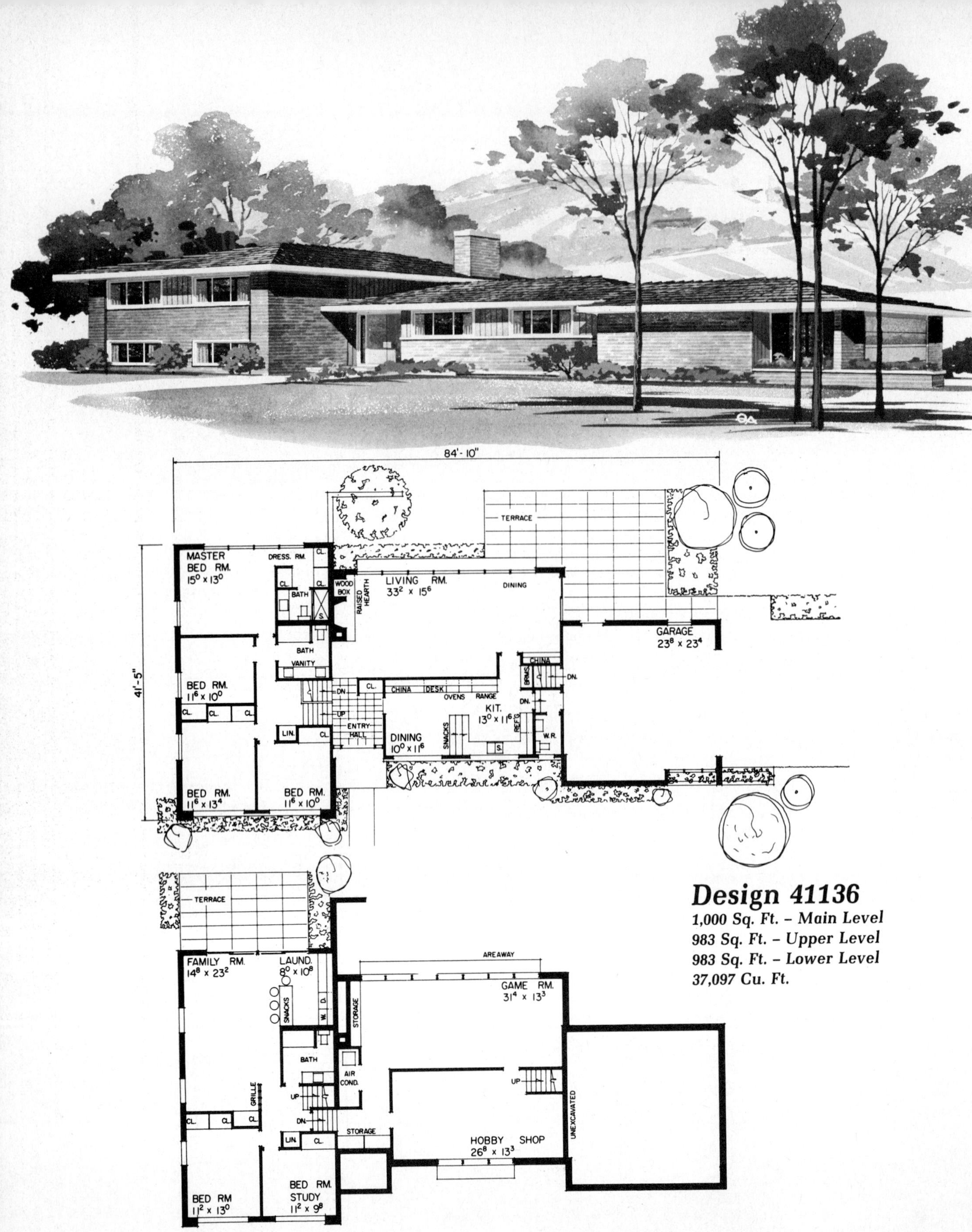

Design 41136

1,000 Sq. Ft. – Main Level
983 Sq. Ft. – Upper Level
983 Sq. Ft. – Lower Level
37,097 Cu. Ft.

● Six bedrooms, three full baths, a family room, a game room, formal and informal areas, a separate laundry, an extra wash room, an outstanding kitchen, a raised hearth fireplace and plenty of storage potential – these are just some of the features to recommend this four level home. And the exterior? Well, one could hardly wish for a more impressive looking home than this. The simple brick masses, the straight-forward window treatment and the wide-overhanging hip roof all add a full measure of eye appeal. Study the exterior carefully for its full appeal.

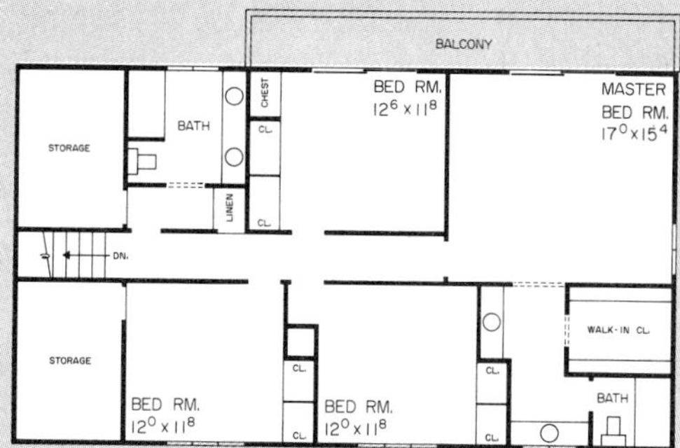

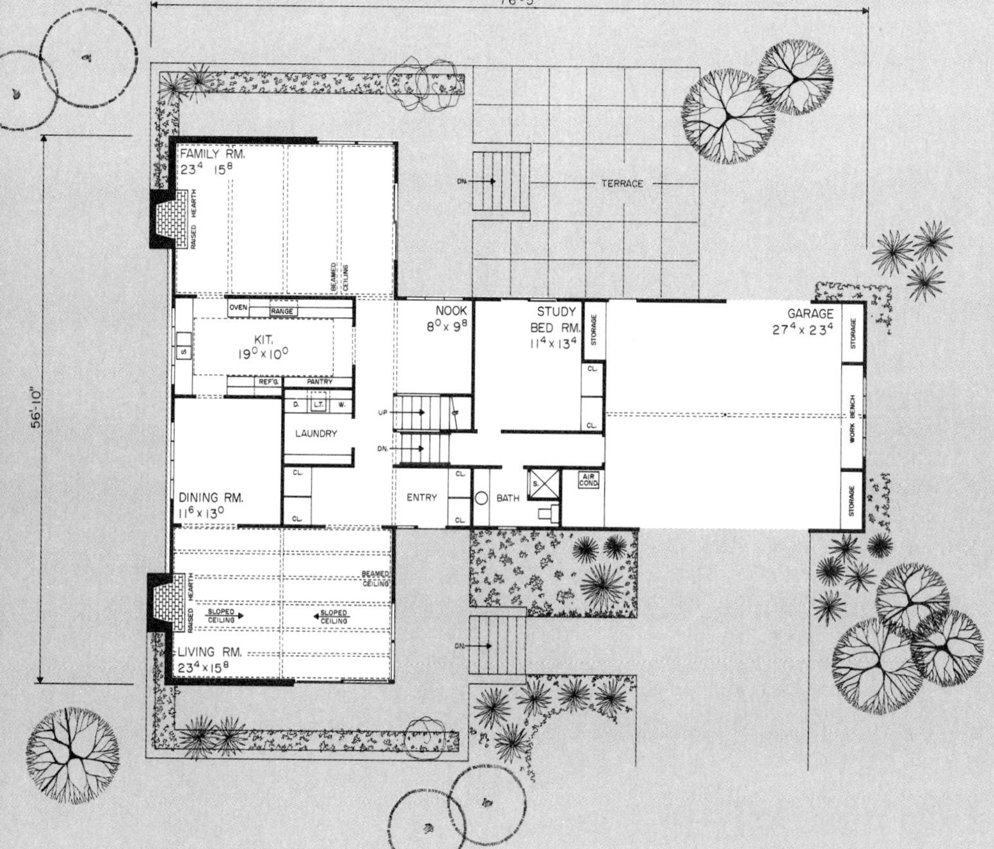

Design 42300

1,579 Sq. Ft. – Main Level
1,176 Sq. Ft. – Upper Level
321 Sq. Ft. – Lower Level
34,820 Cu. Ft.

● A T-shaped contemporary with just loads of livability. You may enter this house on the lower level through the garage, or by ascending the steps to the delightful terrace which leads to the main level front entry. Zoning of the interior is wonderful. Projecting to the front and functioning with the formal dining room is the living room. Projecting to the rear and functioning with the kitchen is the family room. Each of these two living areas features a fireplace, beamed ceiling and sliding glass doors to the outside. Also notice the nook, the laundry and the closet space. On the upper level there are four large bedrooms, two full baths, two storage rooms and an outdoor balcony. The lower level offers that fifth bedroom with a full bath nearby. Don't miss the storage facilities of the garage. Truly fine livability.

A Great, Contemporary Living Experience

● If you have a sloping site and are looking for a multi-level design that will provide you and your family with all kinds of livability, you'd better give this dramatic contemporary a second, and even a third, look. Should your property have a view worth enjoying from the main level living areas, the large expanses of glass would undoubtedly be your most favorite of features. Notice how the glass wall is pushed outward like the prow of a ship. This also permits an extra measure of visiability laterally. Generous glass areas in other parts of the house contribute to the spaciousness. The windows, brick masses and vertical siding set up interesting design contrasts.

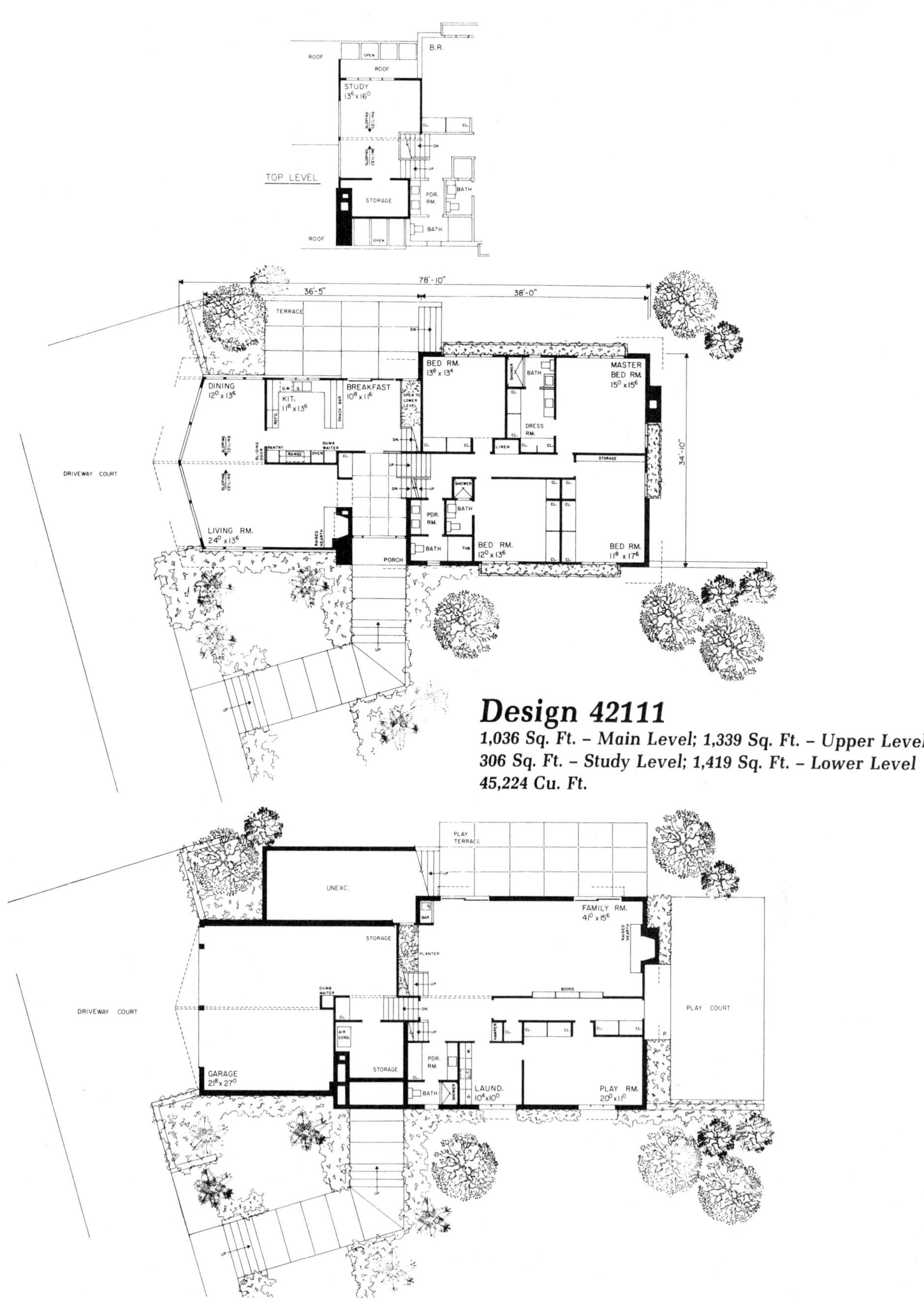

Design 42111

1,036 Sq. Ft. - Main Level; 1,339 Sq. Ft. - Upper Level
306 Sq. Ft. - Study Level; 1,419 Sq. Ft. - Lower Level
45,224 Cu. Ft.

● This home features five distinct level and they all make their contribution to wonderful living patterns. As you enter the front door you are on the main living level where the living, dining, kitchen and breakfast areas are located. A short flight of stairs leads to the four bedroom, three bath sleeping level. Up a few stairs from this level the study is located. This will be a favorite area for retreat from the household commotion. From the front entry on the main level, two short sets of stairs lead to the lower level. Here are the family and play rooms, plus laundry and another bath. The lowest level - the garage - is down a few more steps from this recreation level.

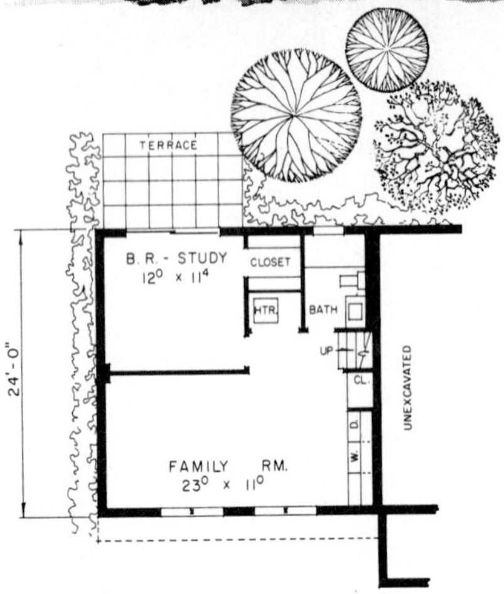

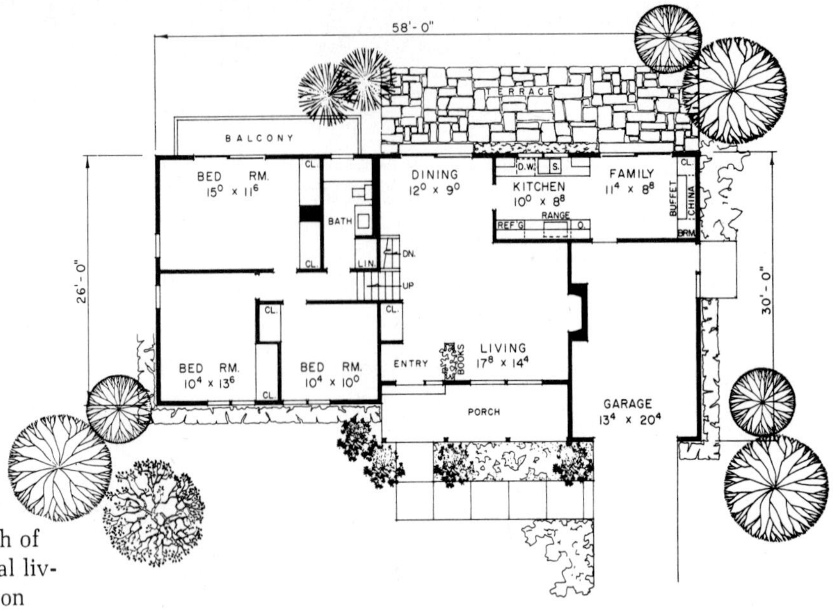

● This is an exciting house. Each of the three levels offers exceptional livability. Count the extra features on each level. The balcony will be a nice feature for the rear bedroom. There is a built-in planter with books below in the living room plus a built-in china cabinet/buffet and closets in the family-dining area.

Design 43217

612 Sq. Ft. – Main Level; 624 Sq. Ft. – Upper Level
576 Sq. Ft. – Lower Level; 17,911 Cu. Ft.

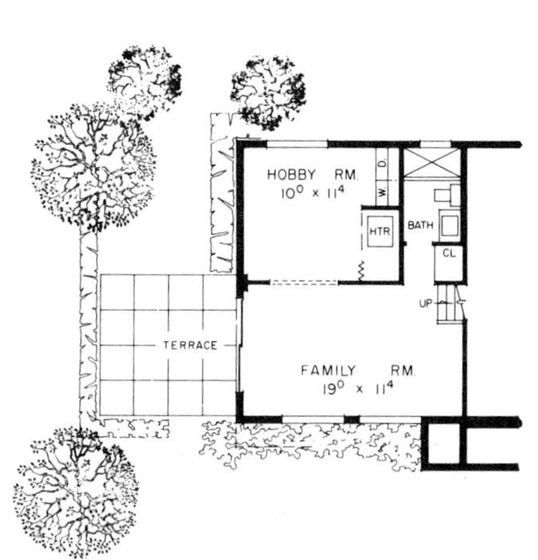

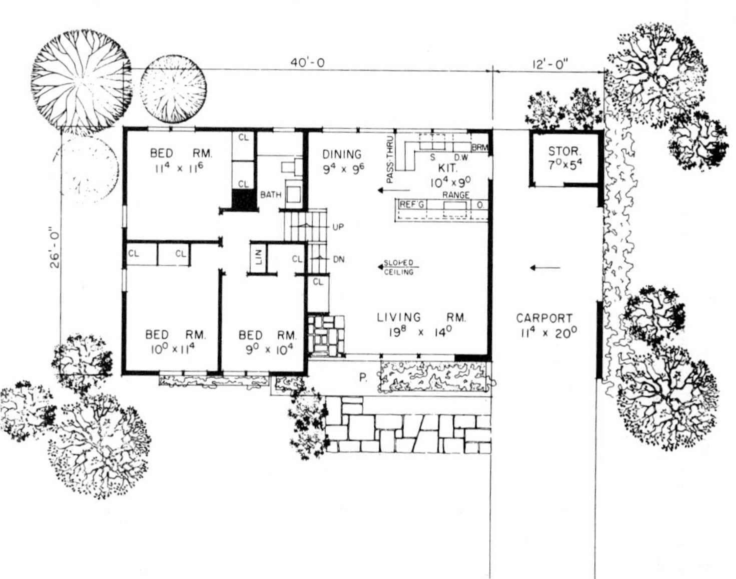

● Contemporary tri-level living that surely will be fun for everyone. Don't miss the lower level family and hobby rooms, plus the extra full bath. The dining room, kitchen and living room will enjoy the openness of sloped ceilings. A convenient storage unit is located in the rear of the carport. Three bedrooms and a full bath are on the upper level.

Design 43216

482 Sq. Ft. – Main Level; 520 Sq. Ft. – Upper Level
480 Sq. Ft. – Lower Level; 15,822 Cu. Ft.

Design 42358

1,007 Sq. Ft. – Main Level; 760 Sq. Ft. – Upper Level
648 Sq. Ft. – Lower Level; 25,685 Cu. Ft.

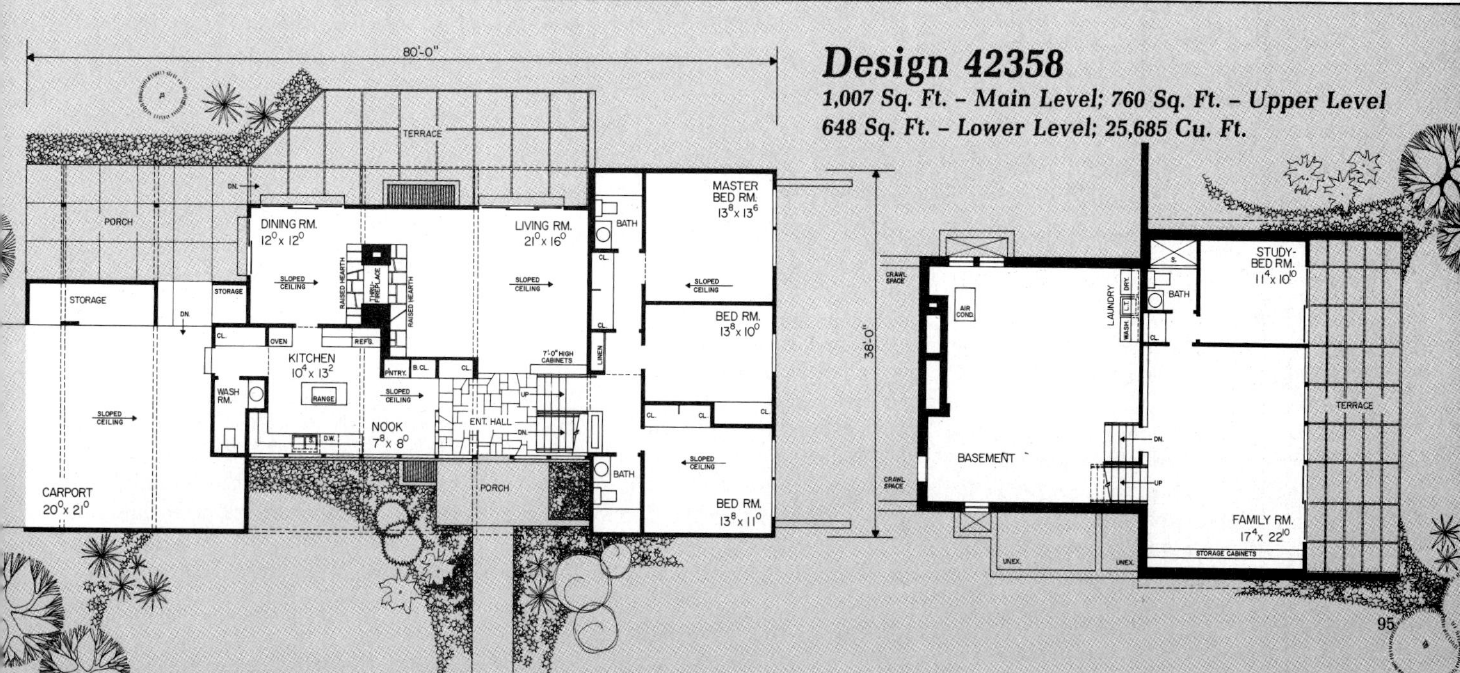

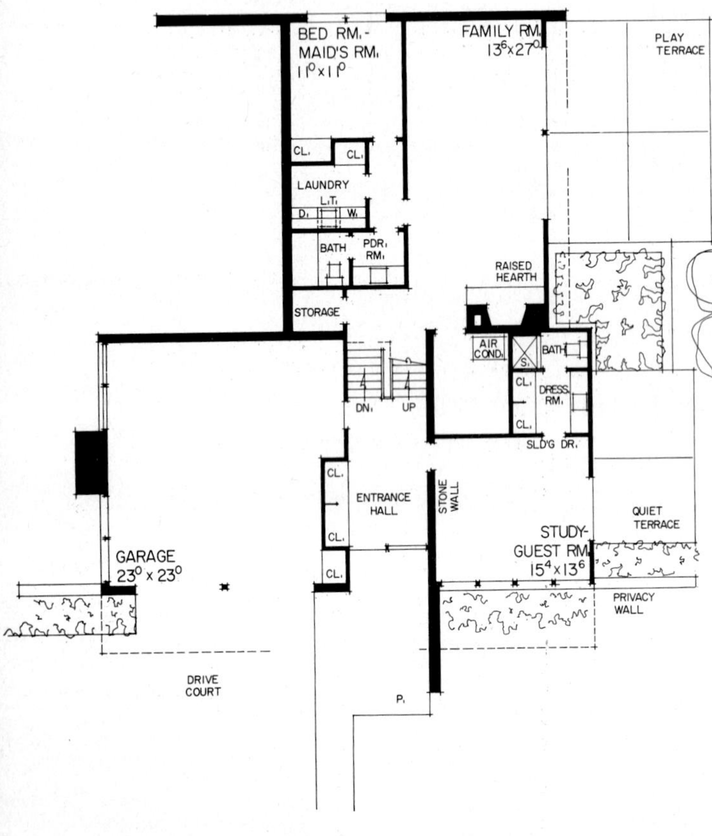

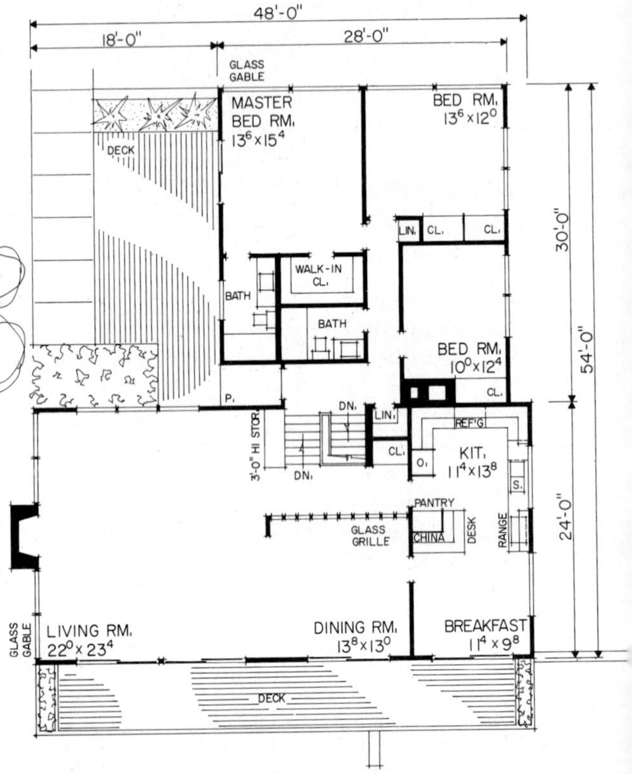

Design 41910

506 Sq. Ft. – Entrance Level; 1,152 Sq. Ft. – Main Level
840 Sq. Ft. – Upper Level; 852 Sq. Ft. – Lower Level; 34,466 Cu. Ft.

● The multi-level design of this house was especially planned to step down a sloping site. Each of the four levels is enhanced and expanded to the outdoors with sliding glass doors leading to decks or terraces. This makes indoor-outdoor living exceptional in this home. The house is also well-suited to a large family. There are three bedrooms on the upper sleeping level, and the two bedrooms with baths on the lower level (which can be used in a diversified way). The lower level family room opens to a play terrace, making it possible for children to run in and out without disturbing the household. There is symmetry of the windows in the large living room.

Design 41840

896 Sq. Ft. - Main Level
952 Sq. Ft. - Upper Level
952 Sq. Ft. - Lower Level
36,113 Cu. Ft.

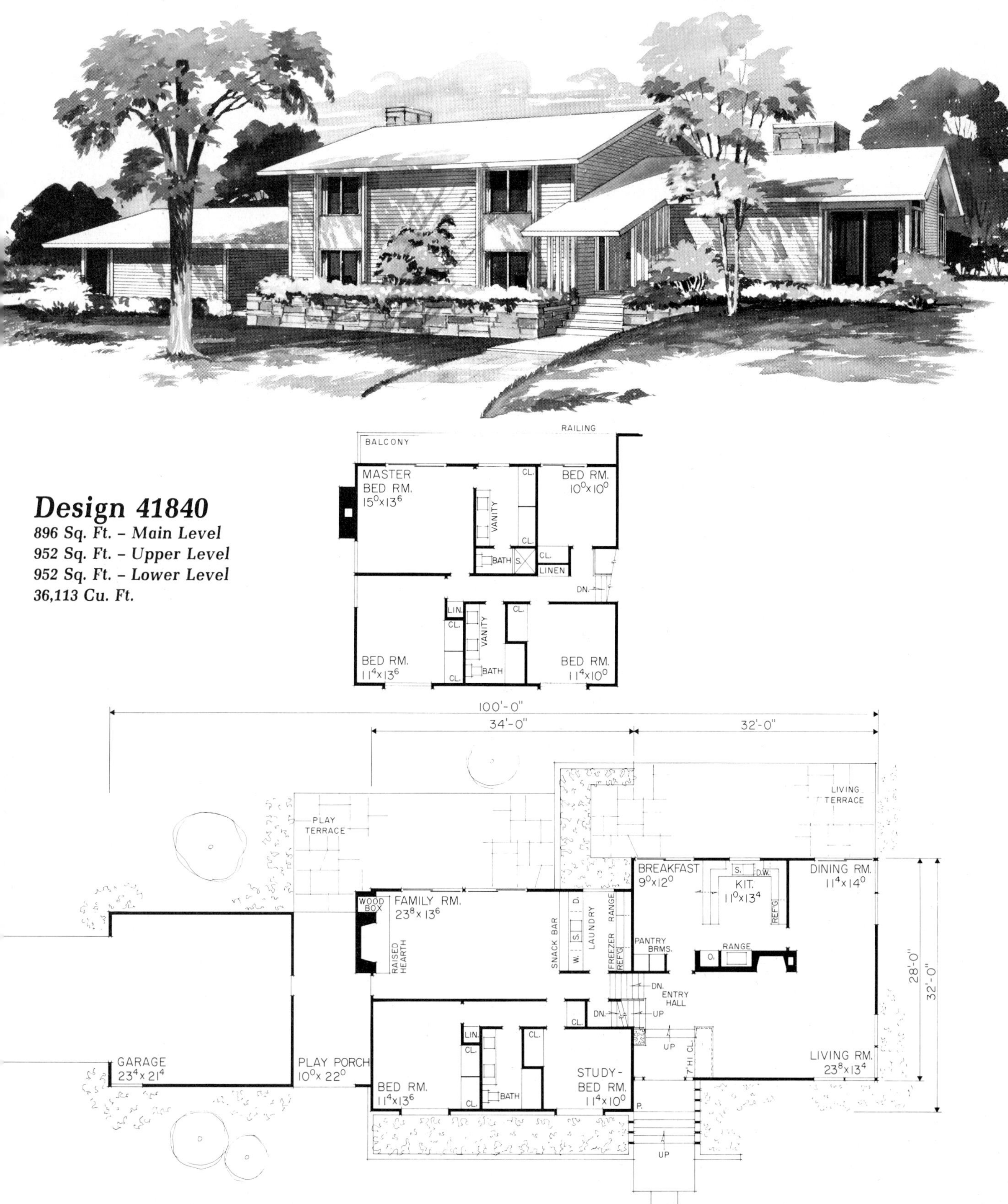

● It takes only a quick look at this plan to determine that it was designed with the large family in mind. The upper level has four bedrooms and two full baths while the lower level has two more bedrooms and another full bath. Imagine, six bedrooms in all! The family will have a lot of fun on the balcony which looks down into the lower terrace. A play porch between the lower level and garage provides a sheltered area for young children. There are two spacious living areas each with a fireplace and plenty of natural light. The excellent kitchen is conveniently flanked by the two eating areas - the informal breakfast room and the formal dining room.

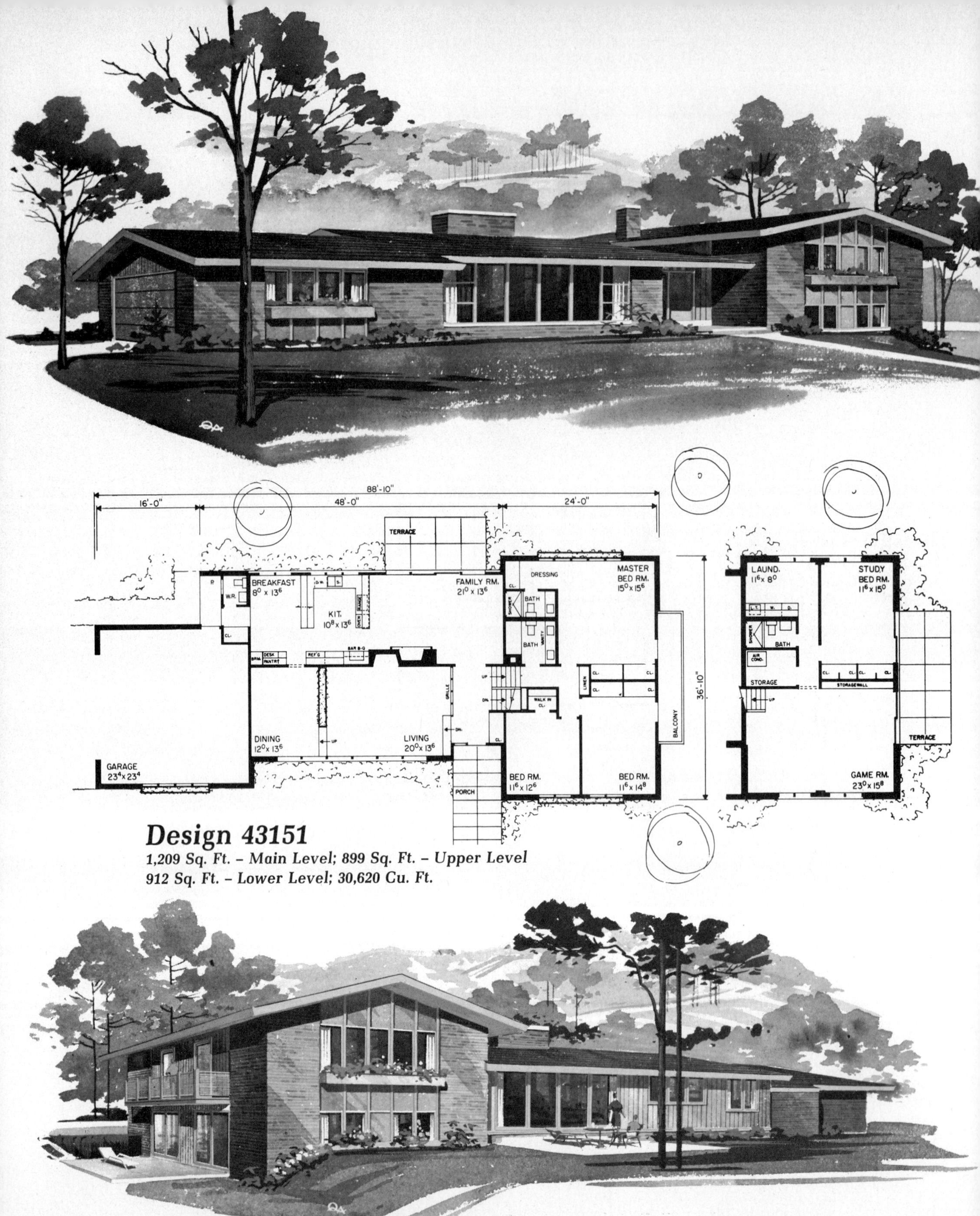

Design 43151

1,209 Sq. Ft. – Main Level; 899 Sq. Ft. – Upper Level
912 Sq. Ft. – Lower Level; 30,620 Cu. Ft.

● Split-level living can be great fun. And it certainly will be for the occupants of this impressive house. First and foremost, you and your family will appreciate the practical zoning. The upper level is the quiet sleeping level. List the features. They are many. The main level is zoned for both formal and informal living. Don't miss the sunken living room or the twin fireplaces. The lower level provides that extra measure of livability for all to enjoy.

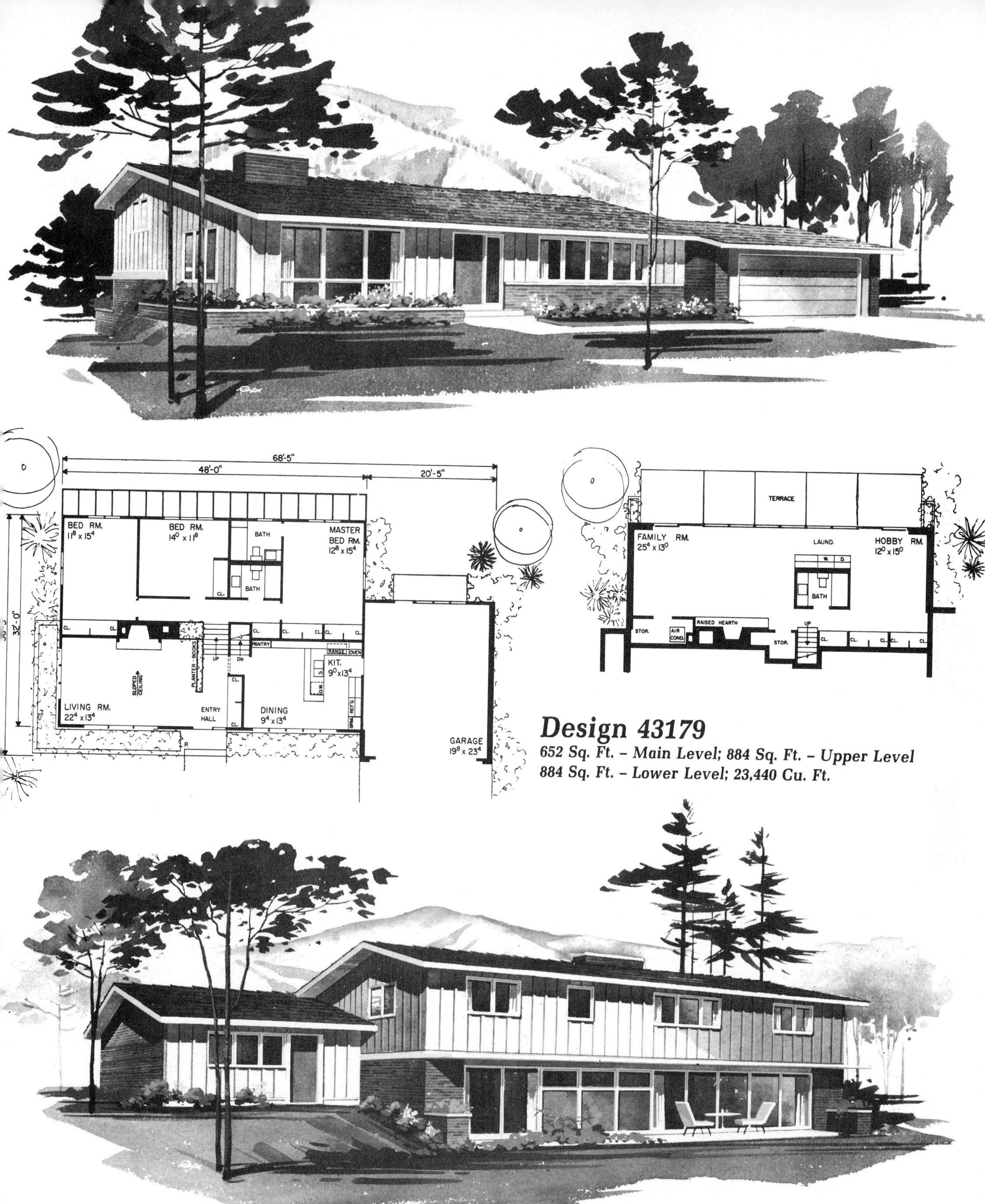

Design 43179

652 Sq. Ft. - Main Level; 884 Sq. Ft. - Upper Level
884 Sq. Ft. - Lower Level; 23,440 Cu. Ft.

● This tri-level home has four distinct areas, each performing its function to perfection. The sleeping area has two full baths, three big bedrooms and plenty of closets. The living area of the main level is spacious, has good light and is free of cross-room traffic. The dining-kitchen is efficient and lends itself to formal and informal dining. The lower level is bright, cheerful, has plenty of space and functions with the outdoors. Note extra full bath.

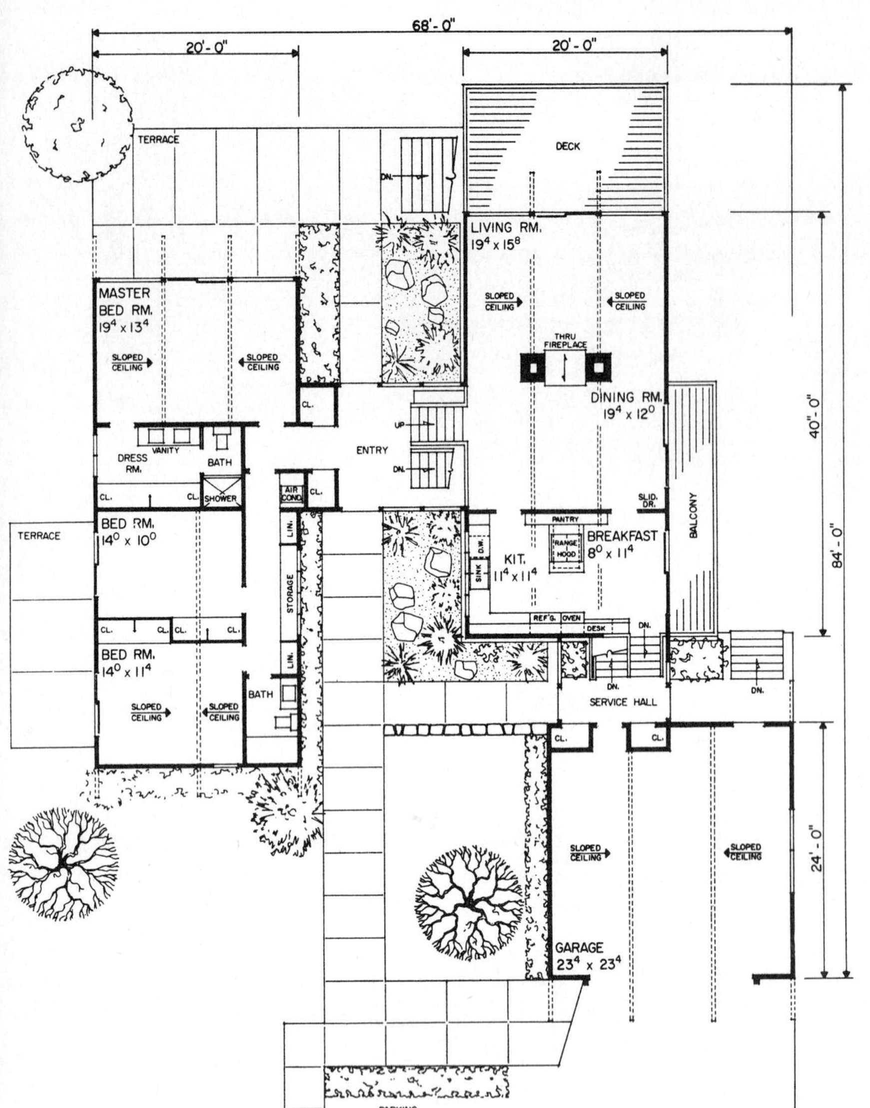

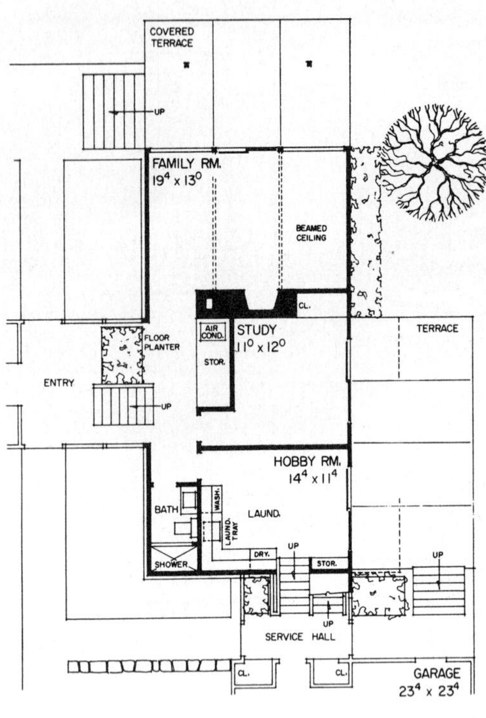

Design 41973

1,112 Sq. Ft. – Sleeping Level
896 Sq. Ft. – Upper Level
820 Sq. Ft. – Lower Level
27,723 Cu. Ft.

● Should you be in the market for something different, without it being outlandish, then don't let this dramatic tri-level get by you. Somewhat similar to the cluster idea, its three distinctly separate low-pitched, wide overhanging roofs are decidedly dramatic. There is, however, nothing fragmented about the delightful living patterns it offers. When bedtime approaches there is the sleeping wing integrated with its own peaceful outdoor terraces. There is most of the upper level for formal family living and dining. Again, outdoor living areas function conveniently. The kitchen is efficient and functions with the breakfast room. Directly below the kitchen on the lower level is the laundry which occupies a corner of the hobby room. A full bath on the lower level is a great convenience.

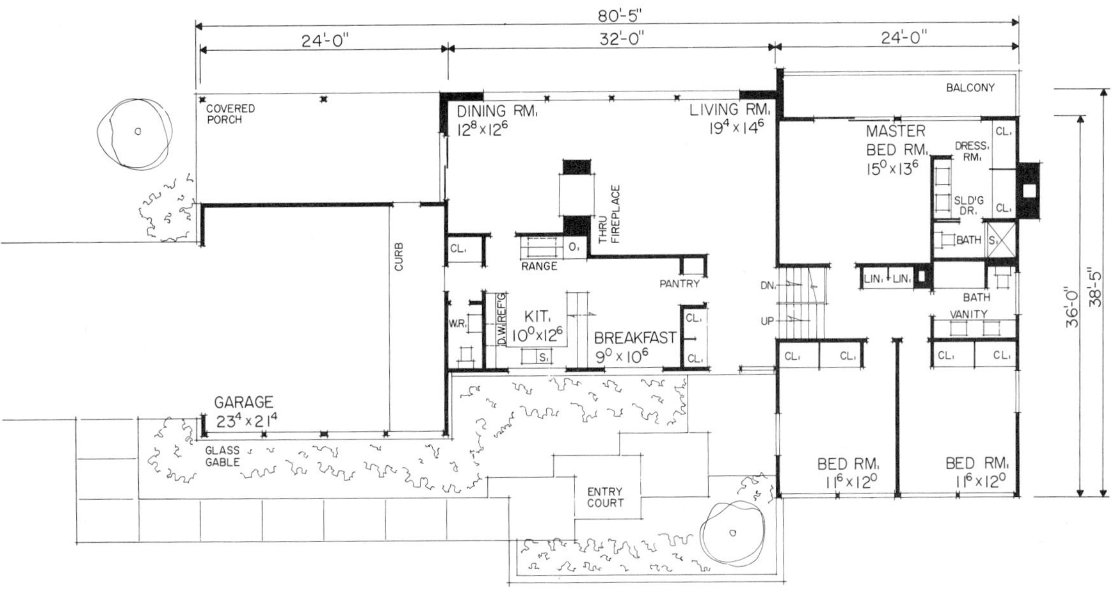

Design 41894

890 Sq. Ft. – Main Level; 864 Sq. Ft. – Upper Level 816 Sq. Ft. – Lower Level; 25,237 Cu. Ft.

● This striking contemporary house incorporates much space-stretching planning into its design. For example in the combination living room-dining room, the eye travels through the two-way fireplace which serves to separate the two areas. The visual adding of the outdoors to the plan by means of large glass areas serves to enlarge the house, too. Other bonuses are the combination study-bedroom, family room with its own terrace and full bath on the lower level. This floor could conveniently form a suite for a live-in relative.

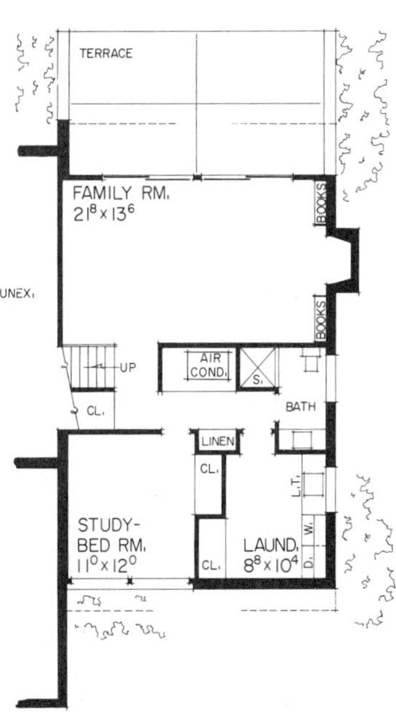

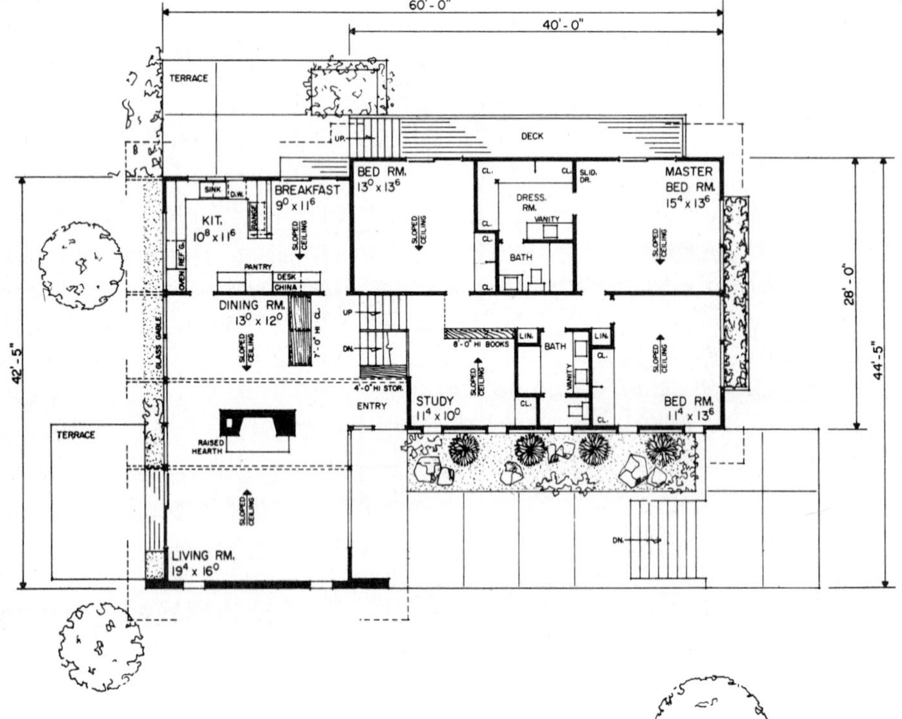

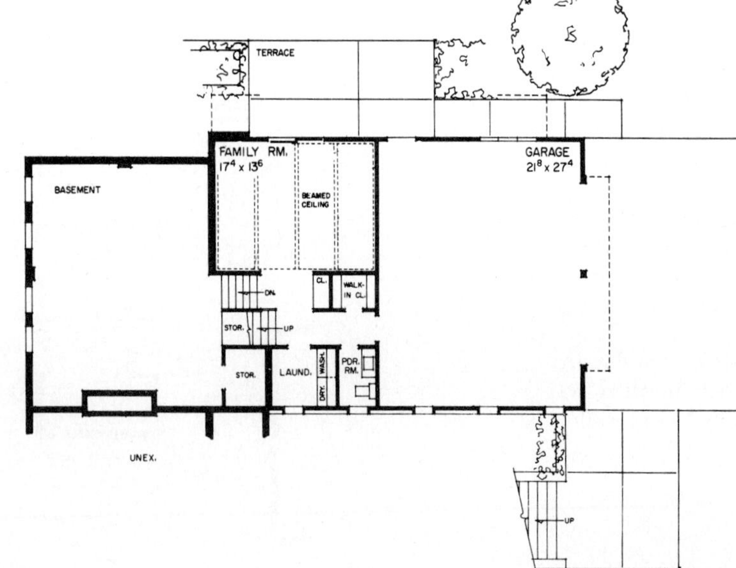

Design 41983

932 Sq. Ft. – Main Level
1,036 Sq. Ft. – Upper Level
466 Sq. Ft. – Lower Level
31,517 Cu. Ft.

● Make no mistake about it, what you do with the "holes" of a house will have a profound effect on its final character. The unique appeal of this contemporary can be traced in large part to the design impact of the windows. When handled with finesse, the various glass areas of the house can be dramatic, indeed. The delightful shadow box effect of the tall vertical glass areas catch the eye immediately. It then shifts to the simple yet appealing detail around the front door. Inside, there is an abundance of space with a fine indoor-outdoor living relationship. The main level with its sloping ceilings and glass gable enjoys a fine feeling of spaciousness. The upper level with its three bedrooms, study and two full baths also has ceilings that slope. The master bedroom features an adjacent dressing room with a vanity and an abundance of wardrobe storage. The lower level offers the family room with beamed ceiling and sliding glass doors to the terrace. Note powder room and laundry. The basement is a large area for the heating and cooling equipment and pursuit of hobbies.

Design 41975

1,424 Sq. Ft. - Main Level
1,064 Sq. Ft. - Upper Level
396 Sq. Ft. - Lower Level
32,591 Cu. Ft.

● Here is a multi-level that provides living that would be difficult to top. Even a quick look at the plan reveals an arrangement that has just got to be enjoyable in which to live. This T-shaped plan is particularly well-zoned. The two large segments are attached by a spacious and dramatic entry. From here traffic flows directly to each area of the house. Notice the size of the living areas with their fireplaces and direct access to outdoor living. Don't miss the breakfast and dining rooms. The upper level features three bedrooms, a study and two full baths. The lower level has a multi-purpose room, a laundry and a powder room. Then there is the basement. There will be a constant awareness of the surrounding environment with all those outdoor living areas. The large terrace services the family, living and dining rooms. The wood deck is accessible from both the family and breakfast rooms. It is also accessible from the balcony which provides the upper level with an exciting, commanding view of the rear yard. The lower level multi-purpose room has access to terrace below balcony.

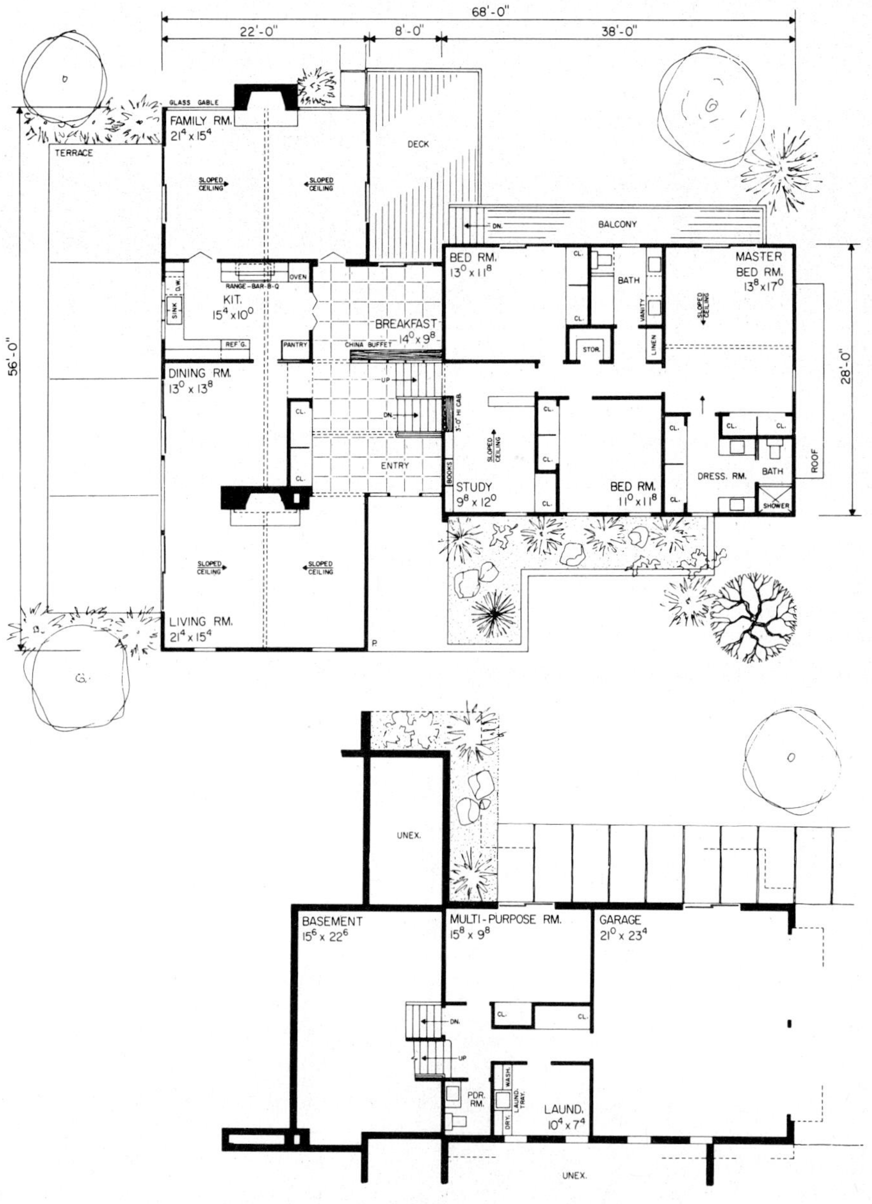

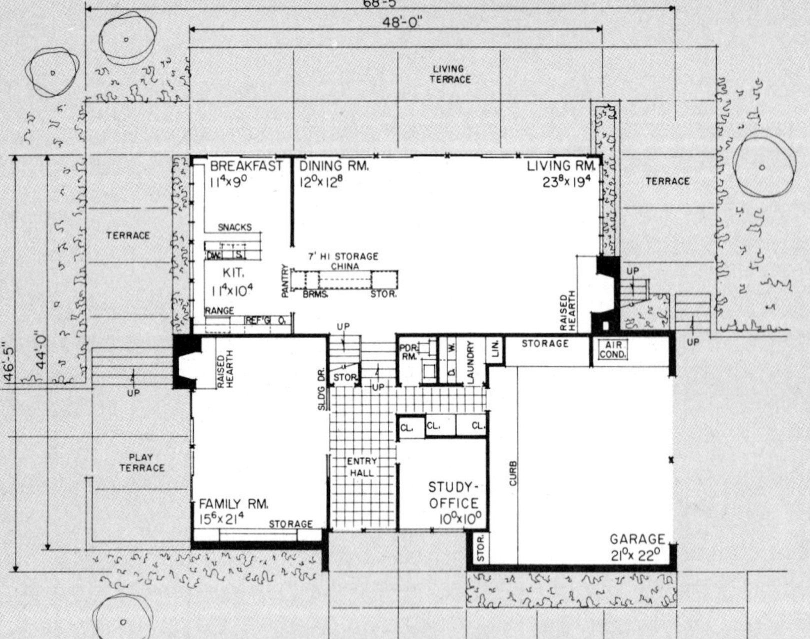

Design 41861

960 Sq. Ft. – Main Level
1,066 Sq. Ft. – Upper Level
794 Sq. Ft. – Lower Level
29,238 Cu. Ft.

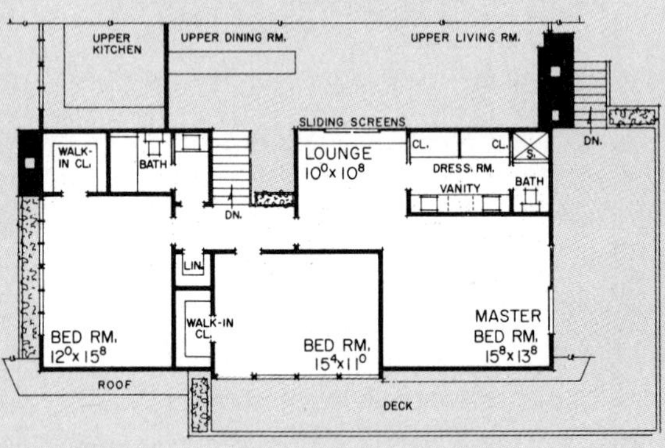

● Here is an exquisitely different design for those with contemporary tastes. It is boldly dramatic, yet pleasing to the eye. Study the floor plans carefully; for the living patterns they offer will be distinctive, indeed. They will be fun, too. Imagine your family occupying the family room and study of the lower, grade level; the spacious living area and efficient kitchen of the main level; the three bedroom, lounge and two baths of the upper level. It will be easy to envision the enjoyment offered by the deck. Notice the sliding screens of the upper level lounge. It is possible to look down into the living area from this spot. The ceilings of the main level are sloping and with all that glass there is a wonderful feeling of spaciousness.

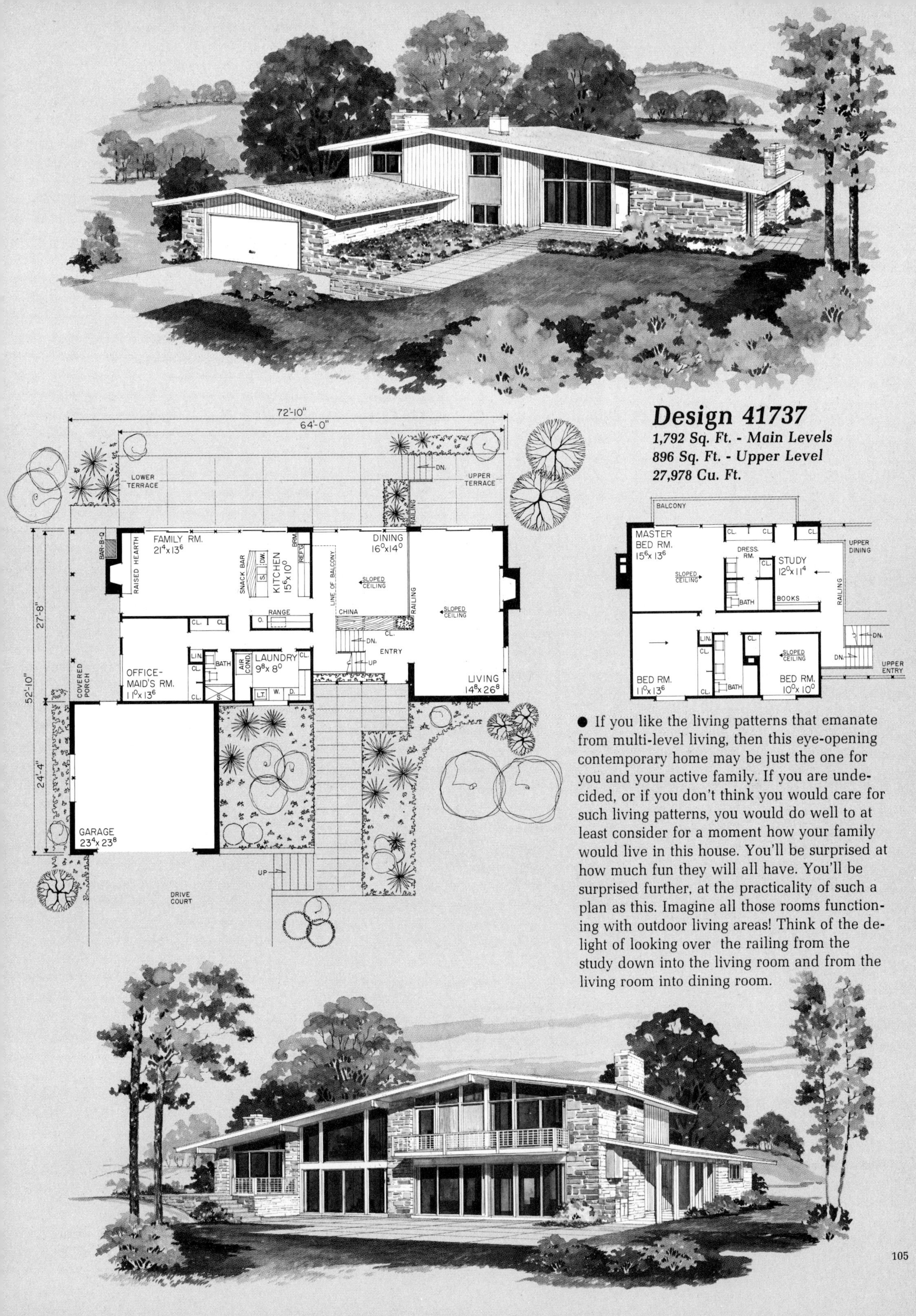

Design 41737

1,792 Sq. Ft. - Main Levels
896 Sq. Ft. - Upper Level
27,978 Cu. Ft.

● If you like the living patterns that emanate from multi-level living, then this eye-opening contemporary home may be just the one for you and your active family. If you are undecided, or if you don't think you would care for such living patterns, you would do well to at least consider for a moment how your family would live in this house. You'll be surprised at how much fun they will all have. You'll be surprised further, at the practicality of such a plan as this. Imagine all those rooms functioning with outdoor living areas! Think of the delight of looking over the railing from the study down into the living room and from the living room into dining room.

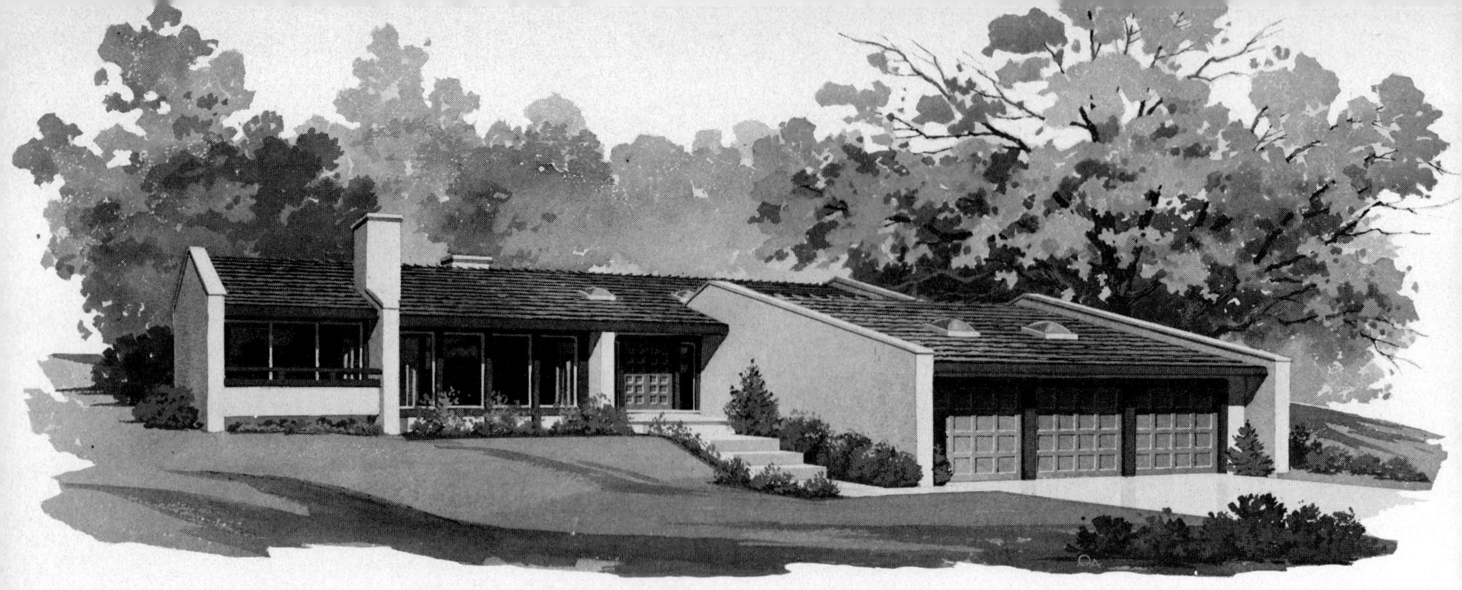

Design 42734 1,626 Sq. Ft. – Main Level; 1,033 Sq. Ft. – Upper Level; 1,273 Sq. Ft. – Lower Level; 47,095 Cu. Ft.

● If you have a desire for something delightfully different that offers unique, yet practical and enjoyable living patterns, then this house deserves careful study by all the members of your family. Having three bedrooms and a study on the upper level and a guest (or hobby) room on the lower level; it offers sleeping flexibility for the growing family. Notice how the living area looks down on the delightful planting area of the lower level. Also it shares a thru-fireplace with the study. Other features of the study include a 7 foot high book shelve, private balcony and separate stairs to the master bedroom. The outstanding U-shaped kitchen is flanked by the family and dining room. In addition to the living room, there is the huge, 32 foot activity room on the lower level. An abundance of storage space will be found in the three-car garage and the basement.

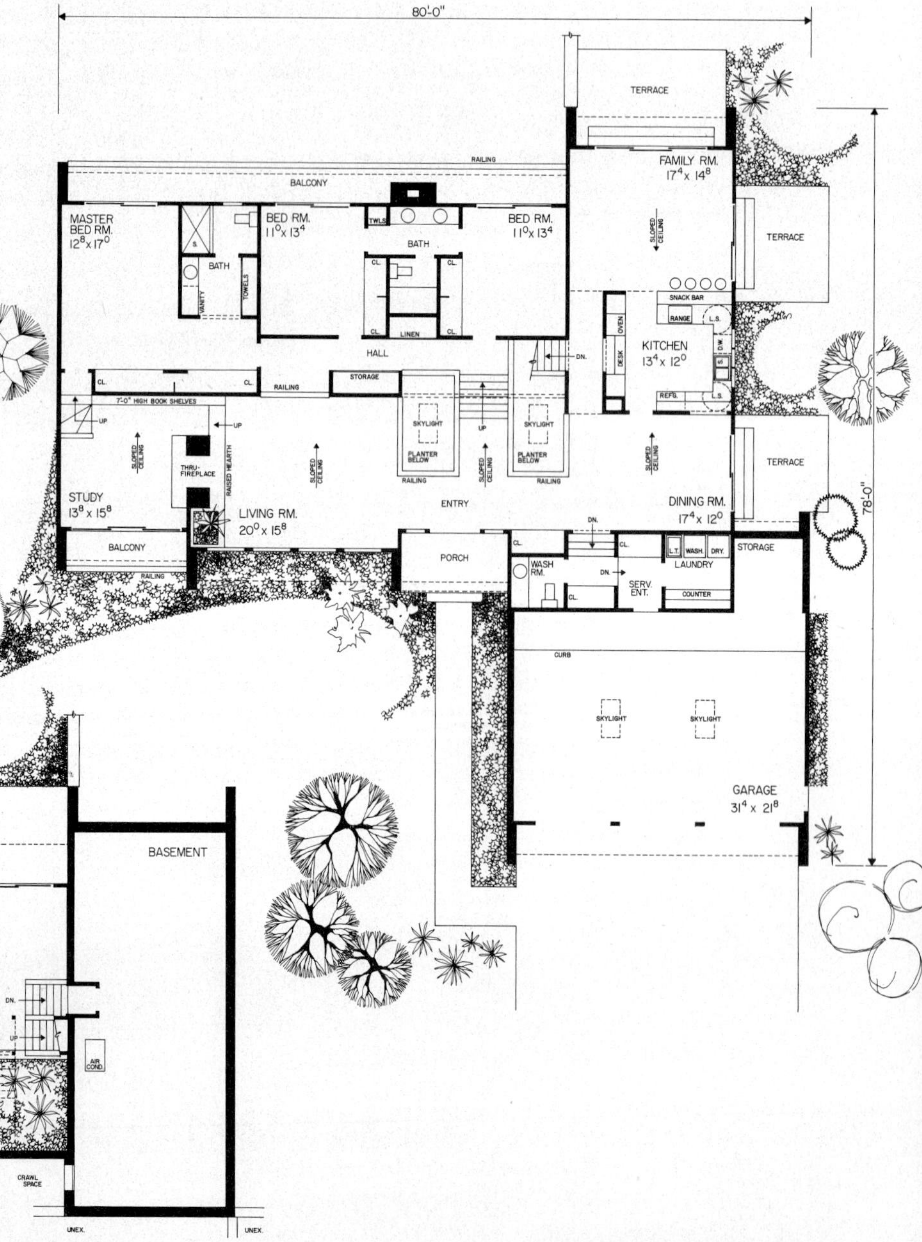

80'-0"
78'-0"
TERRACE
RAILING
BALCONY
MASTER BED RM. 12^8 x 17^0
BATH
BED RM. 11^0 x 13^4
BATH
BED RM. 11^0 x 13^4
FAMILY RM. 17^4 x 14^8
TERRACE
LINEN
HALL
STORAGE
RAILING
7'-0" HIGH BOOK SHELVES
UP
THRU-FIREPLACE
RAISED HEARTH
STUDY 13^8 x 15^8
SKYLIGHT
PLANTER BELOW
KITCHEN 13^4 x 12^0
SNACK BAR
RANGE
REFG.
LIVING RM. 20^0 x 15^8
ENTRY
DINING RM. 17^4 x 12^0
TERRACE
BALCONY
PORCH
WASH RM.
SERV. ENT.
LAUNDRY
COUNTER
STORAGE
CURB
SKYLIGHT
SKYLIGHT
GARAGE 31^4 x 21^8

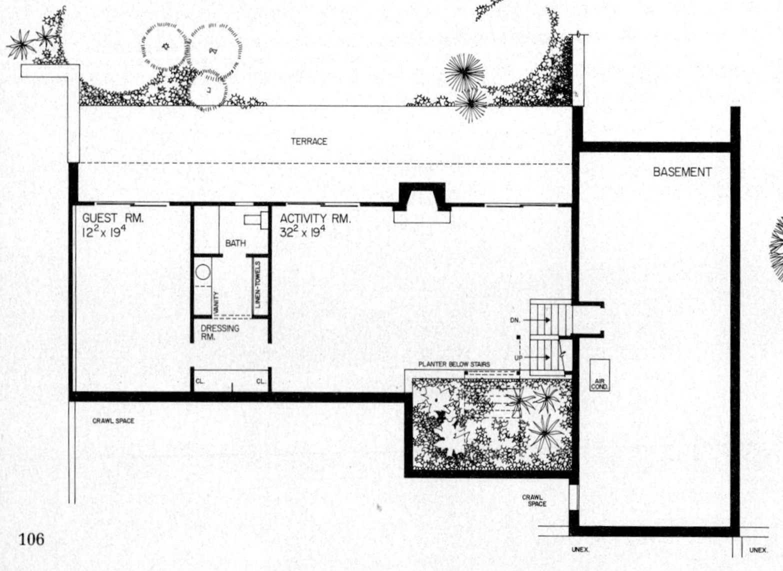

TERRACE
BASEMENT
GUEST RM. 12^2 x 19^4
BATH
ACTIVITY RM. 32^2 x 19^4
DRESSING RM.
CL.
CL.
PLANTER BELOW STAIRS
AIR COND.
CRAWL SPACE
CRAWL SPACE
UNEX.
UNEX.

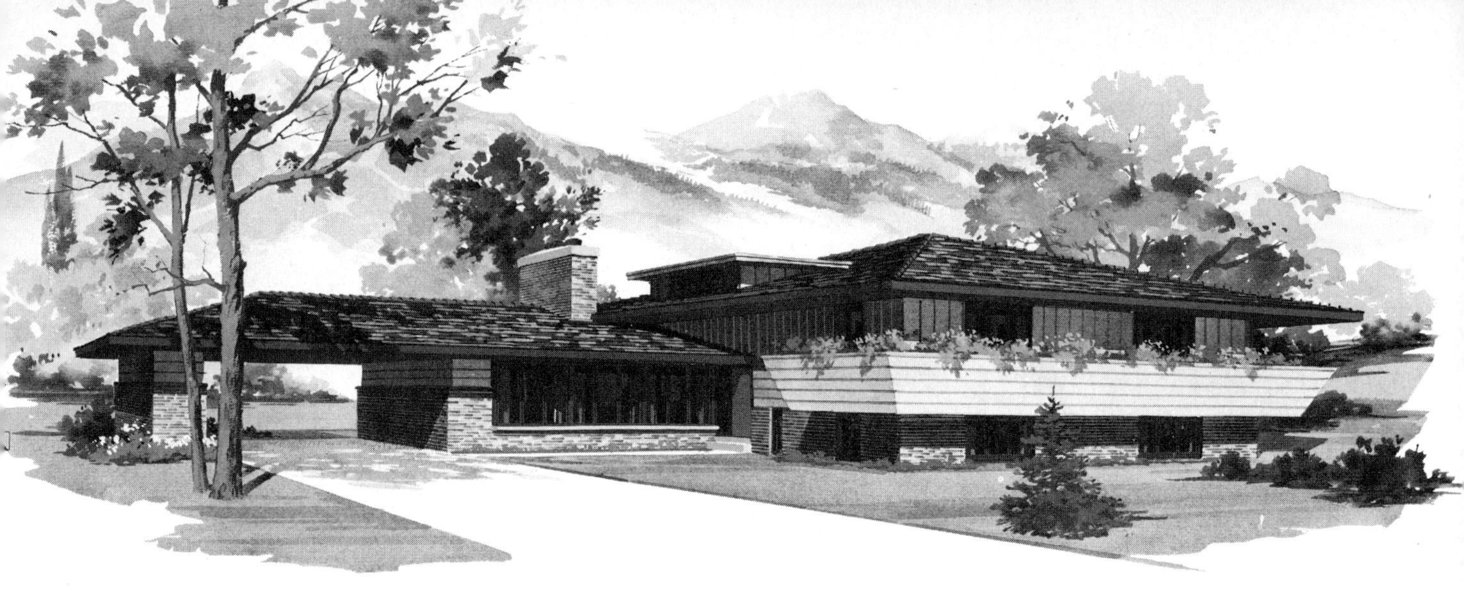

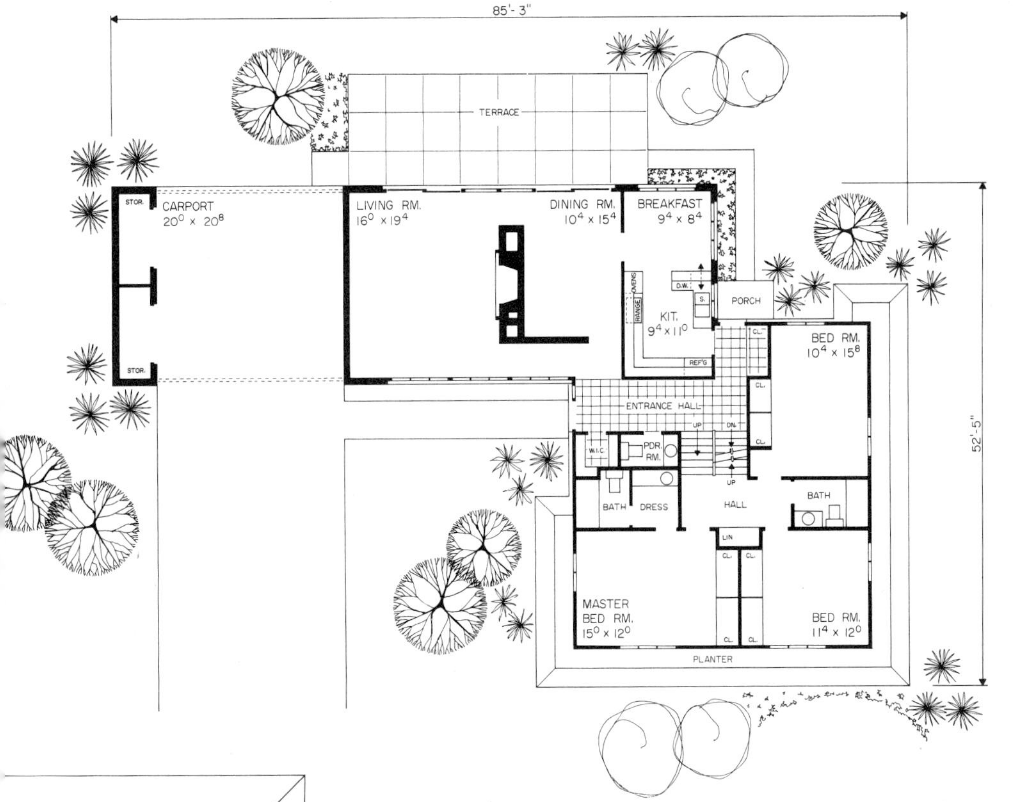
85'- 3"
TERRACE
CARPORT
20^0 x 20^8
STOR.
STOR.
LIVING RM.
16^0 x 19^4
DINING RM.
10^4 x 15^4
BREAKFAST
9^4 x 8^4
D.W.
S.
OVENS
RANGE
KIT.
9^4 x 11^0
REF'G
PORCH
CL.
BED RM.
10^4 x 15^8
ENTRANCE HALL
UP
DN.
W.I.C.
PDR. RM.
BATH
DRESS
HALL
BATH
LIN
CL.
CL.
MASTER
BED RM.
15^0 x 12^0
BED RM.
11^4 x 12^0
CL.
CL.
PLANTER
52'-5"

Design 42296

993 Sq. Ft. – Main Level
810 Sq. Ft. – Upper Level
842 Sq. Ft. – Lower Level
255 Sq. Ft. – Study Level
38,788 Cu. Ft.

● Here is real multi-level living with each level making a fine contribution to convenient living for the whole family. Including the basement, there are five distinct levels. The L-shaped plan gives birth to the distinctive contemporary exterior. The hip-roof has a wide, pleasing overhang. Just below is the unique, overhanging planter. If desired, this refreshing design could function wonderfully as a five bedroom home. The formal dining room and the breakfast room provide excellent eating facilities. The 32 foot lower level family room provides all the space necessary for multi-purpose activities. The basement offers an extra area for hobbies. Don't miss the laundry, three full baths and the powder room.

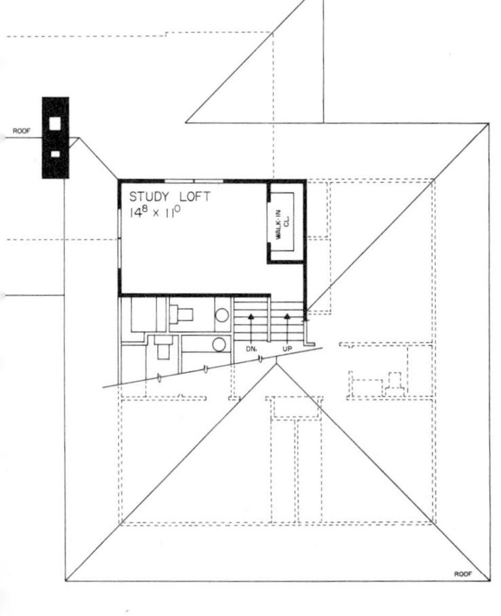
ROOF
STUDY LOFT
14^8 x 11^0
WALK-IN CL.
DN.
UP
ROOF

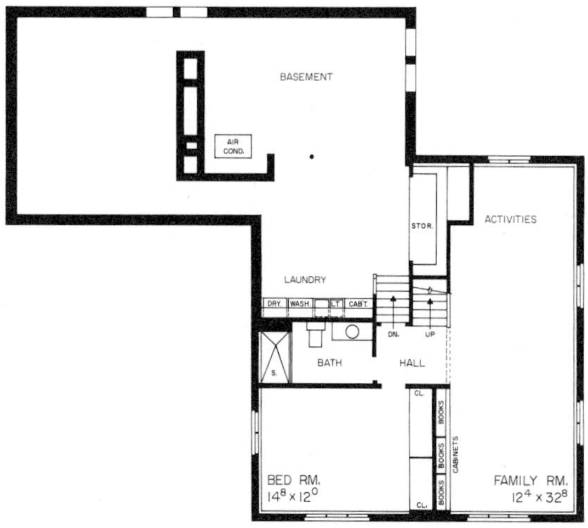
BASEMENT
AIR COND.
STOR.
ACTIVITIES
LAUNDRY
DRY
WASH
L.T.
CABT.
DN.
UP
BATH
HALL
S.
CL.
BOOKS
CABINETS
BOOKS
BOOKS
CL.
BED RM.
14^8 x 12^0
FAMILY RM.
12^4 x 32^8

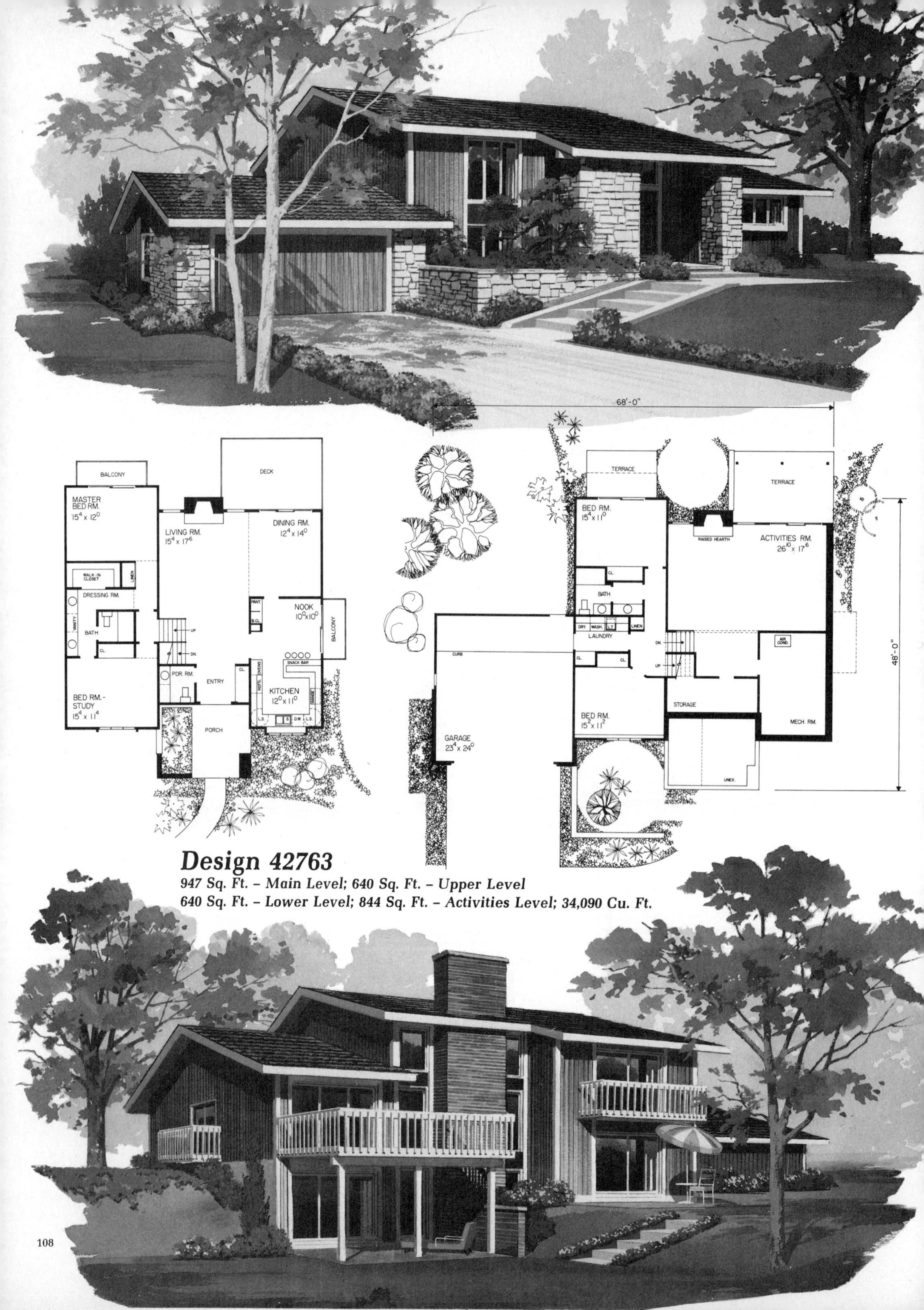

Design 42763

947 Sq. Ft. – Main Level; 640 Sq. Ft. – Upper Level
640 Sq. Ft. – Lower Level; 844 Sq. Ft. – Activities Level; 34,090 Cu. Ft.

Design 42580

1,852 Sq. Ft. - Upper Level
1,297 Sq. Ft. - Lower Level
32,805 Cu. Ft.

● Indoor-outdoor living could hardly ask for more. And here's why. Imagine, five balconies and three terraces! These unique balconies add great beauty to the exterior while adding pleasure to those who utilize them from the interior. And there's more. This home has enough space for all to appreciate. Take note of the size of the gathering room, family room and activity room. There's also a large dining room. Four bedrooms too, for the large or growing family. Or three plus a study. Two fireplaces, one to service each of the two levels in this bi-level design. The rear terrace is accessible thru sliding glass doors from the lower level bedroom and activity room. The side terrace functions with the activity/family room area. The master suite has two walk-in closets and a private bath.

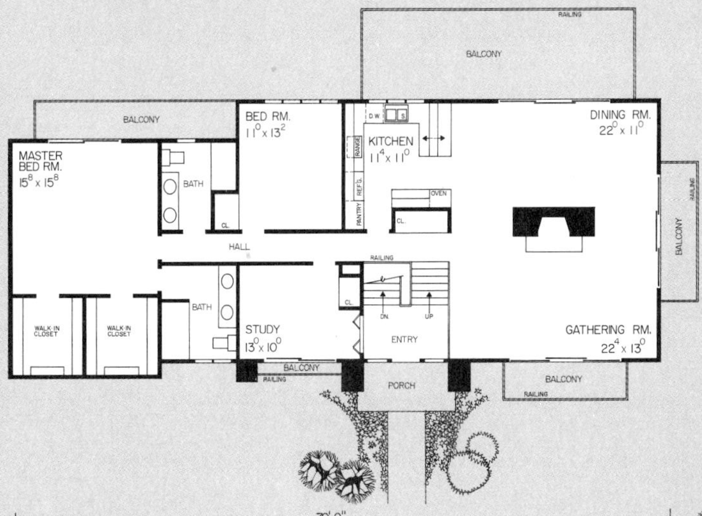

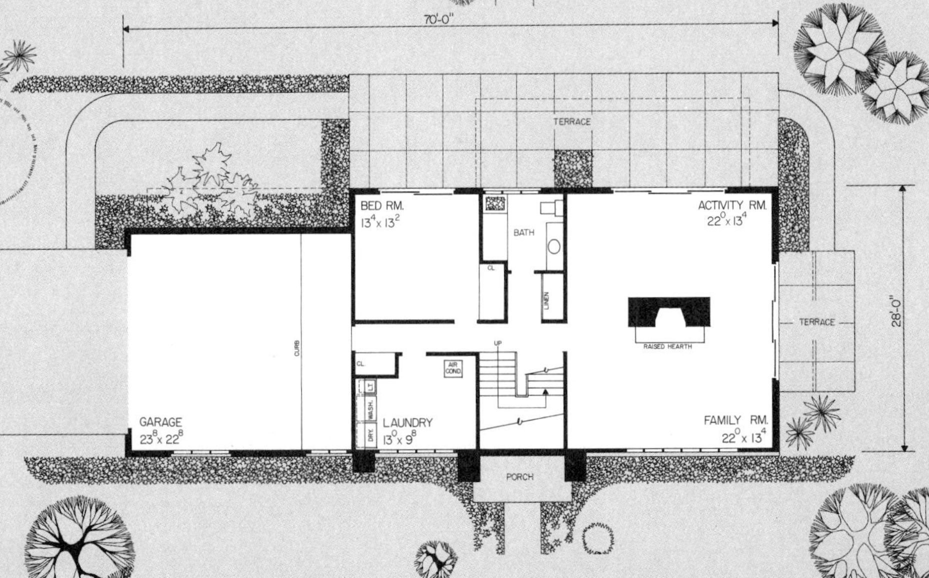

Design 42715

2,299 Sq. Ft. – Upper Level
1,524 Sq. Ft. – Lower Level; 40,700 Cu. Ft.

● A lounge with built-in seating and a thru-fireplace to the gathering room. A delightful attraction to view upon entrance of this home. Plus a formal dining room, a study and a U-shaped kitchen with breakfast nook. That is a lot of room. There's more! A huge activities room with a fireplace, snack bar and adjacent summer kitchen. This is the perfect set-up for teenage parties or family cook-outs on the terrace. The entire family will certainly enjoy the convenience of this area. All this, plus three bedrooms (optional four without the study), including a luxury master suite with all the extras and its own outdoor balcony. The upper level outdoor deck provides partial cover for the lower level terrace. This home offers glorious outdoor living potential on both levels.

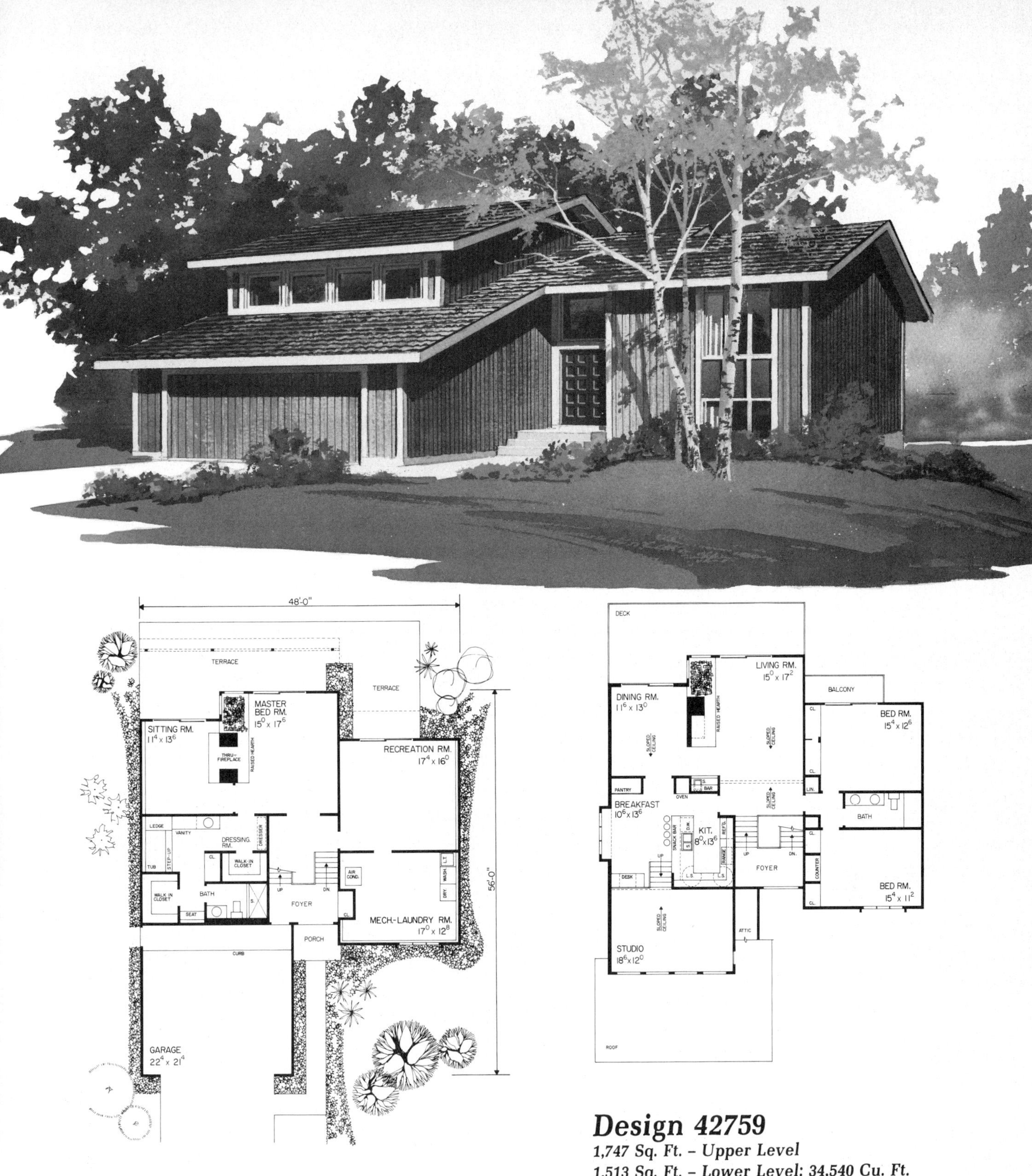

Design 42759

1,747 Sq. Ft. – Upper Level
1,513 Sq. Ft. – Lower Level; 34,540 Cu. Ft.

● A contemporary bi-level with a large bonus room on a third level over the garage. This studio will serve as a great room to be creative in or just to sit back in. The design also provides great indoor/outdoor living relationships with terraces and decks. The formal living/dining area has a sloped ceiling and built-in wet bar. The dramatic beauty of a raised hearth fireplace and built-in planter will be enjoyed by those in the living room. Both have sliding glass doors to the rear deck. The breakfast area will serve as a pleasant eating room with ample space for a table plus the built-in snack bar. The lower level houses the recreation room, laundry and an outstanding master suite. This master suite includes a thru-fireplace, sitting room, tub and shower and more.

Design 42788

1,795 Sq. Ft. – Upper Level
866 Sq. Ft. – Lower Level
34,230 Cu. Ft.

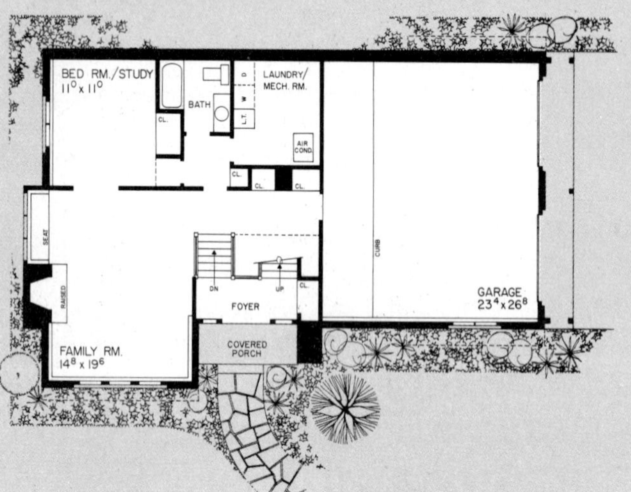

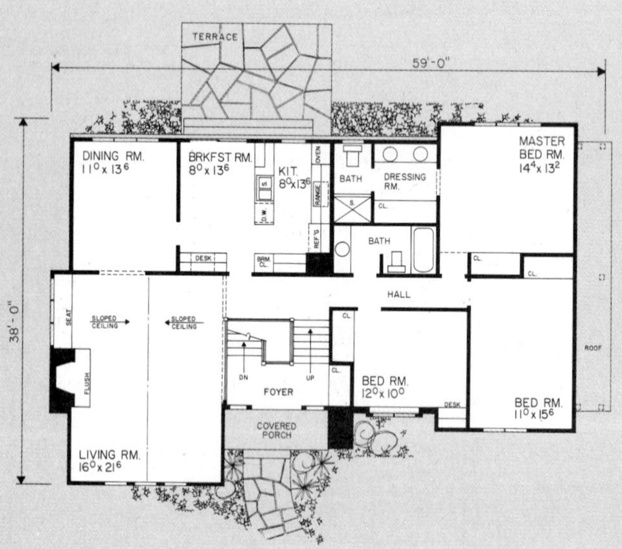

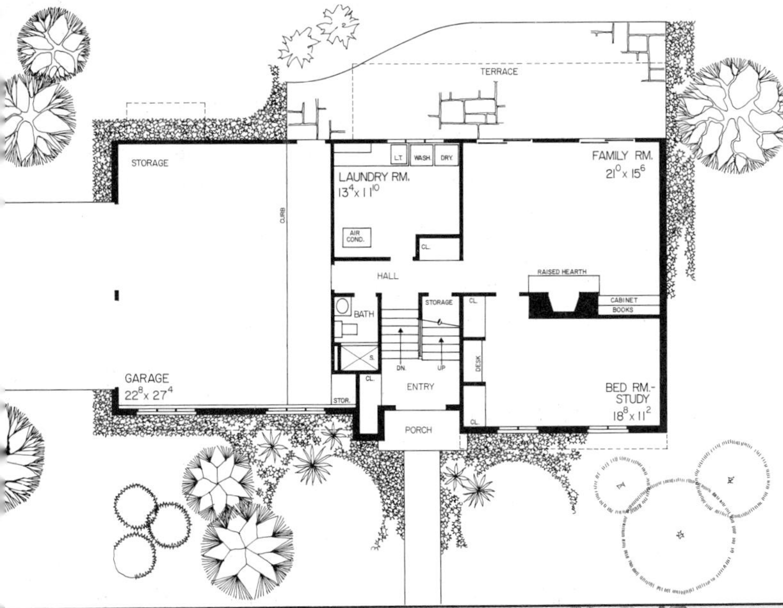

Design 42589

1,801 Sq. Ft. – Upper Level
1,061 Sq. Ft. – Lower Level
32,770 Cu. Ft.

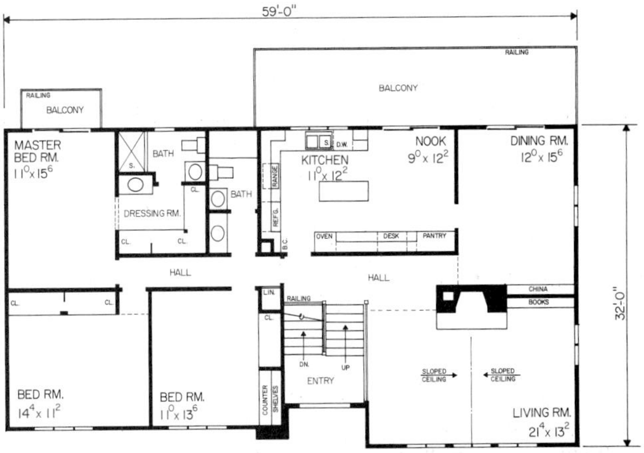

Design 41267

1,114 Sq. Ft. – Upper Level
1,194 Sq. Ft. – Lower Level
23,351 Cu. Ft.

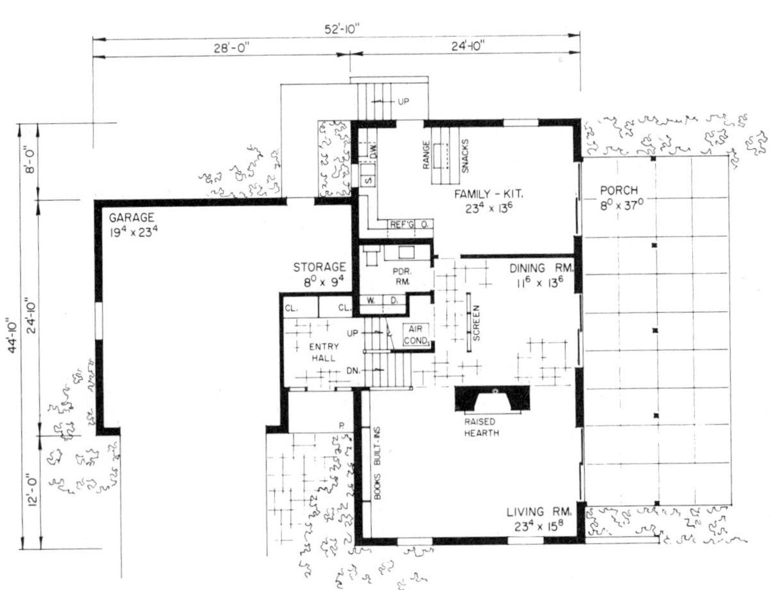

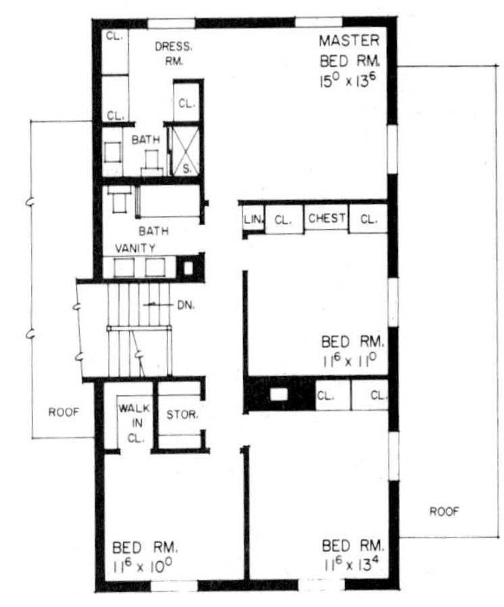

Design 42567 *1,773 Sq. Ft. – Main Level; 1,356 Sq. Ft. – Lower Level; 35,750 Cu. Ft.*

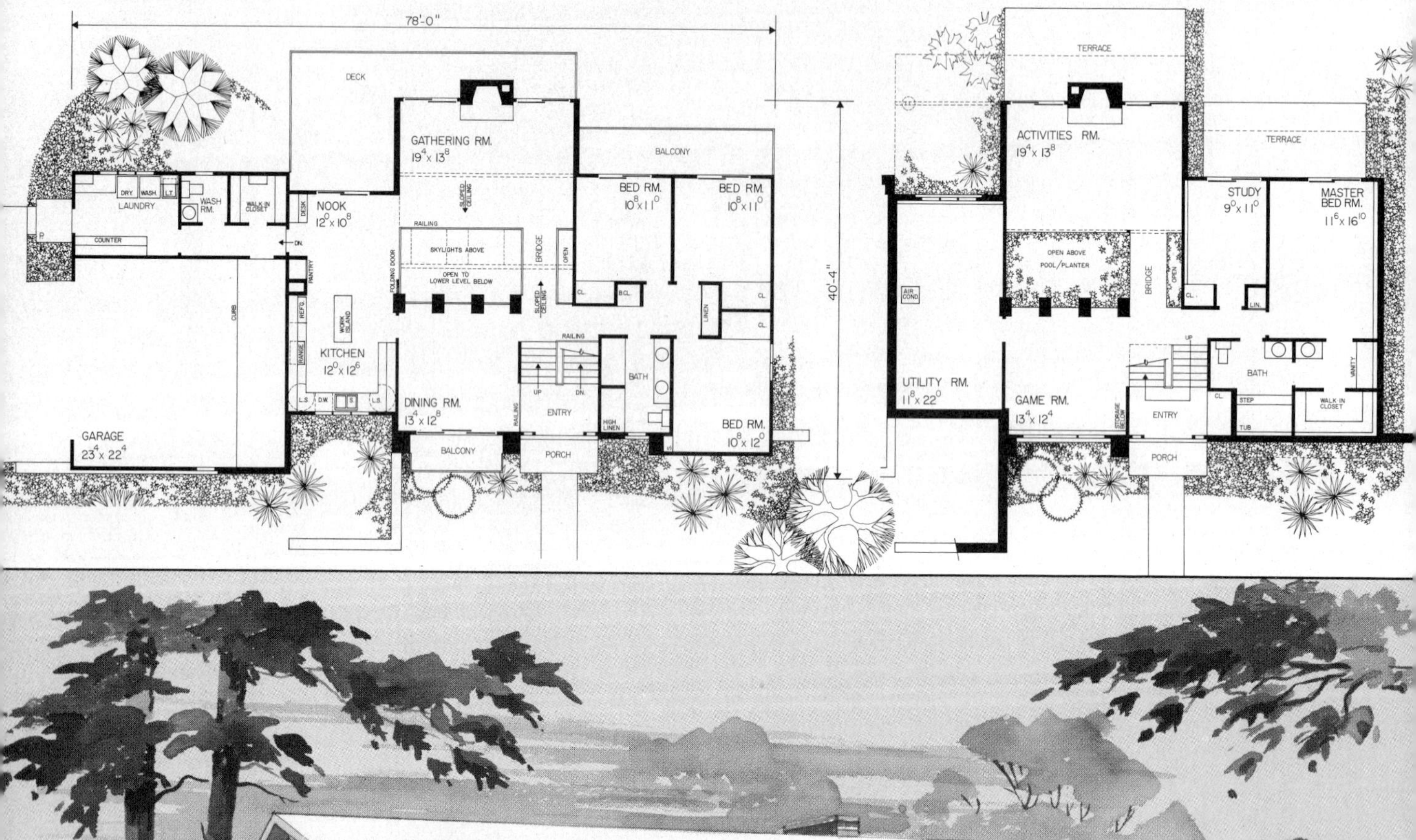

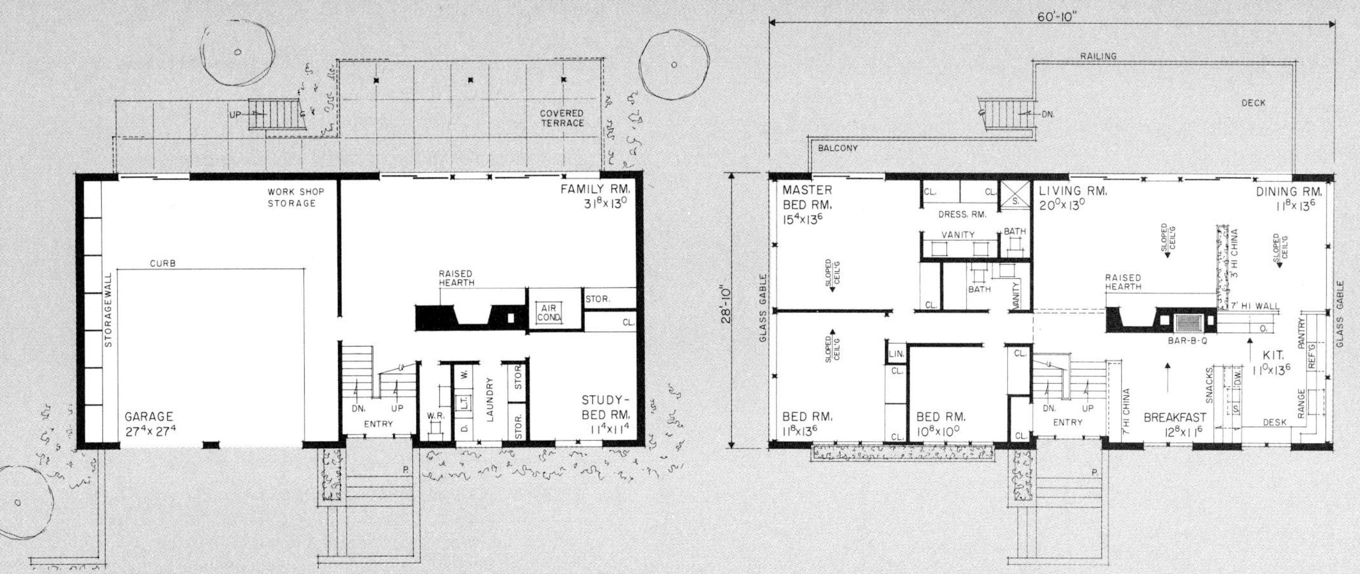

Design 41842 1,747 Sq. Ft. - Upper Level; 937 Sq. Ft. - Lower Level; 27,212 Cu. Ft.

Design 41782 1,920 Sq. Ft. - Upper Level; 1,036 Sq. Ft. - Lower Level; 30,672 Cu. Ft.

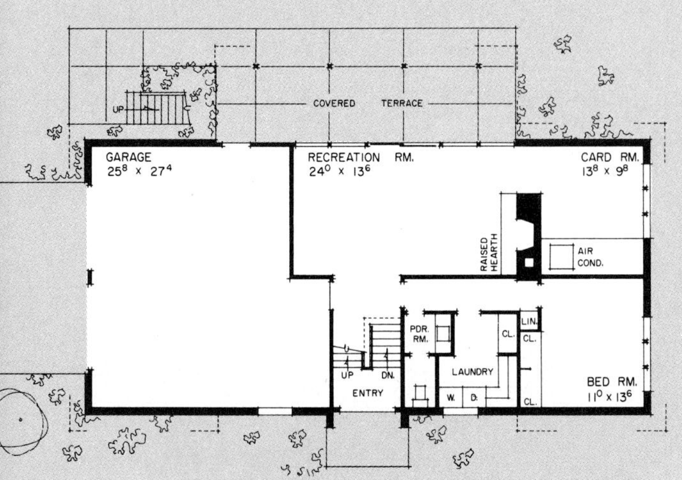

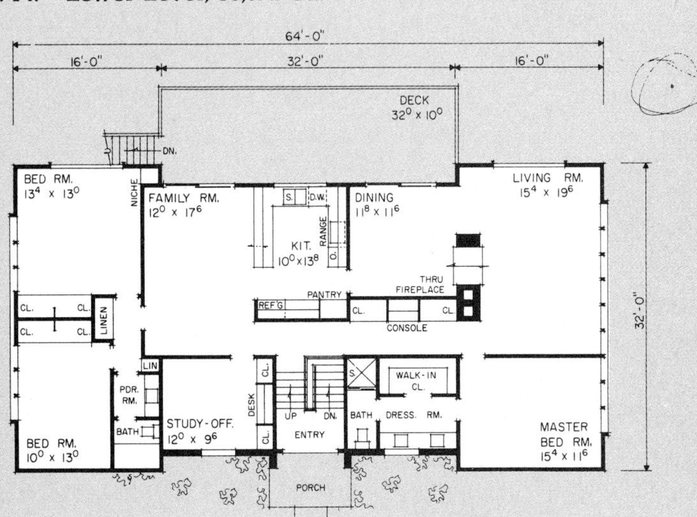

Design 43199

1,120 Sq. Ft. – Upper Level; 728 Sq. Ft. – Lower Level
19,364 Cu. Ft.

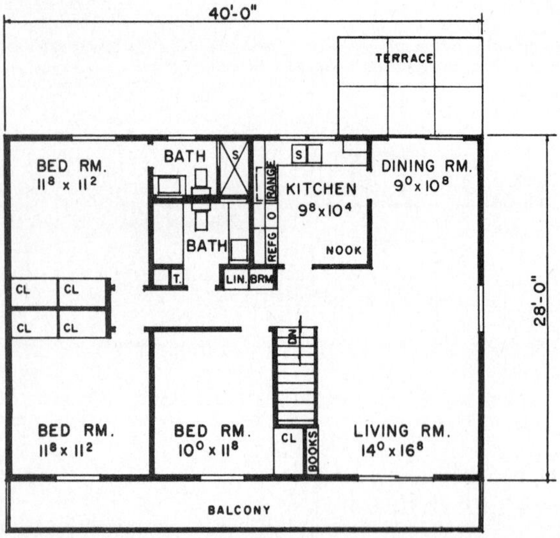

40'-0"
TERRACE
BED RM.
11⁸ x 11²
BATH
KITCHEN
9⁸ x 10⁴
DINING RM.
9⁰ x 10⁸
BATH
NOOK
REFG
O
RANGE
LIN.
BRM
CL
CL
CL
CL
DN
28'-0"
BED RM.
11⁸ x 11²
BED RM.
10⁰ x 11⁸
CL
BOOKS
LIVING RM.
14⁰ x 16⁸
BALCONY

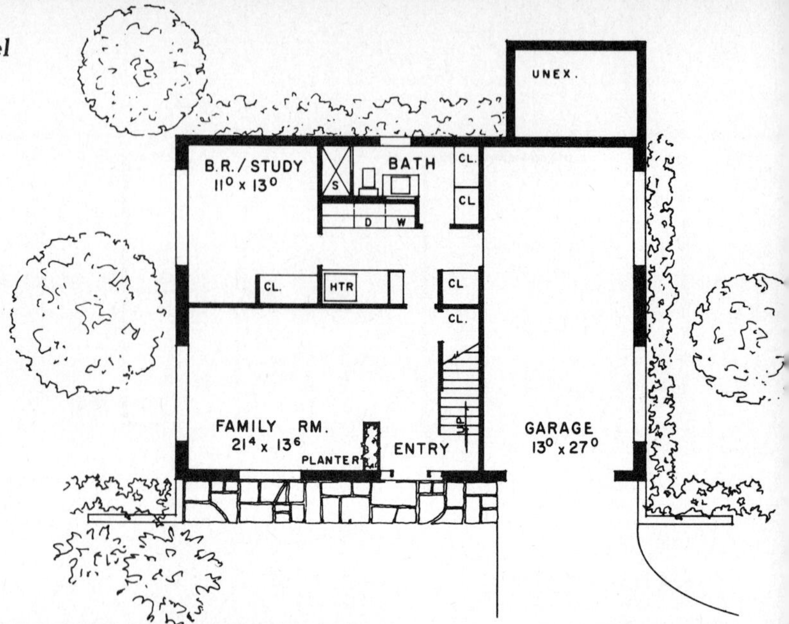

UNEX.
B.R. / STUDY
11⁰ x 13⁰
BATH
CL.
CL
D
W
CL.
HTR
CL
CL.
FAMILY RM.
21⁴ x 13⁶
PLANTER
ENTRY
UP
GARAGE
13⁰ x 27⁰

Design 43198

1,040 Sq. Ft. – Upper Level; 986 Sq. Ft. – Lower Level
20,368 Cu. Ft.

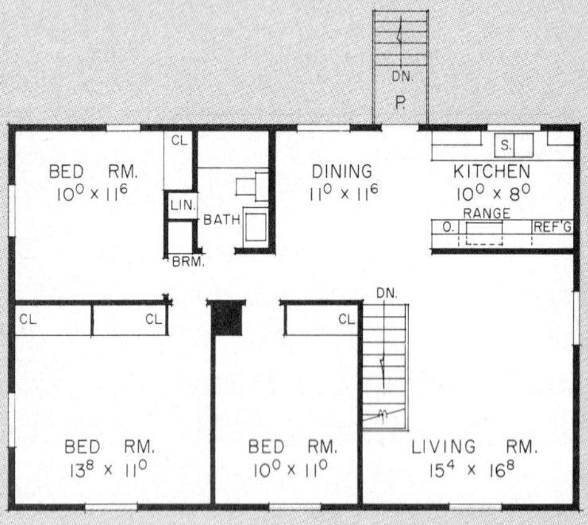

DN.
P.
BED RM.
10⁰ x 11⁶
CL.
LIN.
BATH
BRM.
DINING
11⁰ x 11⁶
KITCHEN
10⁰ x 8⁰
S.
RANGE
O.
REF'G
CL
CL
CL
DN.
BED RM.
13⁸ x 11⁰
BED RM.
10⁰ x 11⁰
LIVING RM.
15⁴ x 16⁸

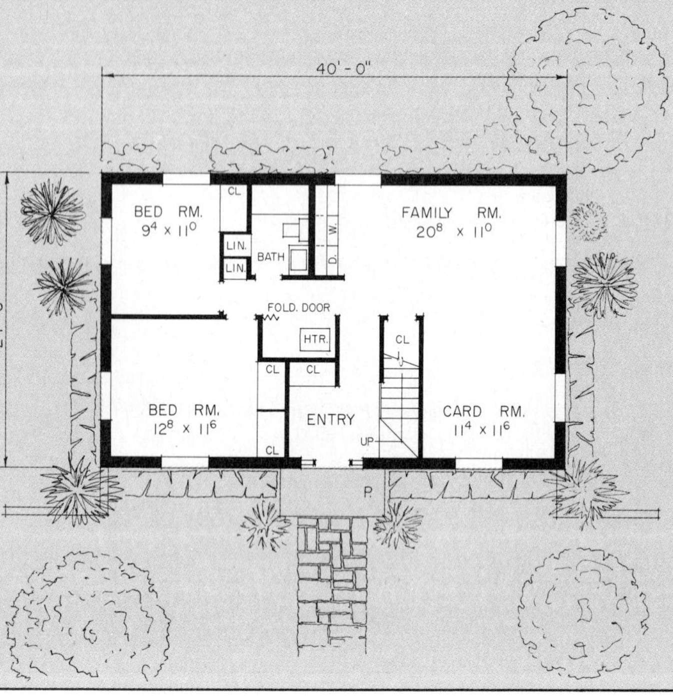

40'-0"
24'-8"
BED RM.
9⁴ x 11⁰
CL
LIN.
LIN.
BATH
D.
W.
FAMILY RM.
20⁸ x 11⁰
FOLD. DOOR
HTR.
CL
CL
CL
BED RM.
12⁸ x 11⁶
CL
ENTRY
UP
CARD RM.
11⁴ x 11⁶
P.

Design 42334

1,694 Sq. Ft. – Upper Level; 1,020 Sq. Ft. – Lower Level
34,259 Cu. Ft.

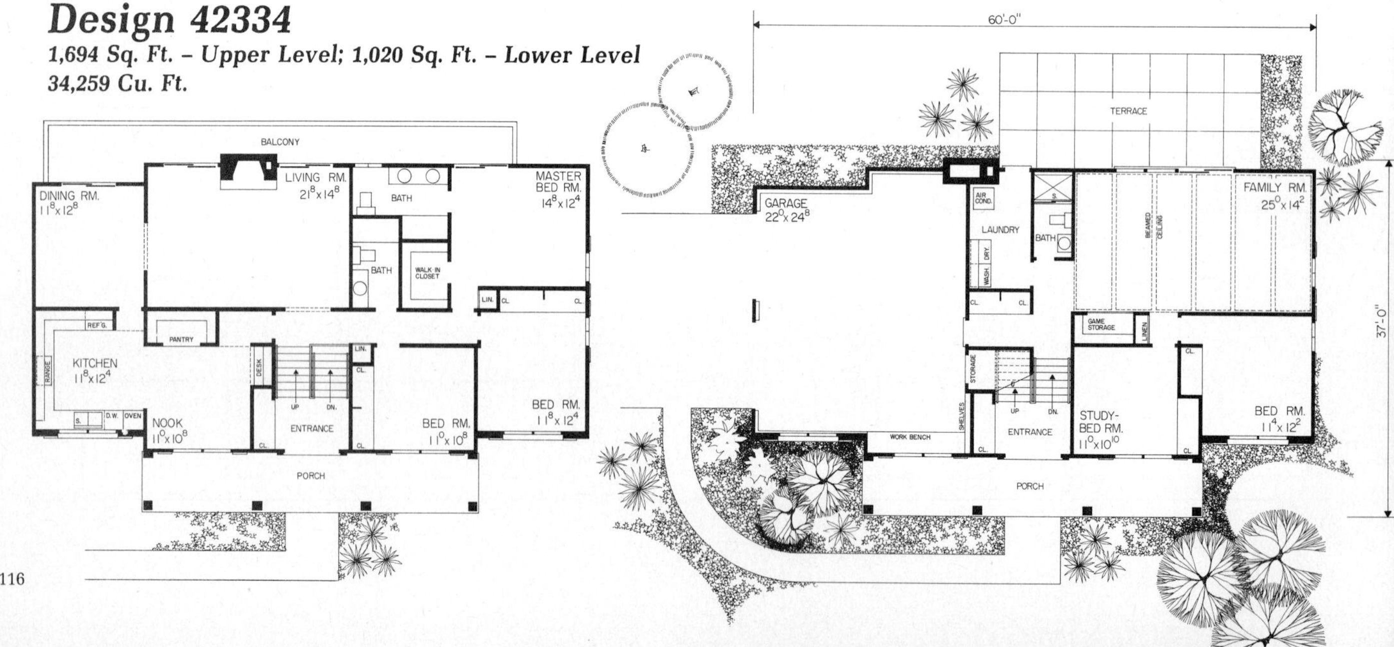

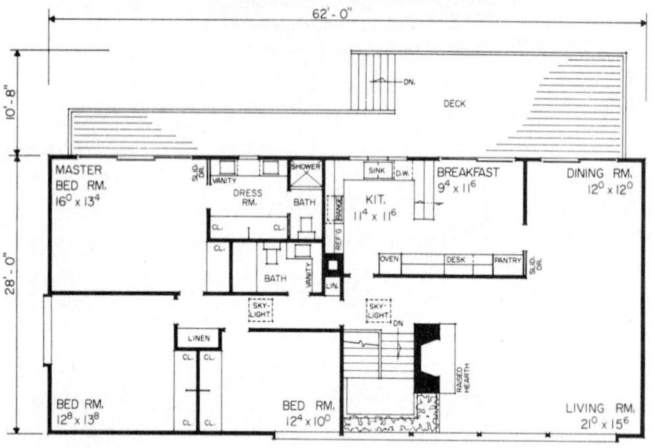

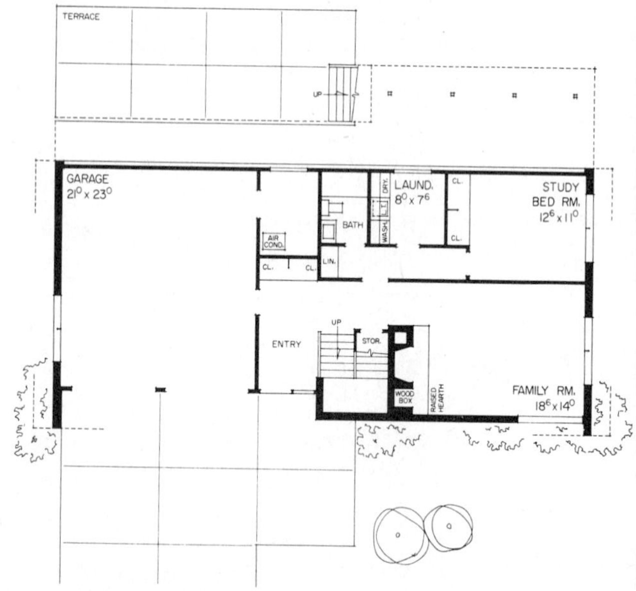

62'-0"
10'-8"
28'-0"
DECK
DN.
MASTER BED RM. 16⁰ x 13⁴
VANITY
DRESS RM.
SHOWER
BATH
SINK
D.W.
BREAKFAST 9⁴ x 11⁶
DINING RM. 12⁰ x 12⁰
KIT. 11⁴ x 11⁶
REF'G.
OVEN
DESK
PANTRY
CL.
BATH
VANITY
LIN.
SKY-LIGHT
LINEN
BED RM. 12⁸ x 13⁸
BED RM. 12⁴ x 10⁰
RAISED HEARTH
LIVING RM. 21⁰ x 15⁶

TERRACE
UP
GARAGE 21⁰ x 23⁰
AIR COND.
BATH
LAUND. 8⁰ x 7⁶
DRY
WASH
STUDY BED RM. 12⁶ x 11⁰
LIN.
ENTRY
STOR.
WOOD BOX
RAISED HEARTH
FAMILY RM. 18⁶ x 14⁰

Design 41968 *1,736 Sq. Ft. – Upper Level; 954 Sq. Ft. – Lower Level; 24,836 Cu. Ft.*

● When a designer works in a contemporary vein, he can create original forms and relationships. Here, a large, airy mass (second floor) is contrasted with a smaller, but heavier first floor base. The house takes advantage of a sloping site with the back wall doubling as foundation and an enclosure of living space. The excellent indoor-outdoor relationship you would expect is accomplished with sliding glass doors in three of the rooms opening to a deck. Two large, raised hearth fireplaces are especially luxurious features in the living and family rooms.

Design 41166 *1,274 Sq. Ft. - Main Level; 1,242 Sq. Ft. - Lower Level; 22,015 Cu. Ft.*

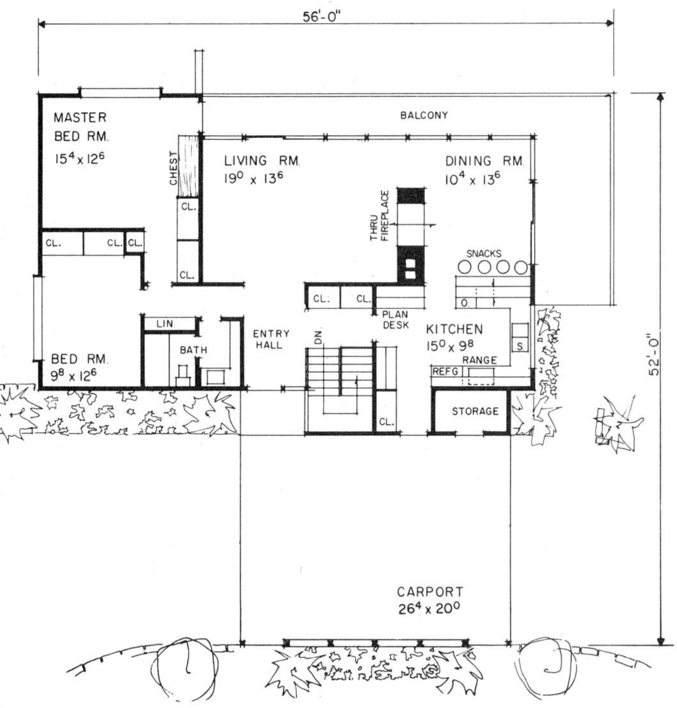

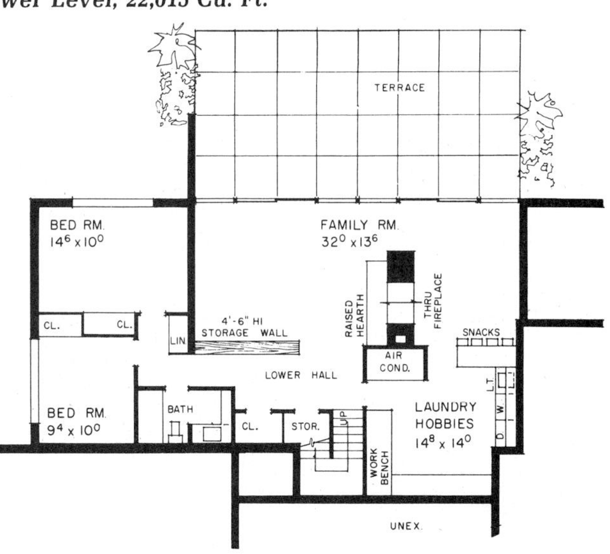

● Two-level living that will be hard to beat. Each level features two bedrooms, a full bath, a thru-fireplace, a spacious living area, delightful glass areas, plenty of storage potential and even a snack bar. Don't miss the efficient kitchen, the laundry/hobby room and the outdoor balcony.

Design 42723

1,748 Sq. Ft. - Main Level
294 Sq. Ft. - Entry Level
743 Sq. Ft. - Upper Level
45,400 Cu. Ft.

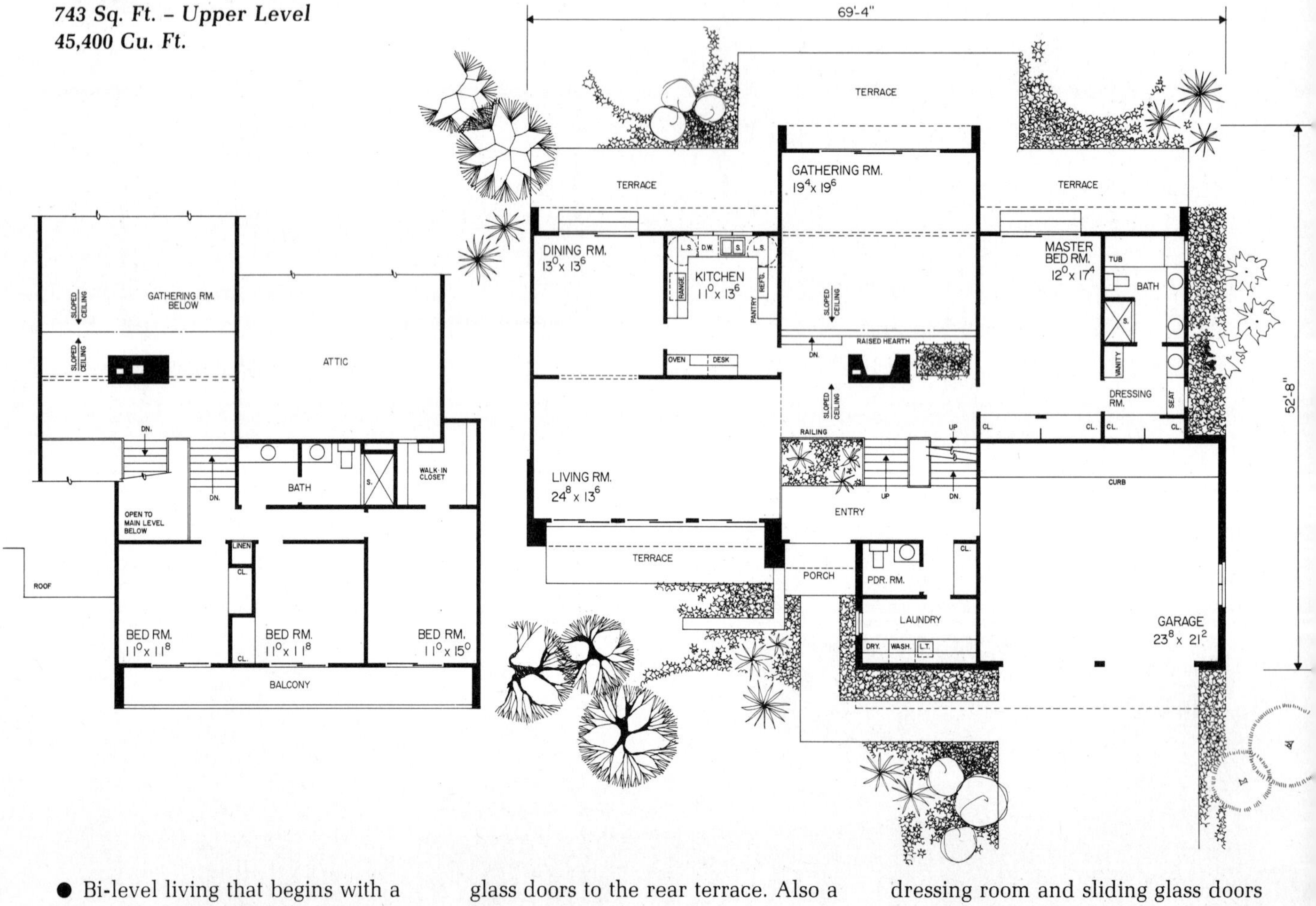

GATHERING RM. BELOW
SLOPED CEILING
ATTIC
DN.
BATH
S.
WALK-IN CLOSET
OPEN TO MAIN LEVEL BELOW
LINEN
CL.
ROOF
BED RM. $11^0 \times 11^8$
BED RM. $11^0 \times 11^8$
BED RM. $11^0 \times 15^0$
BALCONY

69'-4"
TERRACE
TERRACE
GATHERING RM. $19^4 \times 19^6$
TERRACE
DINING RM. $13^0 \times 13^6$
KITCHEN $11^0 \times 13^6$
L.S. D.W. S. L.S.
RANGE REFG. PANTRY
OVEN DESK
SLOPED CEILING
RAISED HEARTH
MASTER BED RM. $12^0 \times 17^4$
TUB
BATH
S.
VANITY
DRESSING RM.
SEAT
CL.
52'-8"
RAILING
UP
DN.
LIVING RM. $24^8 \times 13^6$
ENTRY
CURB
TERRACE
PORCH
PDR. RM.
LAUNDRY
DRY. WASH. L.T.
GARAGE $23^8 \times 21^2$

● Bi-level living that begins with a front entry level which houses a powder room and a laundry. The main level has a sunken gathering room! This area has a sloped ceiling, raised hearth fireplace, built-in planter and sliding glass doors to the rear terrace. Also a living room with an adjacent formal dining room. The U-shaped kitchen is a convenient work center with easy access to all areas. The master bedroom's occupants will enjoy their private bath, dressing room and sliding glass doors to another section of the rear terrace. The upper level is an entire sleeping zone. Three bedrooms and a bath will further serve the family. View the gathering room from the upper level.

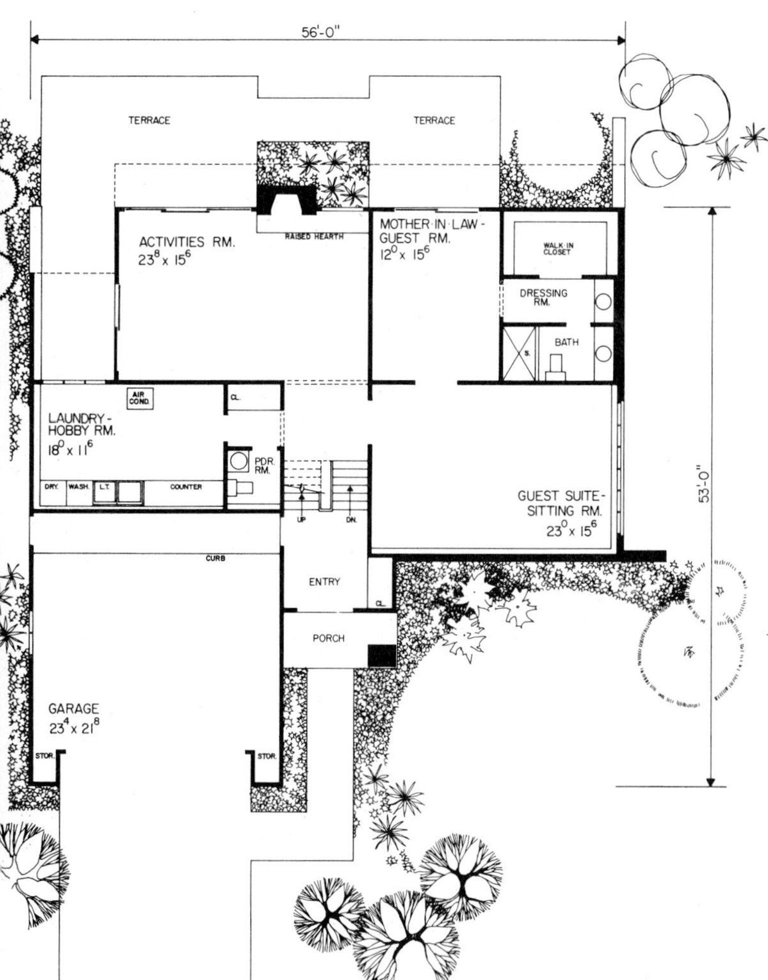

Design 42735

1,545 Sq. Ft. - Upper Level
1,633 Sq. Ft. - Lower Level
33,295 Cu. Ft.

● Whether entering this house through the double front doors, or from the garage, access is gained to the lower level by descending seven stairs. Here, there is a bonus of livability. If desired, this level could be used to accommodate a live-in relative while still providing the family with a fine informal activities room and a separate laundry/hobby room and extra powder room. Up seven risers from the entry is the main living level. It has a large gathering room; a sizable nook which could be called upon to function as a separate dining room; an efficient kitchen with pass-thru to a formal dining area and a two bedroom, two bath and study sleeping zone. Don't miss the balconies and deck.

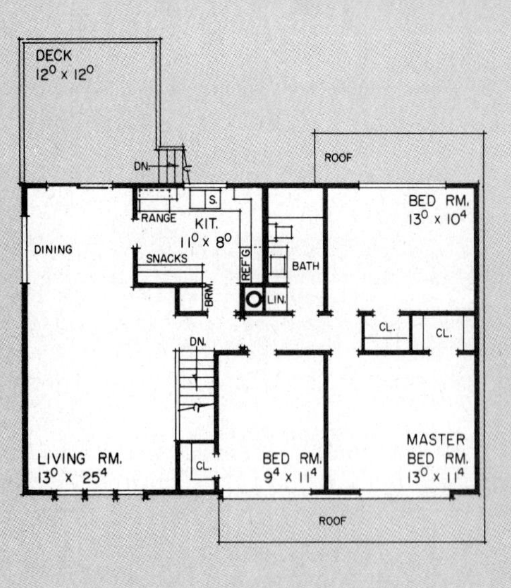
DECK
12⁰ x 12⁰
DN.
ROOF
RANGE
S.
KIT.
11⁰ x 8⁰
BED RM.
13⁰ x 10⁴
DINING
SNACKS
REF'G
BATH
BRM.
LIN.
CL.
CL.
DN.
LIVING RM.
13⁰ x 25⁴
CL.
BED RM.
9⁴ x 11⁴
MASTER
BED RM.
13⁰ x 11⁴
ROOF

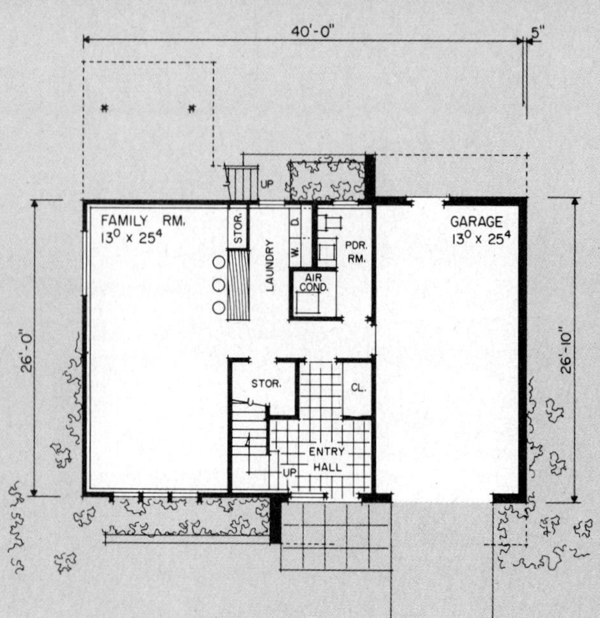
40'-0"
5"
UP
FAMILY RM.
13⁰ x 25⁴
STOR.
W.
D.
LAUNDRY
PDR.
RM.
AIR
COND.
GARAGE
13⁰ x 25⁴
26'-0"
26'-10"
STOR.
CL.
ENTRY
HALL
UP

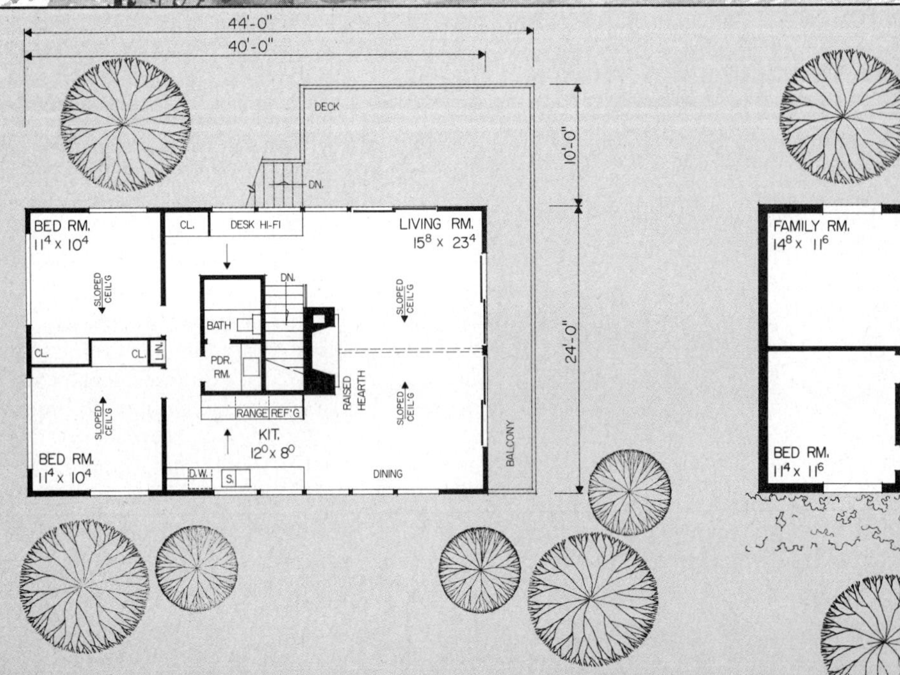
44'-0"
40'-0"
DECK
DN.
10'-0"
24'-0"
BED RM.
11⁴ x 10⁴
CL.
DESK HI-FI
LIVING RM.
15⁸ x 23⁴
SLOPED
CEIL'G
DN.
BATH
CL.
CL.
LIN.
PDR.
RM.
RAISED
HEARTH
RANGE
REF'G
KIT.
12⁰ x 8⁰
BED RM.
11⁴ x 10⁴
D.W.
S.
DINING
BALCONY

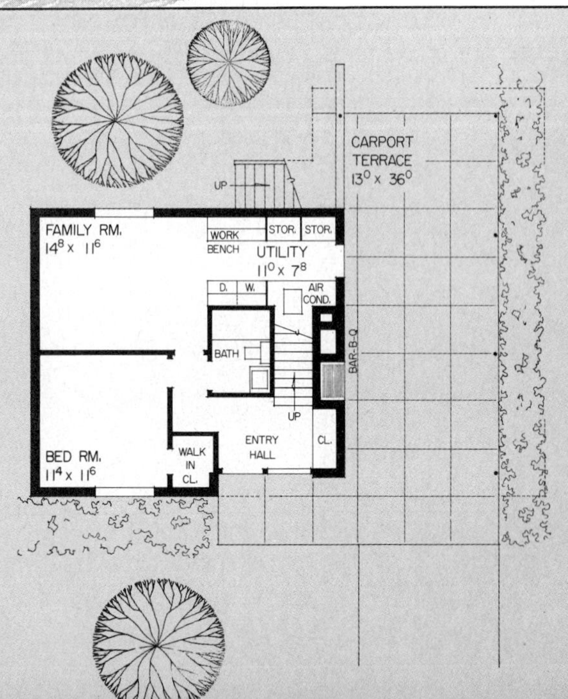
CARPORT
TERRACE
13⁰ x 36⁰
UP
FAMILY RM.
14⁸ x 11⁶
WORK
BENCH
STOR.
STOR.
UTILITY
11⁰ x 7⁸
D.
W.
AIR
COND.
BATH
BAR-B-Q
UP
ENTRY
HALL
CL.
BED RM.
11⁴ x 11⁶
WALK
IN
CL.

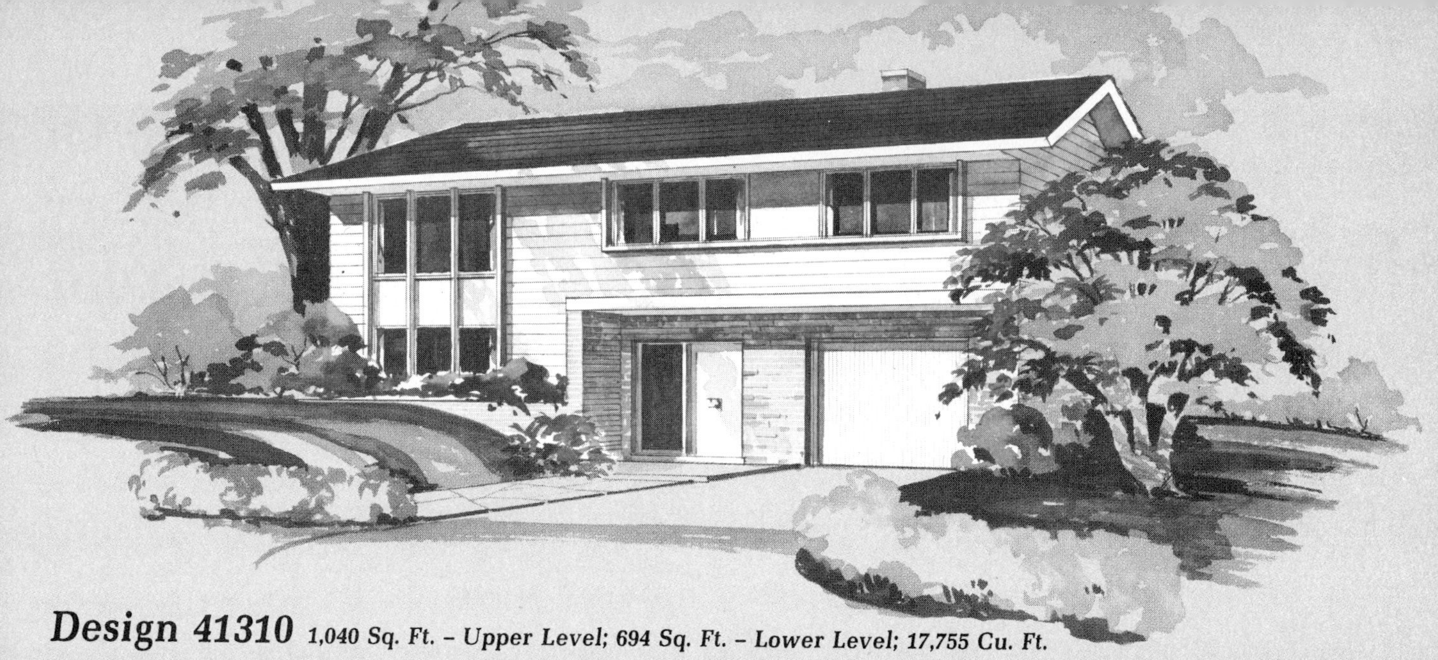

Design 41310 *1,040 Sq. Ft. – Upper Level; 694 Sq. Ft. – Lower Level; 17,755 Cu. Ft.*

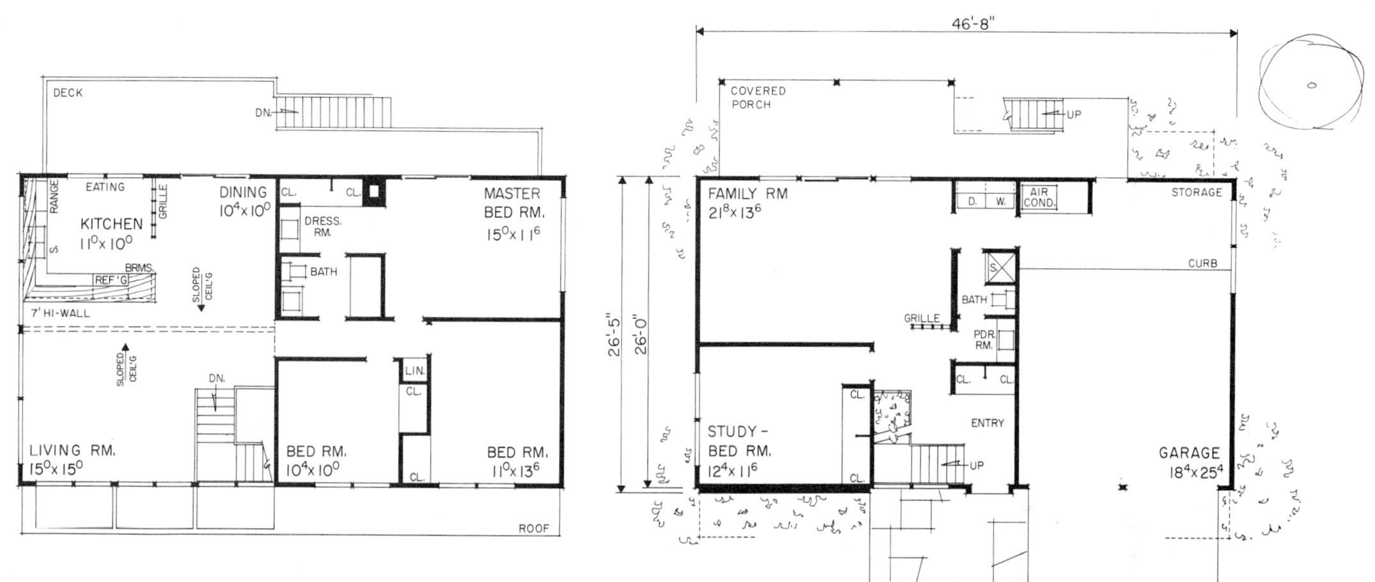

Design 41800 *1,212 Sq. Ft. – Upper Level; 720 Sq. Ft. – Lower Level; 19,480 Cu. Ft.*

Design 41234 *960 Sq. Ft. – Upper Level; 628 Sq. Ft. – Lower Level; 15,304 Cu. Ft.*

Your Choice Of Contemporary - Plus Your 2 Floor Plans

● The bi-level, or split foyer design has become increasingly popular. Here are six alternate elevations - three Traditional and three Contemporary - which may be built with either of two basic floor plans. One plan contains 960 square feet on each level and is 24 feet in depth; the other contains 1,040 square feet on each level and is 26 feet in depth. Plans for traditional and contemporary series include each of the three optional elevations.

Design 41377 *Traditional Exteriors*
24 Foot Depth Plan (18,960 Cu. Ft.)

Design 41375 *Traditional Exteriors*
26 Foot Depth Plan (20,778 Cu. Ft.)

960 Sq. Ft. - Upper Level; 960 Sq. Ft. - Lower Level

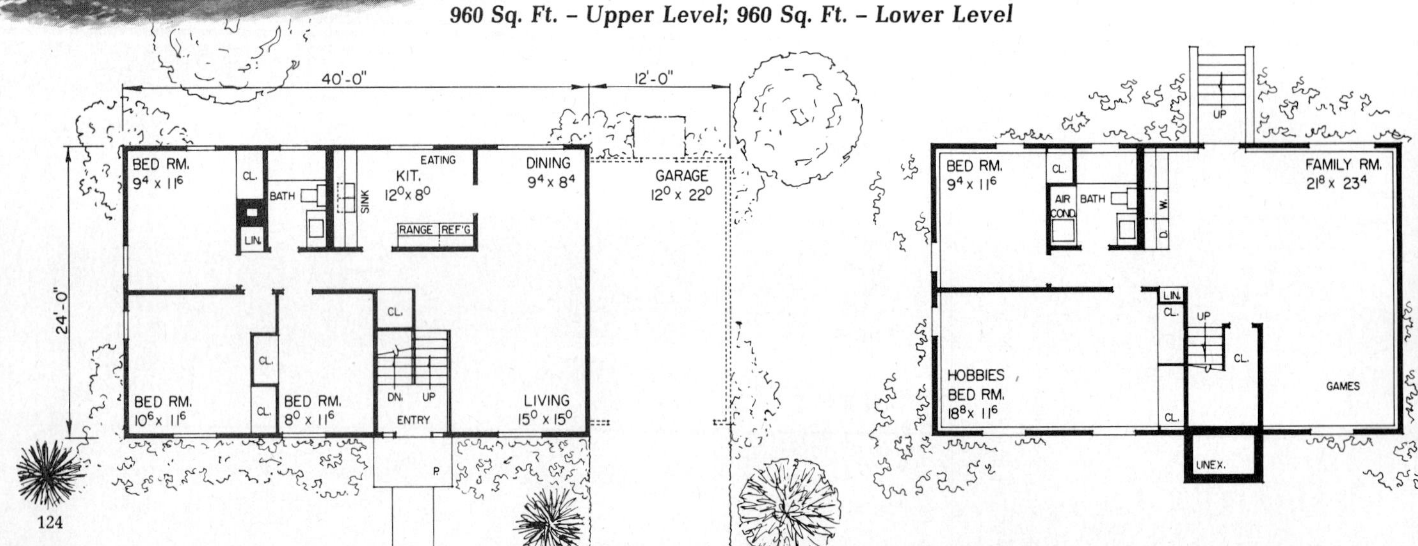

Traditional or Exterior Styling - Choice of For Each Style

● The popularity of the bi-level design can be traced to the tremendous amount of the livable space that such a design provides per construction dollar. While the lower level is partially below grade, it enjoys plenty of natural light and, hence, provides a bright, cheerful atmosphere for total livability. While these two basic floor plans are essentially the same, it is important to note that the larger of the two features a private bath for the master bedroom. Study the plans.

Design 41376 **Contemporary Exteriors**
24 Foot Depth Plan (18,000 Cu. Ft.)

Design 41378 **Contemporary Exteriors**
26 Foot Depth Plan (19,624 Cu. Ft.)

1,040 Sq. Ft. – Upper Level; 1,040 Sq. Ft. – Lower Level

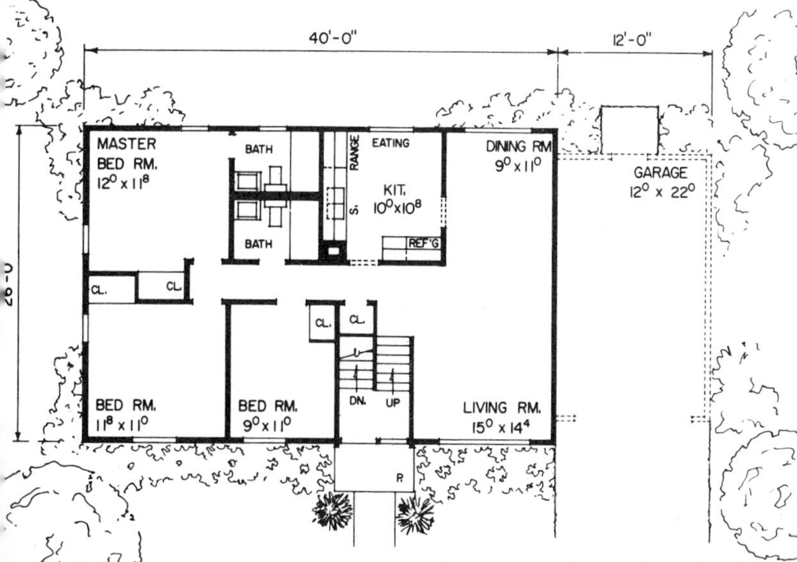

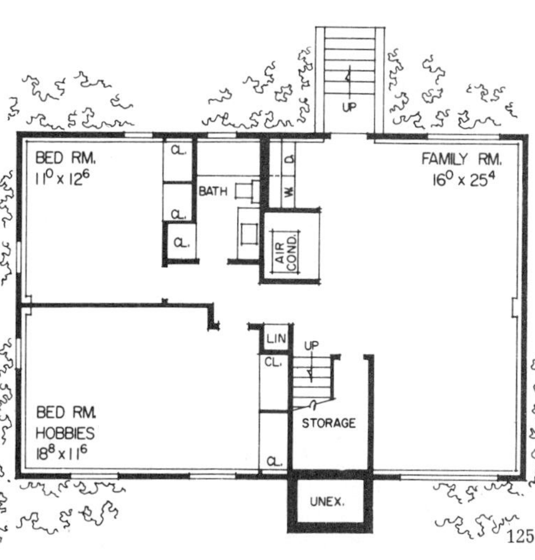

Design 42337

1,228 Sq. Ft. – Upper level
1,812 Sq. Ft. – Lower Level
32,377 Cu. Ft.

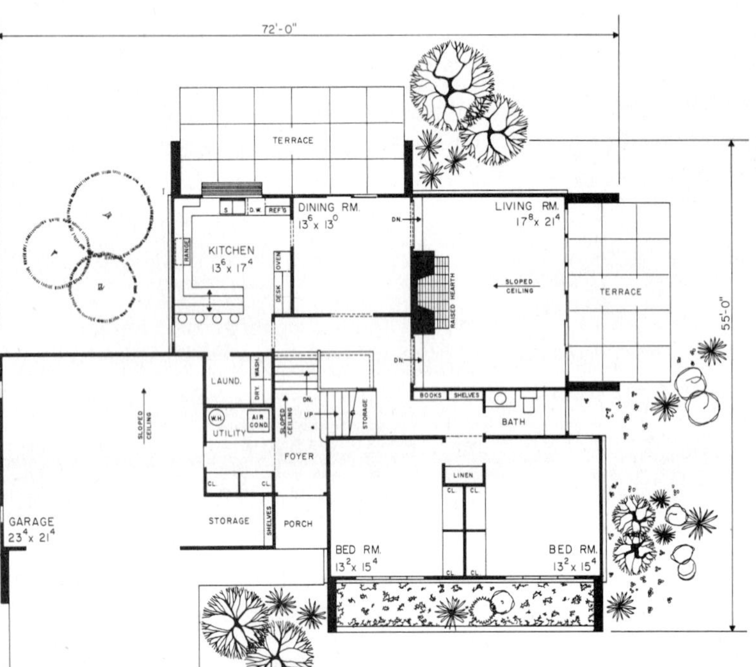

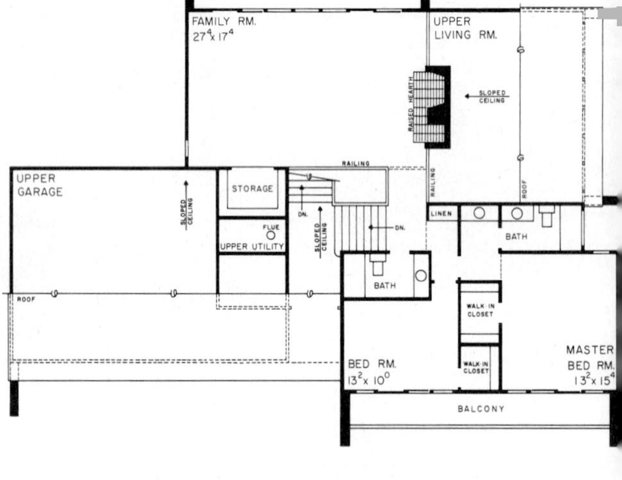

● Distinctive, indeed! And little wonder with its simple, straightforward lines and unadorned masses of stucco, this home will be an eye-catching attraction in any location. Inside, the foyer routes traffic to the two levels of livability. Each level features two bedrooms and two living areas, each with a raised hearth fireplace. For eating facilities the efficient kitchen has a pass-thru to the snack bar and is but a step from the formal dining room. Outdoor balconies, front and rear, provide upper access to outdoor enjoyment. A dramatic interior highlight is the relationship between the upper level family room and the lower level living room. From the family room it is possible to look down into the living room as well as into the foyer. The storage potential is outstanding throughout this design.

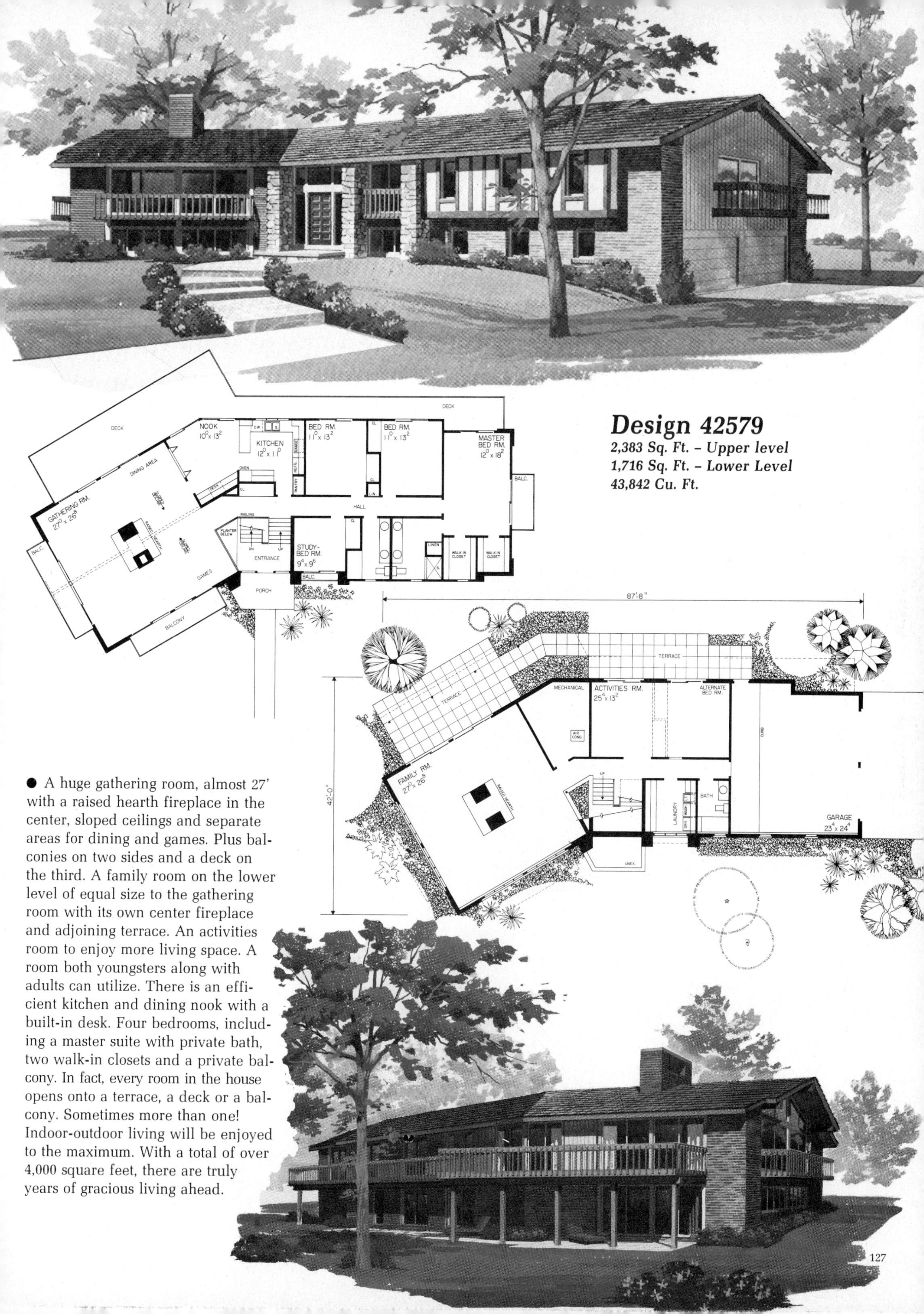

Design 42579

2,383 Sq. Ft. – Upper level
1,716 Sq. Ft. – Lower Level
43,842 Cu. Ft.

● A huge gathering room, almost 27' with a raised hearth fireplace in the center, sloped ceilings and separate areas for dining and games. Plus balconies on two sides and a deck on the third. A family room on the lower level of equal size to the gathering room with its own center fireplace and adjoining terrace. An activities room to enjoy more living space. A room both youngsters along with adults can utilize. There is an efficient kitchen and dining nook with a built-in desk. Four bedrooms, including a master suite with private bath, two walk-in closets and a private balcony. In fact, every room in the house opens onto a terrace, a deck or a balcony. Sometimes more than one! Indoor-outdoor living will be enjoyed to the maximum. With a total of over 4,000 square feet, there are truly years of gracious living ahead.

Design 41850

1,456 Sq. Ft. – Upper Level
728 Sq. Ft. – Lower Level
23,850 Cu. Ft.

● This attractive, traditional bi-level house will surely prove to be an outstanding investment. While it is a perfect rectangle - which leads to economical construction - it has a full measure of eye-appeal. Setting the character of the exterior is the effective window treatment, plus the unique design of the recessed front entrance.

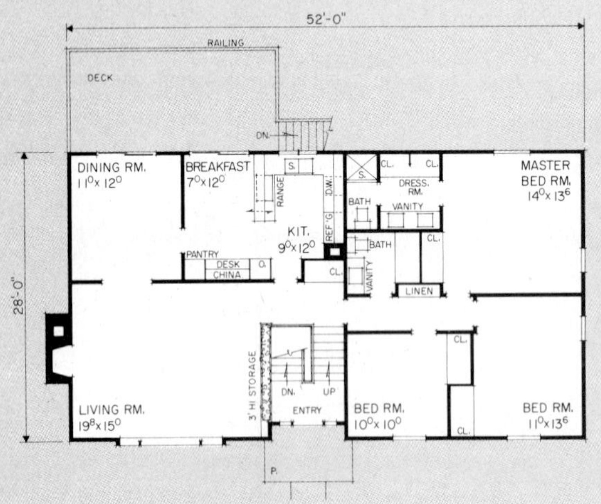

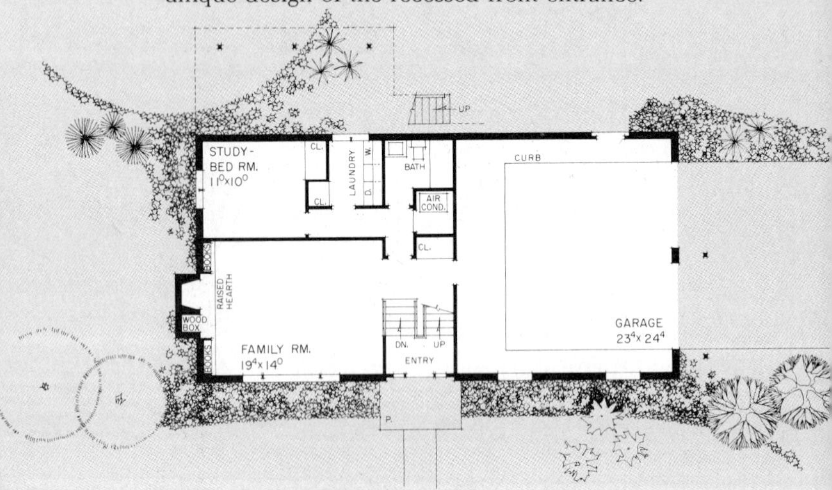

Design 41219

1,456 Sq. Ft. – Upper Level
670 Sq. Ft. – Lower Level
22,772 Cu. Ft.

● This bi-level home has large glass areas to enhance the exterior appeal. They will also enable you to have an exciting and commanding view of your natural surroundings. Two terraces and a balcony to serve the family outdoors. Interior livability will take place on the upper level with the exception of the family room on the lower level.

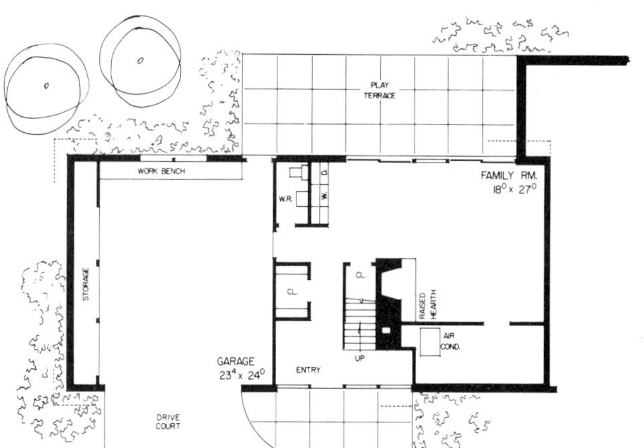

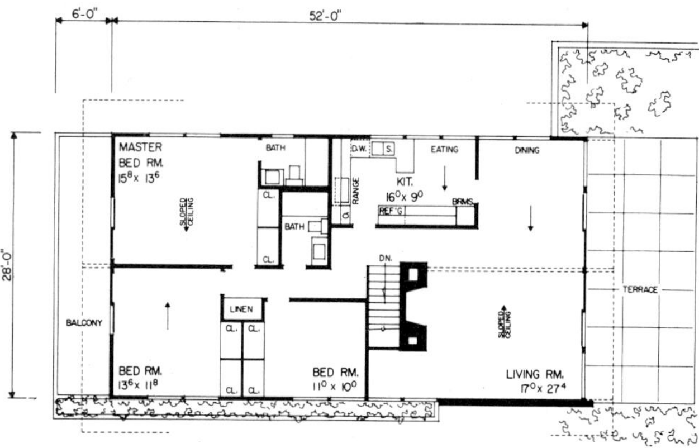

Design 41220

1,456 Sq. Ft. – Upper Level
862 Sq. Ft. – Lower Level
22,563 Cu. Ft.

● This fresh, contemporary exterior sets the stage for exceptional livability. Measuring only 52 across the front, this bi-level home offers the large family outstanding features. Whether called upon to function as a four or five bedroom home, there is plenty of space in which to move around.

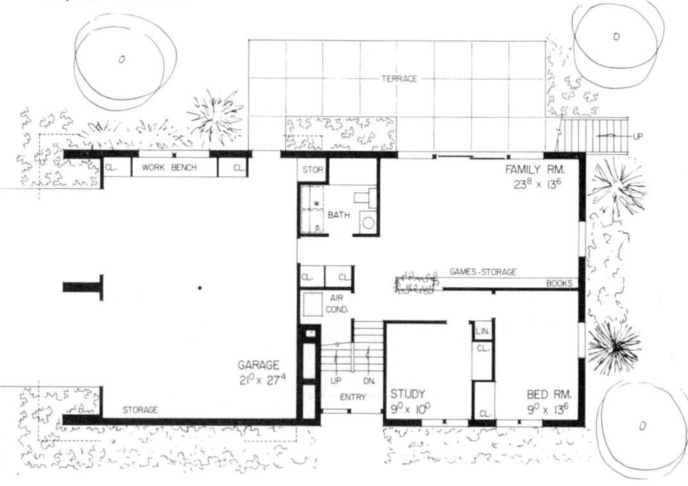

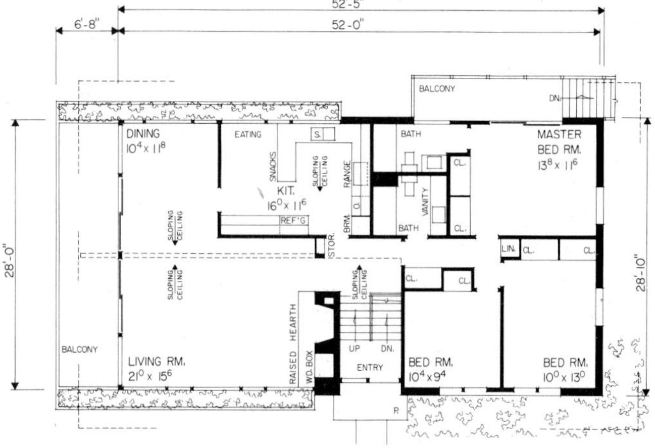

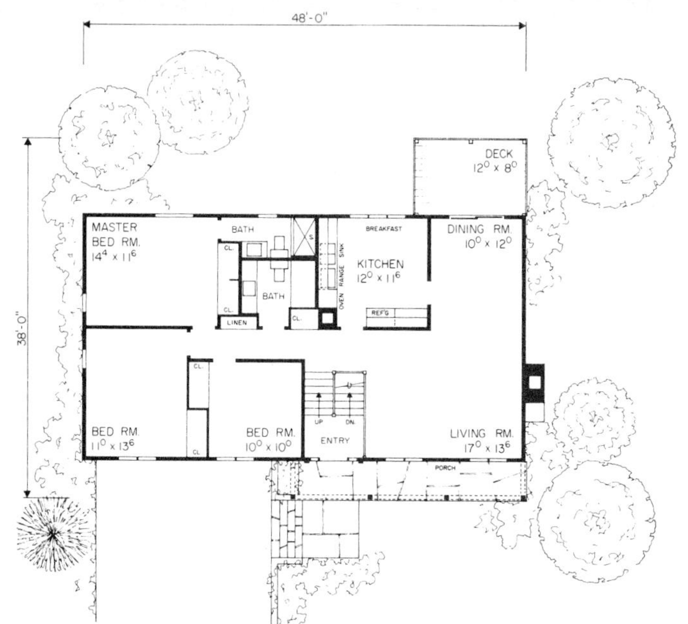

Design 41210

1,248 Sq. Ft. - Upper Level
676 Sq. Ft. - Lower Level
19,812 Cu. Ft.

Design 41386

880 Sq. Ft. - Upper Level
596 Sq. Ft. - Lower Level
14,043 Cu. Ft.

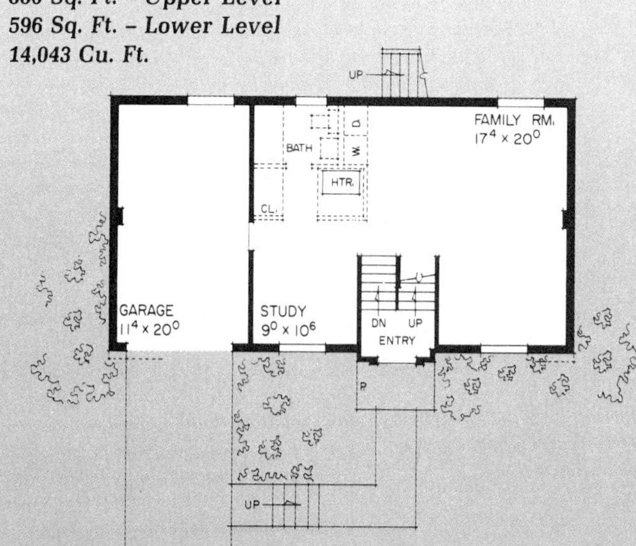

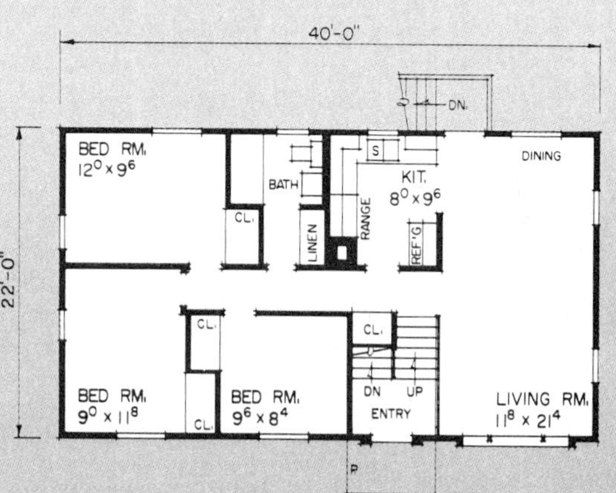

● These three designs feature traditional exterior styling with a split-foyer bi-level living interior. Owners of a bi-level design achieve a great amount of livability from an economical plan without sacrificing any of the fine qualities of a much larger and more expensive plan.

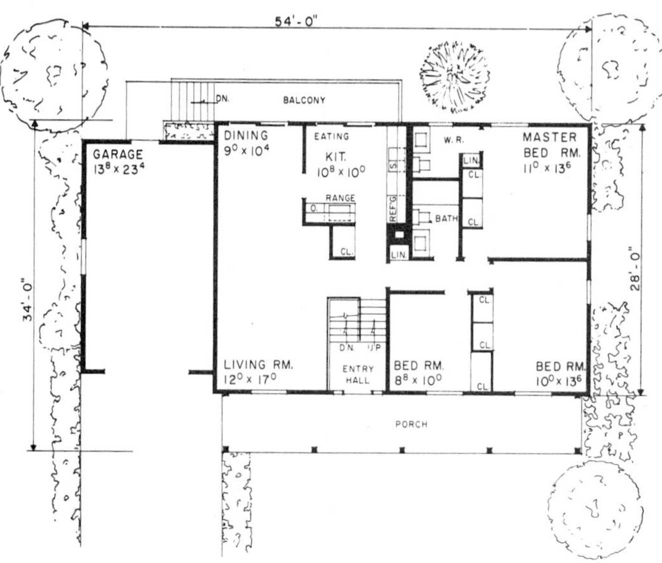

Design 41192

1,120 Sq. Ft. - Upper Level
1,120 Sq. Ft. - Lower Level
24,912 Cu. Ft.

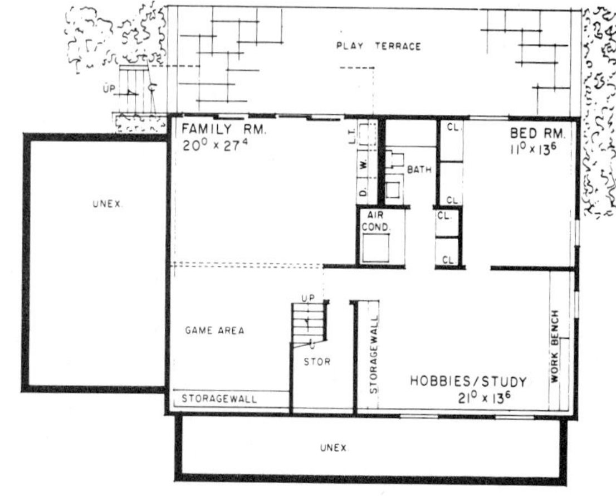

Design 42299

1,281 Sq. Ft. – Upper Level
1,320 Sq. Ft. – Lower Level
30,817 Cu. Ft.

● A dramatically simple, contemporary bi-level, or split-foyer, home. This rather geometric design is as interesting and distinctive on the inside as on the outside. Double front doors open to a well-lighted entry made possible by the large glass panel above. Down six steps is the main living level. Across the back and functioning with the rear terrace are the spacious formal and informal living areas. A massive two-way fireplace may be enjoyed from each room. The kitchen is most efficient and is but a step from the nook. The laundry and wash room are nearby. The study, or TV room, surely will be a popular haven. Up eight steps from the front entry is the four bedroom, two bath upper level. A balcony looks out upon the sloping ceiling and down into the lower level living areas. The master bedroom has a dressing room and is adjacent to a sizable storage area. This will be a handy spot to store all the seasonal paraphernalia a large family falls heir to. Bi-level living will be fun.

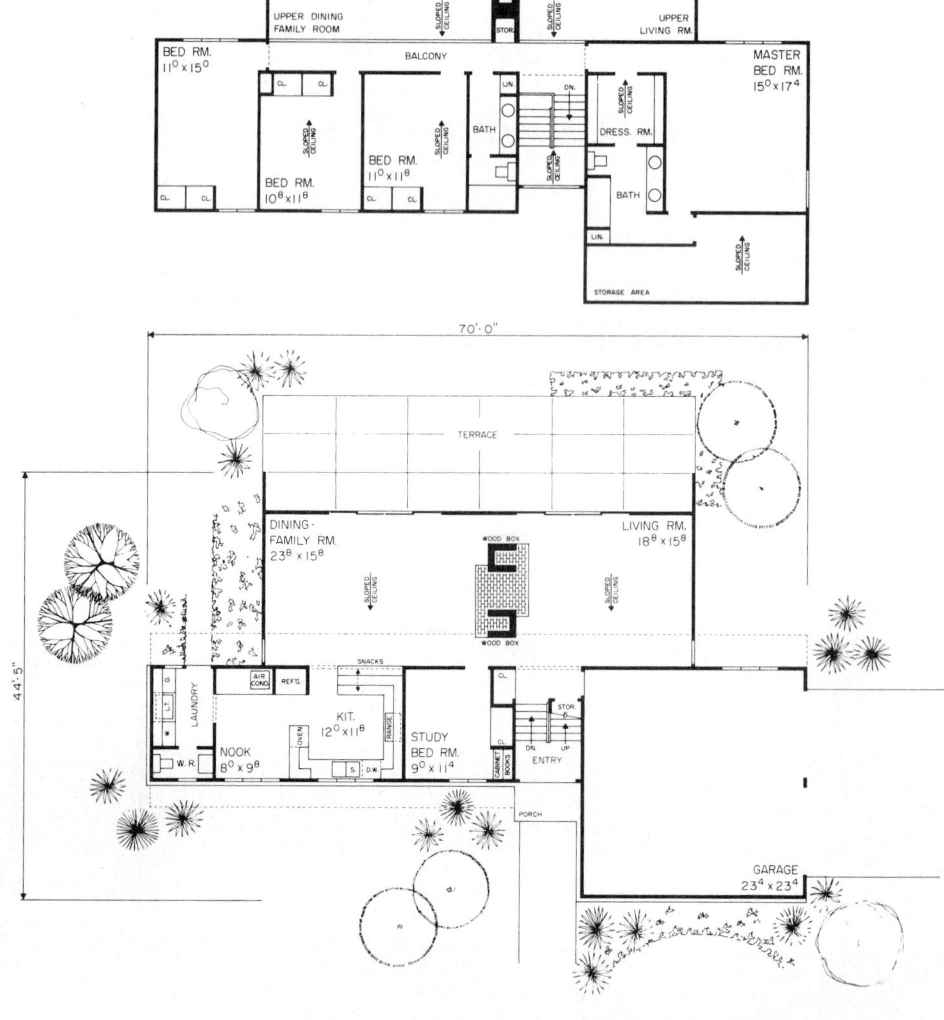

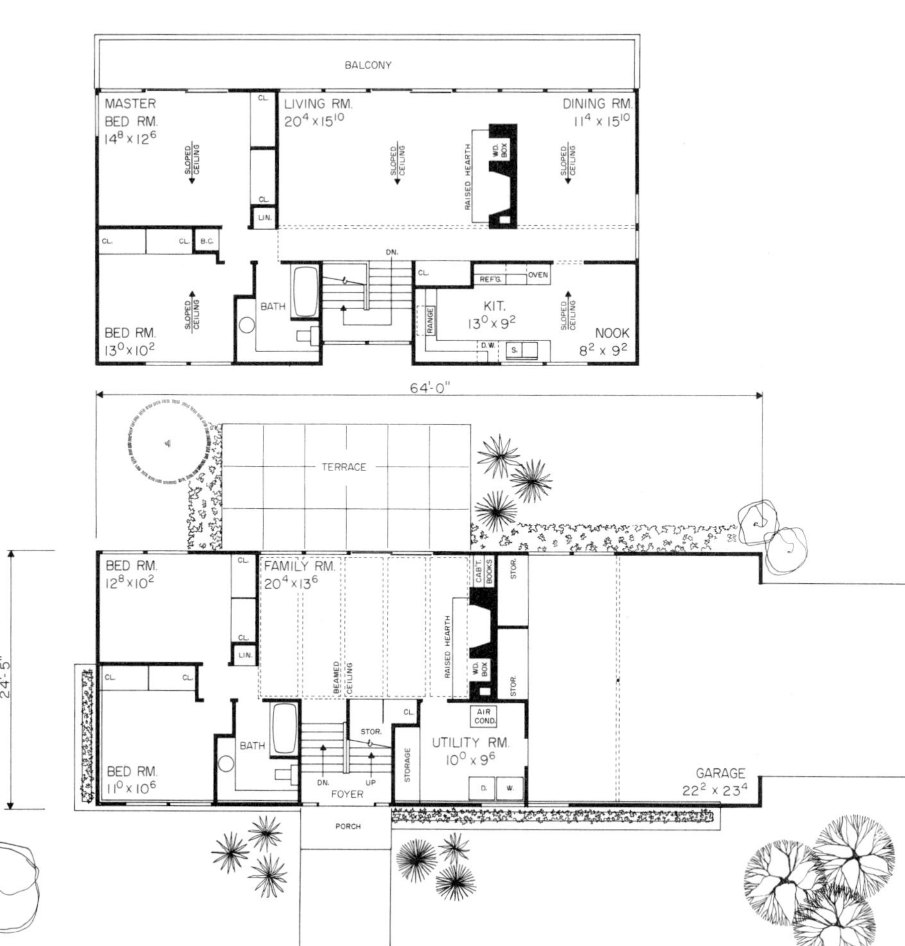

Design 42319

1,343 Sq. Ft. - Upper Level
980 Sq. Ft. - Lower Level
23,290 Cu. Ft.

● This rectangular bi-level home will be most economical to build. The wide overhanging, low-pitched roof enhances the appeal of the exterior. Notice how the upper level in turn overhangs the lower level. Your invested dollar delivers tremendous livability. There are four bedrooms (two on each level); a family activities area, plus a more formal living room with an adjacent dining room; a fine functioning kitchen opening into the breakfast nook; two raised hearth fireplaces; a sizable utility room and two-car garage. Be sure to observe the sloping ceilings of the upper level and the beamed ceiling of the family room. Don't miss the sweeping outdoor balcony which provides the main level with an outdoor living facility.

Design 41355 *879 Sq. Ft. – Upper Level; 879 Sq. Ft. – Lower Level; 16,920 Cu. Ft.*

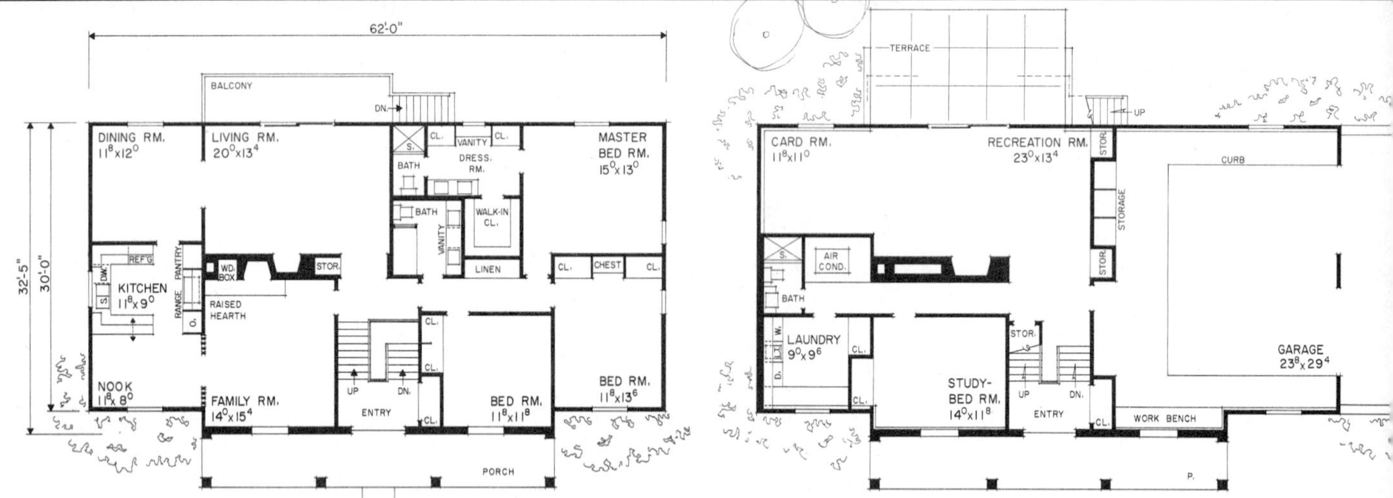

● Here is a unique bi-level. Not only in its delightful exterior appeal, but in its practical planning. The covered porch with its impressive columns, the contrasting use of materials and the traditional window and door detailing, are all features which will surely provoke comment from passers-by. The upper level is a complete living unit of three bedrooms, two baths, separate living, dining and family rooms, a kitchen with an eating area, two fireplaces and an outdoor balcony. The lower level represents extra living space which is bright and cheerful.

Design 41349 *1,132 Sq. Ft. – Upper Level; 1,132 Sq. Ft. – Lower Level; 23,206 Cu. Ft.*

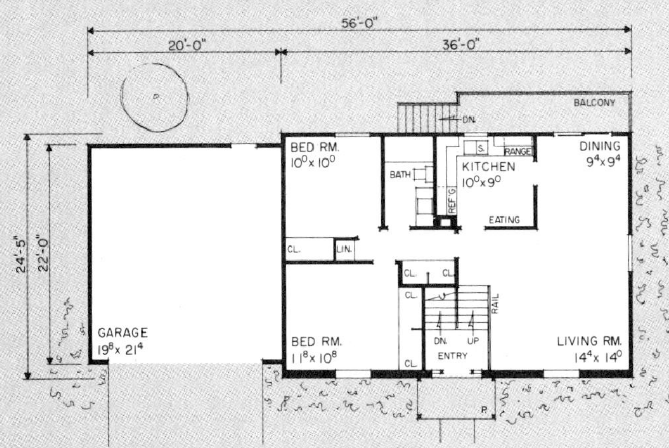

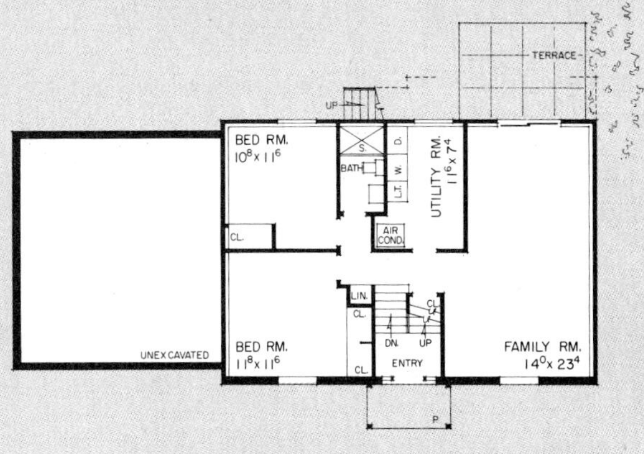

● An economically built Colonial bi-level with an exceptional amount of living space. Here is, perhaps, your best possible return on your construction dollar. Under a relatively small roof and on a similar foundation there's over 1,700 square feet of livability. It is used to provide four bedrooms, two baths, a 23 foot family room, a spacious living and dining area, an L-shaped kitchen with eating space, good closet facilities and a utility room. Everyone will enjoy the balcony off the dining room. There is a terrace off the family room.

Design 41822 *1,836 Sq. Ft. – Upper Level; 1,150 Sq. Ft. – Lower Level; 33,280 Cu. Ft.*

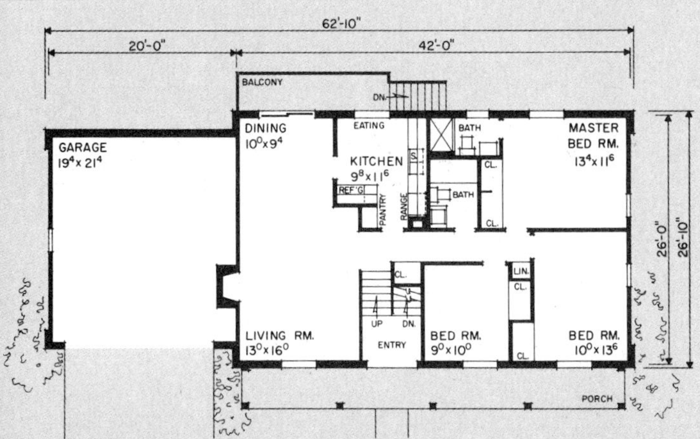

● Just think, six bedrooms! Or five bedrooms and a study, or any other kind of combination you might want. If you wanted to enlarge the master bedroom into a suite of rooms this would be entirely possible. However, if you have a lot of children and want plenty of low-cost space this bi-level should fill the bill. There is kitchen eating space and a formal dining area for those special occasions. Other attractions include a 25 foot family room, a utility room and three full baths. The upper level balcony, accessible from the formal area will be a lot of fun.

Design 42547

1,946 Sq. Ft. – Main Level
1,340 Sq. Ft. – Lower Level; 40,166 Cu. Ft.

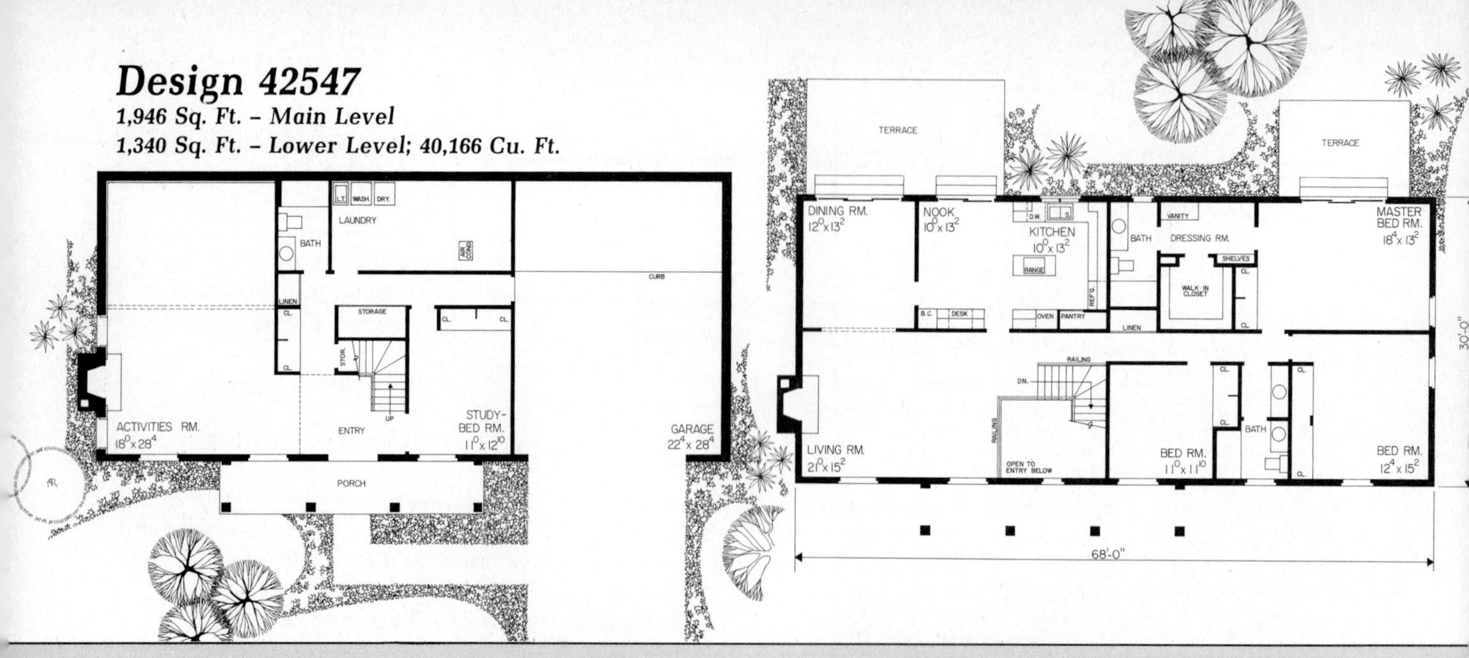

Design 42514

1,713 Sq. Ft. – Upper Level
916 Sq. Ft. – Lower Level; 32,000 Cu. Ft.

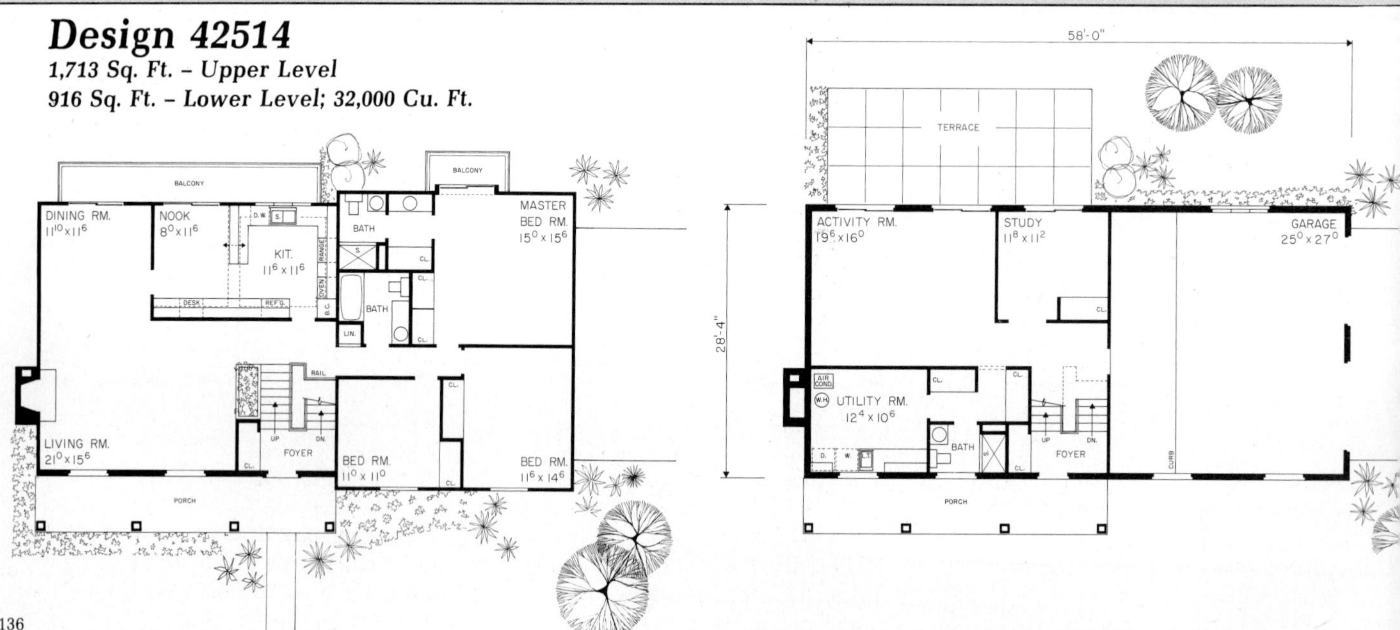

Design 41341 1,248 Sq. Ft. – Upper Level; 676 Sq. Ft. – Lower Level; 19,812 Cu. Ft.

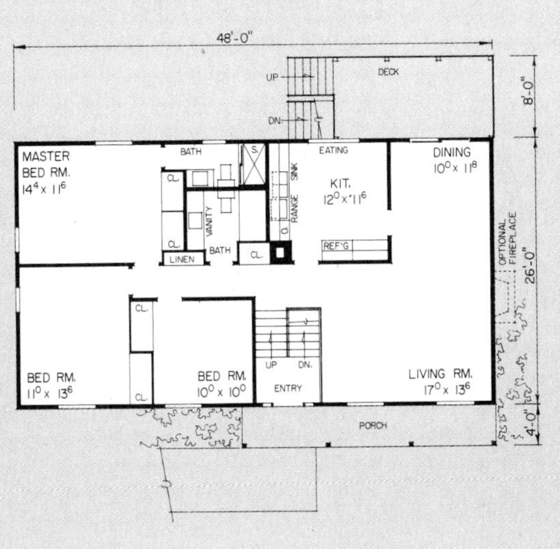

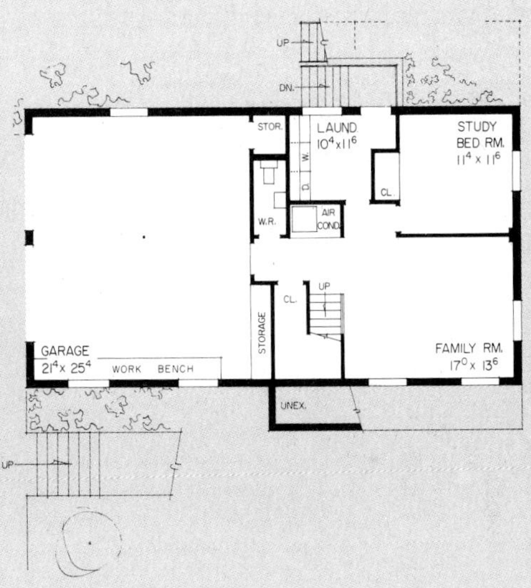

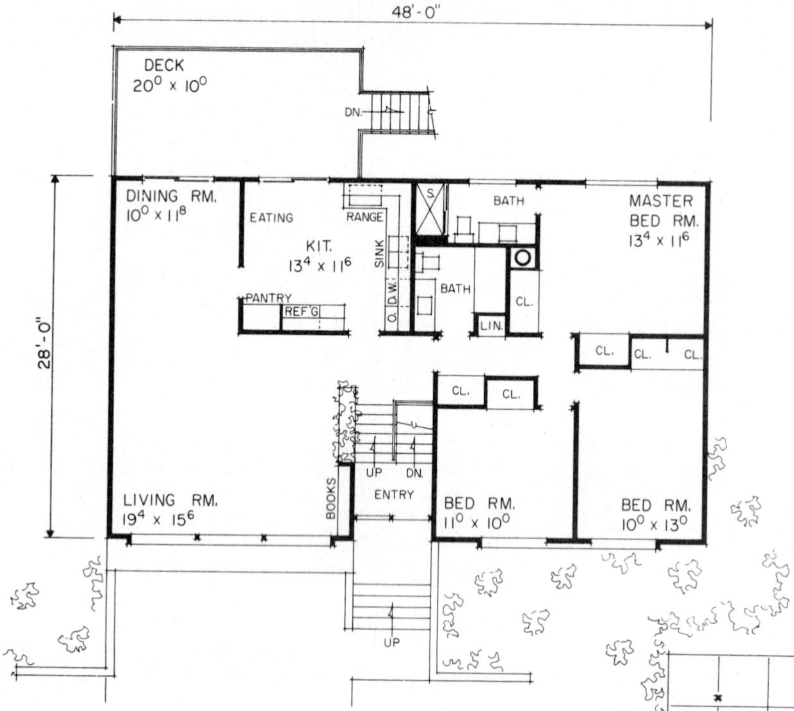

Design 41778

1,344 Sq. Ft. – Upper Level
768 Sq. Ft. – Lower Level
22,266 Cu. Ft.

● Interesting? You bet it is. The low-pitched, wide overhanging roof, the vertical siding and the dramatic glass areas give the facade of this contemporary bi-level house an appearance all its own.

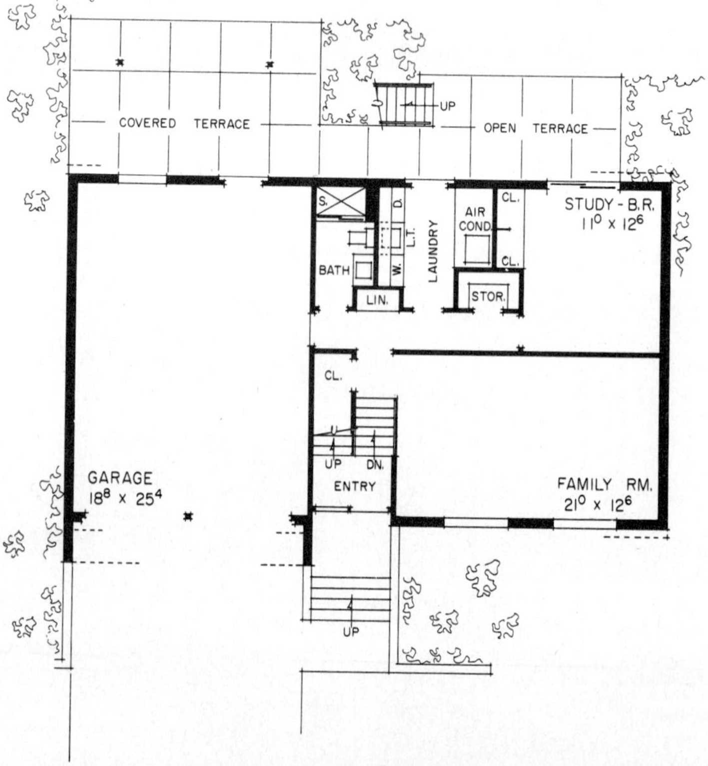

This type of house provides an outstanding return on your construction dollar. By raising the main, upper level off the ground, what would normally be a depressed full basement area also is raised. This results in a well-lighted, completely livable lower level. True, living patterns will be different. But, they will be delightfully so. The lower level is wonderfully functional and even houses the two-car garage. The upper level is a complete living unit. Note baths, eating space and deck.

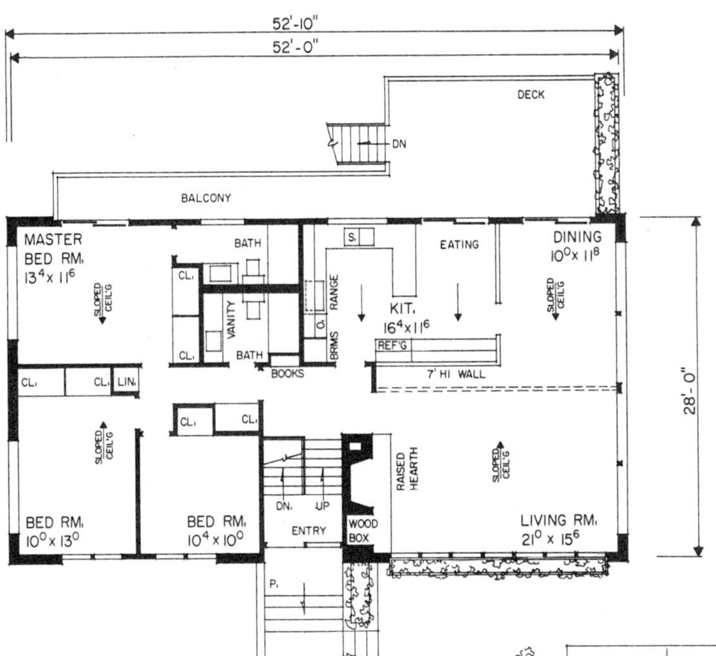

Design 41704

1,498 Sq. Ft. - Upper Level
870 Sq. Ft. - Lower Level
23,882 Cu. Ft.

● The bi-level concept of living has become popular. This is understandable, for it represents a fine way in which to gain a maximum amount of extra livable area beneath the basic floor plan.

In the case of this pleasant contemporary, the lower level not only represents an extra living area in the form of a family room, a fourth bedroom and a study; but there is also the two-car garage which because of its size can serve the dual role of work shop and car storage. The large glass areas and the sloping ceilings of the upper level living area make this a cheerful and spacious area for family use. Each level has outdoor living. The family's favorite spot will be the spacious deck.

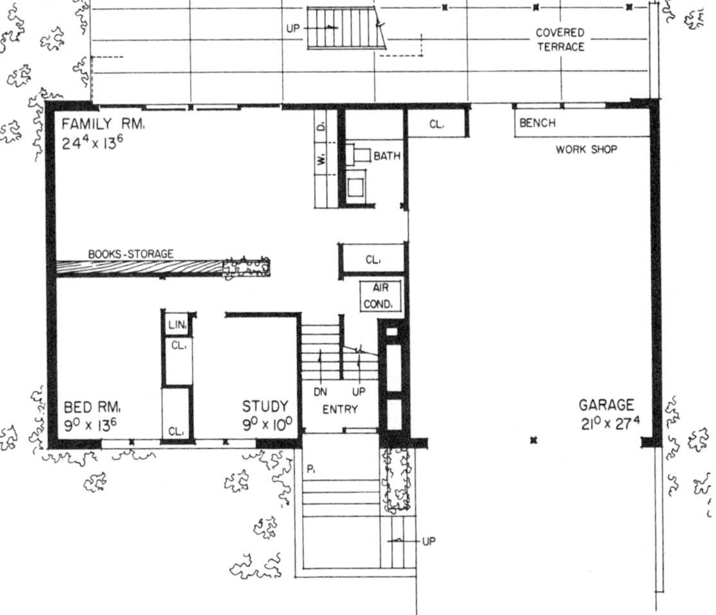

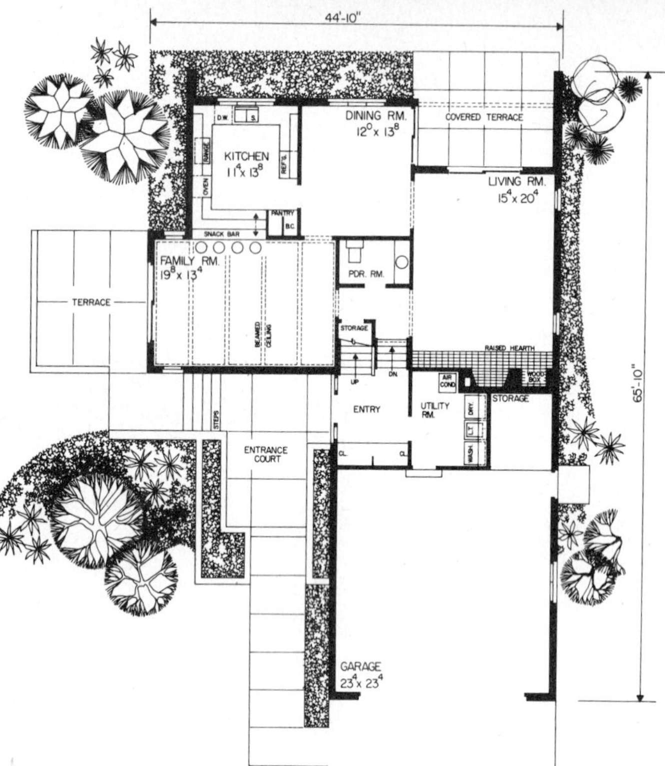

Design 42394

180 Sq. Ft. – Entry Level; 1,098 Sq. Ft. – Upper Level 1,145 Sq. Ft. – Lower Level; 24,660 Cu. Ft.

● This handsome traditional adaptation with a Mansard roof is dramatic. Its appeal is to be found in its straight forward simplicity. The inviting entrance court with its raised planters ties in with the side terrace. Other outdoor living areas include the covered terrace for the large living room and the sweeping balcony overlooking the rear yard for upstairs bedrooms. Certainly, fine indoor-outdoor living relationships.

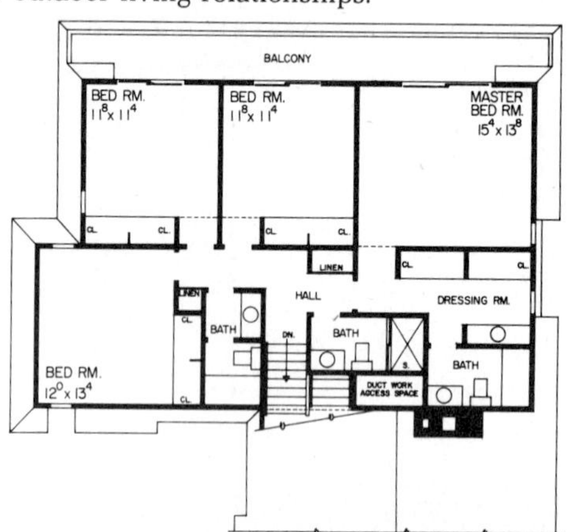

Design 41743

1,580 Sq. Ft. – Upper Level 950 Sq. Ft. – Lower Level; 25,888 Cu. Ft.

● Stately, and a delight to behold. A tailored hip roof and pleasing cornice work cap an exterior whose graciousness emanates from the contrast of the masses of brick and the delightfully delicate detailing of the windows.

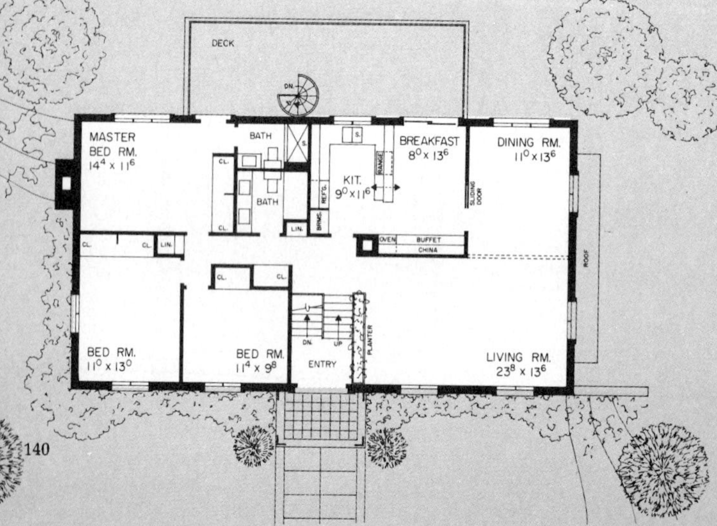

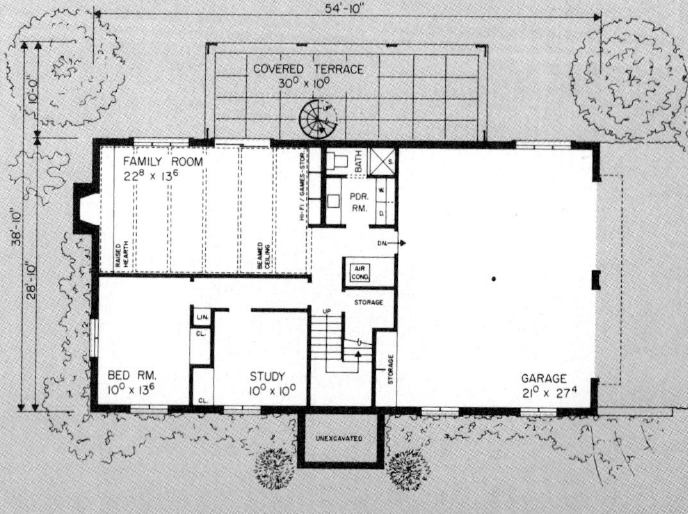

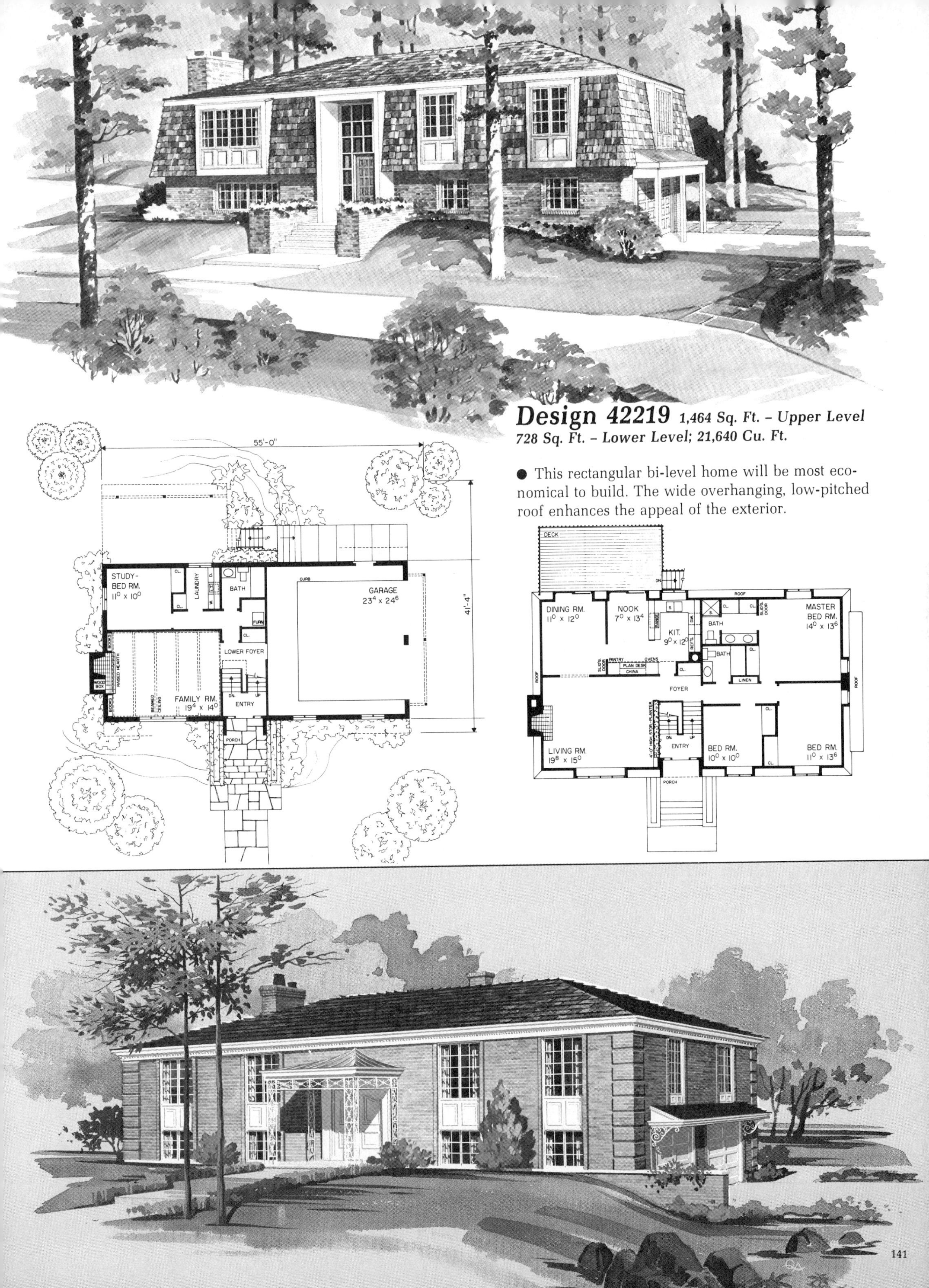

Design 42219 *1,464 Sq. Ft. – Upper Level 728 Sq. Ft. – Lower Level; 21,640 Cu. Ft.*

● This rectangular bi-level home will be most economical to build. The wide overhanging, low-pitched roof enhances the appeal of the exterior.

Design 42830 *1,795 Sq. Ft. – Main Level; 1,546 Sq. Ft. – Lower Level; 49,900 Cu. Ft.*

● Outstanding contemporary design! This home has been created with the advantages of passive solar heating in mind. For optimum energy savings, this delightful design combines passive solar devices, the solarium, with optional active collectors. Included with the purchase of this design are four plot plans to assure that the solar collectors will face the south. The garage in each plan acts as a buffer against cold northern winds. Schematic details for solar application also are included. Along with being energy-efficient, this design has excellent living patterns. Three bedrooms, the master bedroom on the main level and two others on the lower level at each side of the solarium. The living area of the main level will be able to enjoy the delightful view of the solarium and sunken garden.

SUNKEN GARDEN
BED RM. $11^6 \times 14^2$
SOLARIUM $16^0 \times 11^4$
BALCONY ABOVE
BED RM. $11^6 \times 14^2$
LOUNGE $16^0 \times 16^0$
CL. LINEN BATH
LINEN CL. BATH
AIR COND.
BAR
SUMMER KITCHEN $11^6 \times 5^0$
BASEMENT / MECHANICAL (SOLAR EQUIPMENT)
UP
ACTIVITIES RM. $24^4 \times 10^{10}$

58'-4"
SUNKEN GARDEN
ROOF GLASS GLASS GLASS GLASS ROOF
STUDY $12^0 \times 16^4$
OPEN TO SOLARIUM BELOW
RAILING
DINING RM. $12^0 \times 12^0$
COVERED PORCH
GATHERING RM. $16^0 \times 16^0$
MASTER BED RM. $12^0 \times 18^0$
BATH
KITCHEN $12^0 \times 9^6$
WALK-IN CLOSET
VESTIBULE (AIR LOCK)
LAUND.
CL.
WALK-IN CLOSET
DRESSING
BATH
BRKFST. RM. $9^8 \times 11^6$
COVERED PORCH
73'-4"
GARAGE $22^0 \times 25^4$
STORAGE $9^4 \times 7^0$

ALL the "TOOLS" you and your builder need. . .

. . . to, first select an exterior and a floor plan for your new house that satisfy your tastes and your family's living patterns . . .
. . . then, to review the blueprints in great detail and obtain a construction cost figure . . . also, to price out the structural materials required to build . . . and, finally, to review and decide upon the specifications to which your home is to be built. Truly, an invaluable set of "tools" to launch your home planning and building programs.

1. THE PLAN BOOKS

Home Planners' unique Design Category Series makes it easy to look at and study only the types of designs for which you and your family have an interest. Each of five plan books features a specific type of home, namely: 1½ and 2-Story, One-Story Over 2000 Sq. Ft., One-Story Under 2000 Sq. Ft., Multi-Levels and Vacation Homes. In addition to the convenient Design Category Series, there is an impressive selection of other current titles. While the home plans featured in these books are also to be found in the Design Category Series, they, too, are edited for those with special tastes and requirements. Your family will spend many enjoyable hours reviewing the delightfully designed exteriors and the practical floor plans. Surely your home or office library should include a selection of these popular plan books. Your complete satisfaction is guaranteed.

2. THE CONSTRUCTION BLUEPRINTS

There are blueprints available for each of the designs published in Home Planners' current plan books. Depending upon the size, the style and the type of home, each set of blueprints consists of from five to ten large sheets. Only by studying the blueprints is it possible to give complete and final consideration to the proper selection of a design for your next home. The blueprints provide the opportunity for all family members to familiarize themselves with the features of all exterior elevations, interior elevations and details, all dimensions, special built-in features and effects. They also provide a full understanding of the materials to be used and/or selected. The low-cost of our blueprints makes it possible and indeed, practical, to study in detail a number of different sets of blueprints before deciding upon which design to build.

3. THE MATERIAL LIST

A list of materials is an integral part of the plan package. It comprises the last sheet of each set of blueprints and serves as a handy reference during the period of construction. Of course, at the pricing and the material ordering stages, it is indispensable.

4. THE SPECIFICATION OUTLINE

Each order for blueprints is accompanied by one Specification Outline. You and your builder will find this a time-saving tool when deciding upon your own individual specifications. An important reference document should you wish to write your own specifications.

A new design category series– 5 great specially edited plan books:

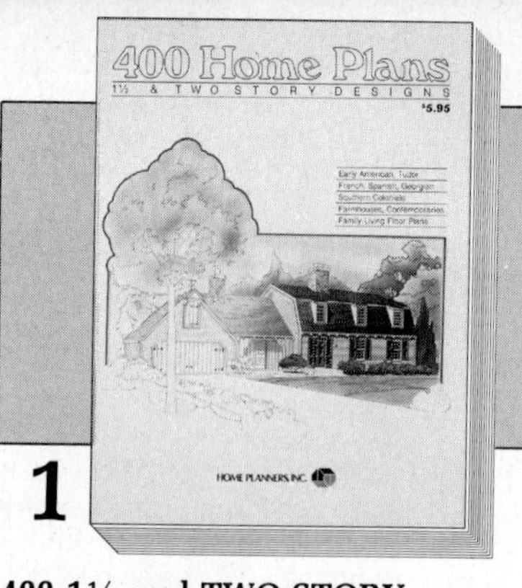

1

400 1½ and TWO-STORY HOME PLANS

Those interested in studying a wide variety of 1½ and two-story exteriors and floor plans need look no further. Here are New England Gambrels, Salt Boxes, Tudors, French Mansards, Georgians, Southern Colonials, Cape Cods, Virginia Tidewater exteriors, variations of the Farmhouse theme, Contemporary and even Multi-Family designs. Homes of varying sizes for all budgets. For family living with two to six bedrooms.
320 Pages, $5.95.

2

210 ONE-STORY HOME PLANS - Over 2,000 Square Feet

Designs for those who prefer one-story living and all the convenience that goes with it. A selection of homes with varying exterior styles - Traditional, Tudor, Spanish, French, Contemporary; with practical and efficient family living floor plans. Family rooms, separate dining rooms, atriums, efficient kitchens, laundries. Sloping beamed ceilings, sunken living rooms, libraries. An excellent design selection for unrestricted budgets.
192 Pages, $3.95.

3

350 ONE-STORY HOME PLANS - Under 2,000 Square Feet

A wide selection of one-story homes for the mod building budgets. Delightfully Traditional exte ors as well as exciting Contemporaries. Fi functioning floor plans for both the small and la family. Gathering rooms, Family room, formal a informal dining facilities, efficient work cente excellent indoor-outdoor relationships. Hon with noteworthy storage, bath and sleeping fa lities. Plans with optional elevations.
256 Pages, $4.95.

Other current titles:

Three Plan Books. . . edited by exterior styling.

6

120 EARLY AMERICAN PLANS

and Other Colonial Adaptations is an outstanding and unique plan book for the home and professional library. Devoted exclusively to Early American architectural interpretations adapted for today's living patterns. Exquisitely detailed exteriors retain all the charm of a proud heritage. One-story, 1½ and two-story and multi-level designs for varying budgets.
96 Pages, $2.25.

7

125 CONTEMPORARY HOME PLANS

Here is an exciting book featuring a wide variety of home designs for the 1980's and far beyond. The exteriors of these delightfully illustrated houses are refreshing with their practical and progressive "new look". The floor plans offer new dimensions in living highlighting such features as gathering rooms, cathedral ceilings and interior balconies. House designs of all sizes and types.
112 Pages, $2.25.

8

135 ENGLISH TUDOR HOMES

and other Popular Family Plans is a favorite many. The current popularity of the English Tud home design is phenomenal. Here is a book whi is loaded with Tudors for all budgets. There a one-story, 1½ and two-story designs, plus mul levels and hillsides from 1,176 to 3,849 squa feet. There is a special 20 page section of Ear American Adaptations.
104 Pages, $2.50.

Plans for varied budgets–

11

102 HOME PLANS

An excellent selection of home designs featuring a wide variety of exterior styles for your review. There are Early American, Tudor, Spanish, French and Contemporary facades. Outstanding, practical family living floor plans with formal and informal livability. Vacation homes for leisure living. Homes for varying budgets. A special 16 page feature section in full color.
96 Pages, $1.75.

An exquisite book in full color–

12

116 TRADITIONAL and CONTEMPORARY PLANS

A beautifully illustrated home plan book in complete, full color. Tudor, French, Spanish, Early American and Contemporary exteriors with efficient, step-saving family living floor plans. One, 1½, two-story and split-level designs for varied budgets. Also, vacation homes. Designs for flat and hillside sites. An ideal gift item.
96 Pages, $4.95.

A selection of most popular designs–

13

166 MOST POPULAR HOMES

A book of best-selling house plans containing ove 400 illustrations. Houses range in size from 1,05 to 3,874 square feet. Tudor, Early American Spanish and French exteriors. Plus Contemporar elevations and floor plans. Homes featuring atri ums, balconies, decks, sloping beamed ceilings exposed lower levels and much more. Truly ar outstanding plan book.
112 Pages, $1.95.

4
5

)5 MULTI-LEVEL HOME PLANS

r those who wish to experience new dimensions total livability. This fine collection includes split yer bi-levels and tri-levels for flat and sloping es. Also, homes with exposed lower levels. A de variety of Traditional and Contemporary terior styles. Homes for all budgets and family zes. Indoor-outdoor living patterns with decks, rraces and balconies. Solar oriented livability. ficient, practical floor plans featuring hobby eas and lounges for the active family.
)2 Pages, $3.95.

223 VACATION HOMES

A popularly acclaimed vacation and leisure-living book of exteriors and floor plans. Designs for all budgets from 480 to 2,928 square feet. Over 600 illustrations. A-Frames, Chalets, Hexagons and other interesting shapes with decks, balconies and terraces. Oriented for flat and sloping sites. Vacation houses with spacious, open planning for lakeshore or woodland living. Sleeping accommodations for two to 22. Lodges for summer and winter enjoyment. 96 exciting full color pages.
176 Pages, $4.25.

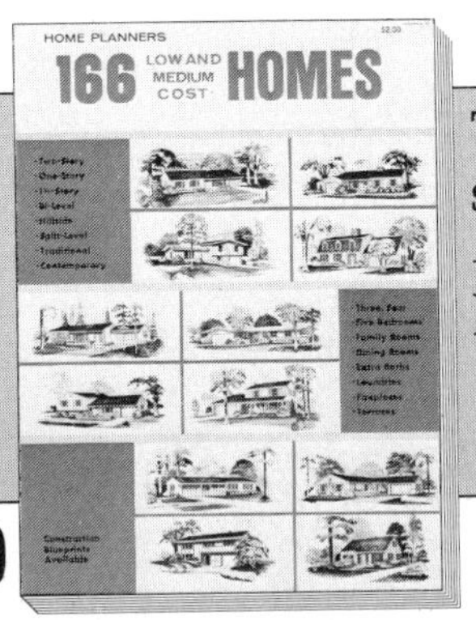

The budget series . . . –Low cost homes Affordable homes–

9
10

66 LOW & MEDIUM COST HOMES

special selection of home designs for the modest restricted building budget. An excellent variety f Traditional and Contemporary designs. One-ory, 1½ and two-story and split-level homes. hree, four and five bedrooms. Family rooms, xtra baths, formal and informal dining rooms. asement and non-basement designs. Attached arages and covered porches.
6 Pages, $2.50.

165 AFFORDABLE HOME PLANS

This book was specially edited with a wide selection of houses and plans for those with a medium budget. While none of these designs are considered low-cost; neither do they require an unlimited budget to build. Square footages range from 1,428. Exteriors of Tudor, French, Early American, Spanish and Contemporary are included.
112 Pages, $2.50

Contemporary living patterns–

14
15

112 TRADITIONAL and CONTEMPORARY FAMILY HOMES

A delightful collection of designs for traditional and contemporary tastes. All sizes and types of designs for family living. Over 300 exterior and floor plan illustrations. Designs from 1,176 to 4,431 square feet. Two to six bedrooms, formal and informal living areas, libraries, efficient kitchens, mud rooms, extra baths.
96 Pages, $1.50.

100 HOUSE PLANS FOR CONTEMPORARY LIVING

This appealing book is packed with floor plans and exteriors reflecting today's popular preferences for contemporary living. There are feature sections showing one, 1½ and two-story homes, plus multi-levels including bi-levels, tri-levels and hillside designs. Traditional as well as refreshing contemporary exteriors.
96 Pages, $1.75.

The Plan Books

. . . are a most valuable tool for anyone planning to build a new home. A study of the hundreds of delightfully designed exteriors and the practical, efficient floor plans will be a great learning and fun-oriented family experience. You will be able to select your preferred styling from among Early American, Tudor, French, Spanish and Contemporary adaptations. Your ideas about floor planning and interior livability will expand. And, of course, after you have selected an appealing home design that satisfies your long list of living requirements, you can order the blueprints for further study of your favorite design in greater detail. Surely the hours spent studying the portfolio of Home Planners' designs will be both enjoyable and rewarding ones.

Kindly note before ordering: The Design Category series contains all of the designs found in the "Other current titles."

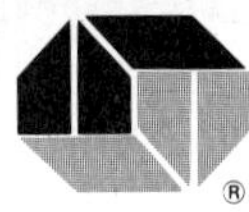

HOME PLANNERS, INC.
23761 Research Drive, Dept. BK
Farmington Hills, Michigan 48024

Please mail to me the following selection of home plan books:

1. ___400 1½ & Two-Story Home Plans @ $5.95 ea. . . . . $____
2. ___210 One-Story - Over 2,000 Sq. Ft. @ $3.95 ea. . . . . . . $____
3. ___350 One-Story - Under 2,000 Sq. Ft. @ $4.95 ea. . . . . . $____
4. ___205 Multi-Level Home Plans @ $3.95 ea. . . . . $____
5. ___223 Vacation Homes @ $4.25 ea. . . . . $____
6. ___120 Early American Home Plans @ $2.25 ea. . . . . $____
7. ___125 Contemporary Home Plans @ $2.25 ea. . . . . $____
8. ___135 English Tudor Homes @ $2.50 ea. . . . . $____
9. ___166 Low & Medium Cost Homes @ $2.50 ea. . . . . $____
10. ___165 Affordable Home Plans @ $2.50 ea. . . . . $____
11. ___102 Home Plans for Varied Budgets @ $1.75 ea. . . . . $____
12. ___116 Traditional & Contemporary Plans @ $4.95 ea. . . . $____
13. ___166 Most Popular Homes @ $1.95 ea. . . . . $____
14. ___112 Trad. & Cont. Family Homes @ $1.50 ea. . . . . $____
15. ___100 House Plans for Contemporary Living @ $1.75 ea. $____

Sub Total $____

SATISFACTION GUARANTEED!

MAIL TODAY!
Your order will be processed & shipped within 48 hours.

Michigan Residents add 4% sales tax $____

TOTAL in U.S.A. funds $____

Please Print
Name ____
Address ____
City ____
State ____ Zip ____

In Canada Mail To: Home Planners, Inc. 772 King St. W. Kitchener, Ontario N2G 1E8

CV4

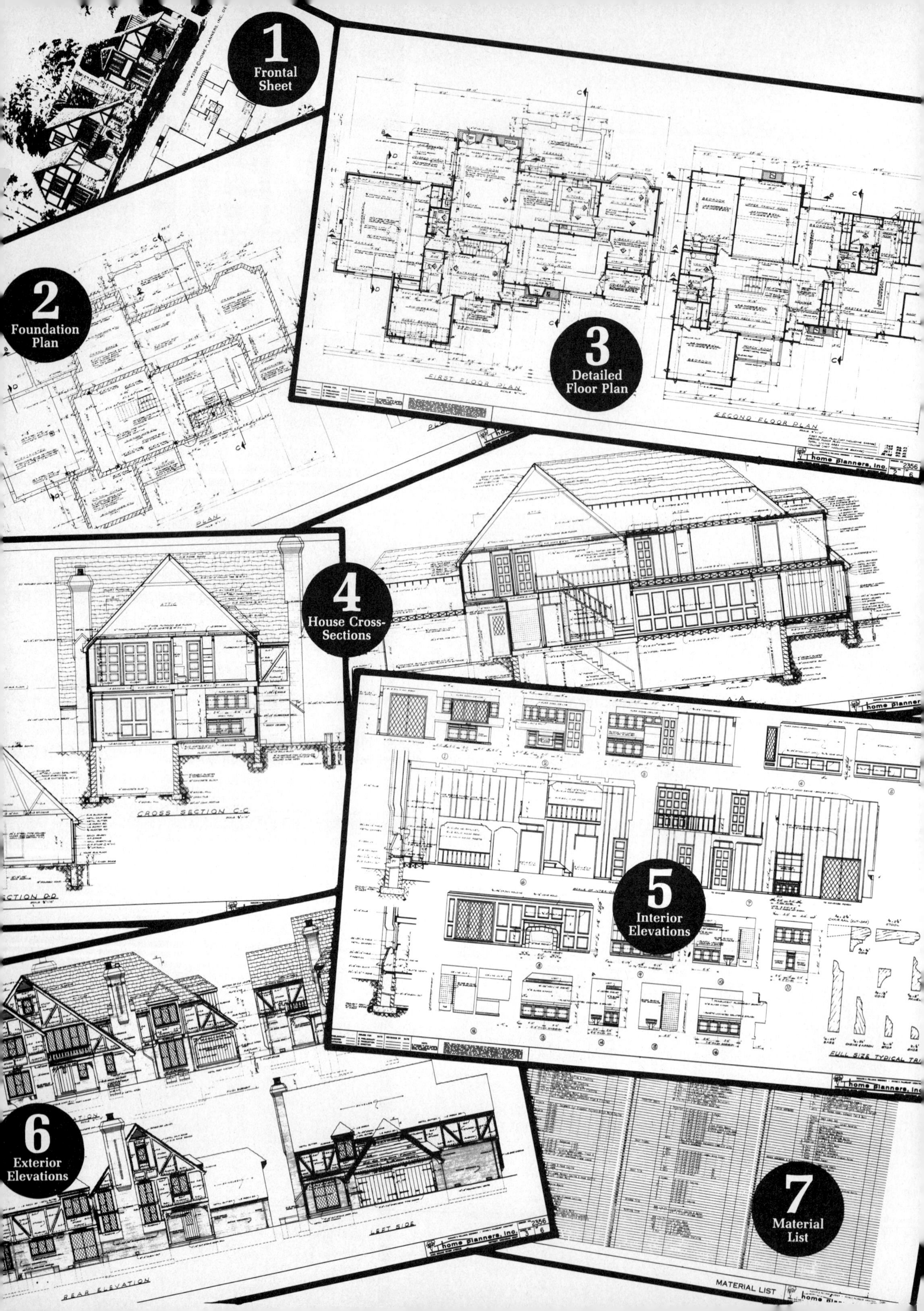

1
Frontal Sheet
2
Foundation Plan
3
Detailed Floor Plan
4
House Cross-Sections
5
Interior Elevations
6
Exterior Elevations
7
Material List
FIRST FLOOR PLAN
SECOND FLOOR PLAN
home planners, inc.
CROSS SECTION C-C
LEFT SIDE
REAR ELEVATION
MATERIAL LIST

The Blueprints. . .

1. FRONTAL SHEET.

Artist's landscaped sketch of the exterior and ink-line floor plans are on the frontal sheet of each set of blueprints.

2. FOUNDATION PLAN.

¼" Scale basement and foundation plan. All necessary notations and dimensions. Plot plan diagram for locating house on building site.

3. DETAILED FLOOR PLAN.

¼" Scale first and second floor plans with complete dimensions. Cross-section detail keys. Diagrammatic layout of electrical outlets and switches.

4. HOUSE CROSS-SECTIONS.

Large scale sections of foundation, interior and exterior walls, floors and roof details for design and construction control.

5. INTERIOR ELEVATIONS.

Large scale interior details of the complete kitchen cabinet design, bathrooms, powder room, laundry, fireplaces, paneling, beam ceilings, built-in cabinets, etc.

6. EXTERIOR ELEVATIONS.

¼" Scale exterior elevation drawings of front, rear, and both sides of the house. All exterior materials and details are shown to indicate the complete design and proportions of the house.

7. MATERIAL LIST.

Complete lists of all materials required for the construction of the house as designed are included in each set of blueprints.

THIS BLUEPRINT PACKAGE

will help you and your family take a major step forward in the final appraisal and planning of your new home. Only by spending many enjoyable and informative hours studying the numerous details included in the complete package, will you feel sure of, and comfortable with, your commitment to build your new home. To assure successful and productive consultation with your builder and/or architect, reference to the various elements of the blueprint package is a must. The blueprints, material list and specification outline will save much consultation time and expense. Don't be without them.

The Material List. . .

With each set of blueprints you order you will receive a material list. Each list shows you the quantity, type and size of the non-mechanical materials required to build your home. It also tells you where these materials are used. This makes the blueprints easy to understand.

Influencing the mechanical requirements are geographical differences in availability of materials, local codes, methods of installation and individual preferences. Because of these factors, your local heating, plumbing and electrical contractors can supply you with necessary material take-offs for their particular trades.

Material lists simplify your material ordering and enable you to get quicker price quotations from your builder and material dealer. Because the material list is an integral part of each set of blueprints, it is not available separately.

Among the materials listed:

• Masonry, Veneer & Fireplace • Framing Lumber • Roofing & Sheet Metal • Windows & Door Frames • Exterior Trim & Insulation • Tile Work, Finish Floors • Interior Trim, Kitchen Cabinets • Rough & Finish Hardware

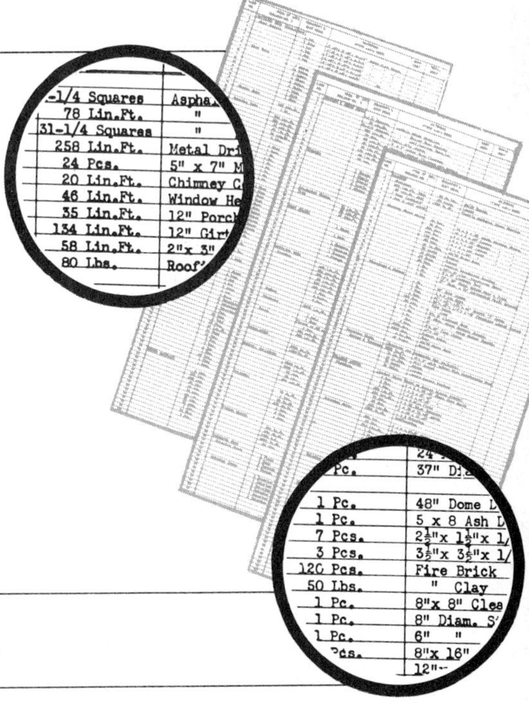

The Specification Outline. . .

This fill-in type specification lists over 150 phases of home construction from excavating to painting and includes wiring, plumbing, heating and air-conditioning. It consists of 16 pages and will prove invaluable for specifying to your builder the exact materials, equipment and methods of construction you want in your new home. One Specification Outline is included free with each order for blueprints. Additional Specification Outlines are available at $3.00 each.

CONTENTS

• General Instructions, Suggestions and Information • Excavating and Grading • Masonry and Concrete Work • Sheet Metal Work • Carpentry, Millwork, Roofing, and Miscellaneous Items • Lath and Plaster or Drywall Wallboard • Schedule for Room Finishes • Painting and Finishing • Tile Work • Electrical Work • Plumbing • Heating and Air-Conditioning

Before you order

1. STUDY THE DESIGNS . . . found in Home Planners current publications. As you review these delightful custom homes, you should keep in mind the total living requirements of your family — both indoors and outdoors. Although we do not make changes in plans, many minor changes can be made prior to the period of construction. If major changes are involved to satisfy your personal requirements, you should consider ordering one set of blueprints and having them redrawn locally. Consultation with your architect is strongly advised when contemplating major changes.

2. HOW TO ORDER BLUEPRINTS . . . After you have chosen the design that satisfies your requirements, or if you have selected one that you wish to study in more detail, simply clip the accompanying order blank and mail with your remittance. However, if it is not convenient for you to send a check or money order, you can use your credit card, or merely indicate C.O.D. shipment. Postman will collect all charges, including postage and C.O.D. fee. C.O.D. shipments are not permitted to Canada or foreign countries. Should time be of essence, as it sometimes is with many of our customers, your telephone order usually can be processed and shipped in the next day's mail. Simply call toll free 1-800-521-6797, (Michigan residents call collect 0-313-477-1854).

3. OUR SERVICE . . . Home Planners makes every effort to process and ship each order for blueprints and books within 48 hours. Because of this, we have deemed it unnecessary to acknowledge receipt of our customers orders. See order coupon below for the postage and handling charges for surface mail, air mail or foreign mail.

4. A NOTE REGARDING REVERSE BLUEPRINTS . . . As a special service to those wishing to build in reverse of the plan as shown, we do include an extra set of reversed blueprints for only $20.00 additional with each order. Even though the lettering and dimensions appear backward on reversed blueprints, they make a handy reference because they show the house just as it's being built in reverse from the standard blueprints — thereby helping you visualize the home better.

5. OUR EXCHANGE POLICY . . . Since blueprints are printed up in specific response to your individual order, we cannot honor requests for refunds. However, the first set of blueprints in any order (or the one set in a single set order) for a given design may be exchanged for a set of another design at a fee of $10.00 plus $2.00 for postage and handling via surface mail; $3.00 via air mail.

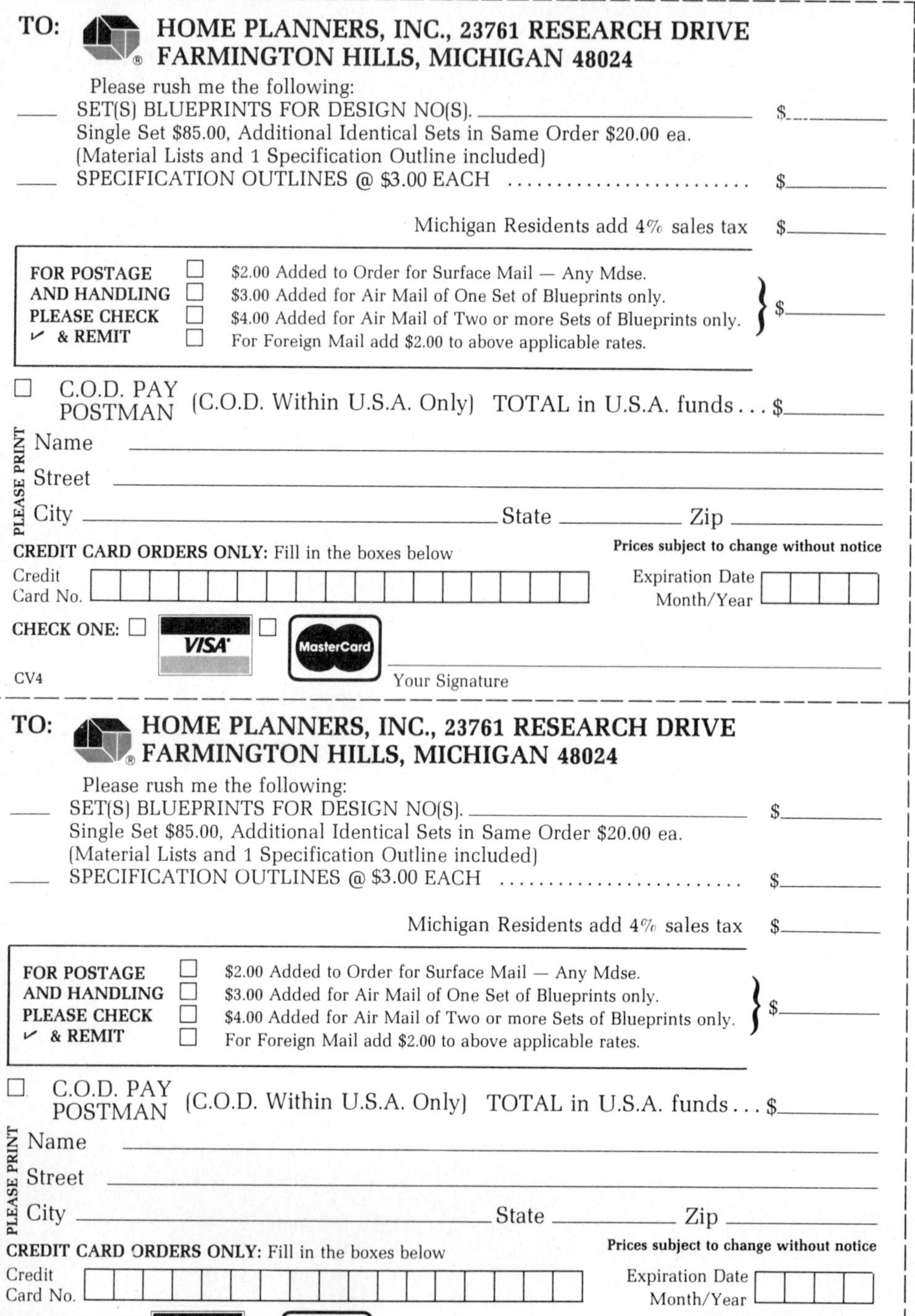

TO: HOME PLANNERS, INC., 23761 RESEARCH DRIVE
FARMINGTON HILLS, MICHIGAN 48024

Please rush me the following:

___ SET(S) BLUEPRINTS FOR DESIGN NO(S). ________ $______
Single Set $85.00, Additional Identical Sets in Same Order $20.00 ea.
(Material Lists and 1 Specification Outline included)

___ SPECIFICATION OUTLINES @ $3.00 EACH . . . . $______

Michigan Residents add 4% sales tax $______

FOR POSTAGE AND HANDLING PLEASE CHECK ✓ & REMIT
☐ $2.00 Added to Order for Surface Mail — Any Mdse.
☐ $3.00 Added for Air Mail of One Set of Blueprints only.
☐ $4.00 Added for Air Mail of Two or more Sets of Blueprints only.
☐ For Foreign Mail add $2.00 to above applicable rates.
$______

☐ C.O.D. PAY POSTMAN (C.O.D. Within U.S.A. Only) TOTAL in U.S.A. funds . . . $______

PLEASE PRINT
Name ________
Street ________
City ________ State ______ Zip ______

CREDIT CARD ORDERS ONLY: Fill in the boxes below

Prices subject to change without notice

Credit Card No. ________ Expiration Date Month/Year ______

CHECK ONE: ☐ VISA ☐ MasterCard

Your Signature ________

CV4

TO: HOME PLANNERS, INC., 23761 RESEARCH DRIVE
FARMINGTON HILLS, MICHIGAN 48024

Please rush me the following:

___ SET(S) BLUEPRINTS FOR DESIGN NO(S). ________ $______
Single Set $85.00, Additional Identical Sets in Same Order $20.00 ea.
(Material Lists and 1 Specification Outline included)

___ SPECIFICATION OUTLINES @ $3.00 EACH . . . . $______

Michigan Residents add 4% sales tax $______

FOR POSTAGE AND HANDLING PLEASE CHECK ✓ & REMIT
☐ $2.00 Added to Order for Surface Mail — Any Mdse.
☐ $3.00 Added for Air Mail of One Set of Blueprints only.
☐ $4.00 Added for Air Mail of Two or more Sets of Blueprints only.
☐ For Foreign Mail add $2.00 to above applicable rates.
$______

☐ C.O.D. PAY POSTMAN (C.O.D. Within U.S.A. Only) TOTAL in U.S.A. funds . . . $______

PLEASE PRINT
Name ________
Street ________
City ________ State ______ Zip ______

CREDIT CARD ORDERS ONLY: Fill in the boxes below

Prices subject to change without notice

Credit Card No. ________ Expiration Date Month/Year ______

CHECK ONE: ☐ VISA ☐ MasterCard

Your Signature ________

CV4

How many sets of blueprints should be ordered?

This question is often asked. The answer can range anywhere from 1 to 8 sets, depending upon circumstances. For instance, a single set of blueprints of your favorite design is sufficient to study the house in greater detail. On the other hand, if you are planning to get cost estimates, or if you are planning to build, you may need as many as eight sets of blueprints. Because the first set of blueprints in each order is $85.00, and because additional sets of the same design in each order are only $20.00 each, you save considerably by ordering your total requirements now. To help you determine the exact number of sets, please refer to the handy check list below.

HOW MANY BLUEPRINTS DO YOU NEED?

___**OWNER'S SET**

___**BUILDER** (Usually requires at least 3 sets: 1 as legal document; 1 for inspection; and at least 1 for tradesmen — usually more.)

___**BUILDING PERMIT** (Sometimes 2 sets are required.)

___**MORTGAGE SOURCE** (Usually 1 set for a conventional mortgage; 3 sets for F.H.A. or V.A. type mortgages.)

___**SUBDIVISION COMMITTEE** (If any.)

___**TOTAL NO. SETS REQUIRED**

BLUEPRINTS SHIPPED WITHIN 48 HOURS!

BLUEPRINT ORDERING HOTLINE

Phone toll free: 1-800-521-6797.

Orders received by 11 a.m. (Detroit time) can usually be processed and shipped to you the same day. Use of this line restricted to blueprint ordering only.

In Canada Mail To:
Home Planners, Inc., 772 King St.W.
Kitchener, Ontario N2G 1E8

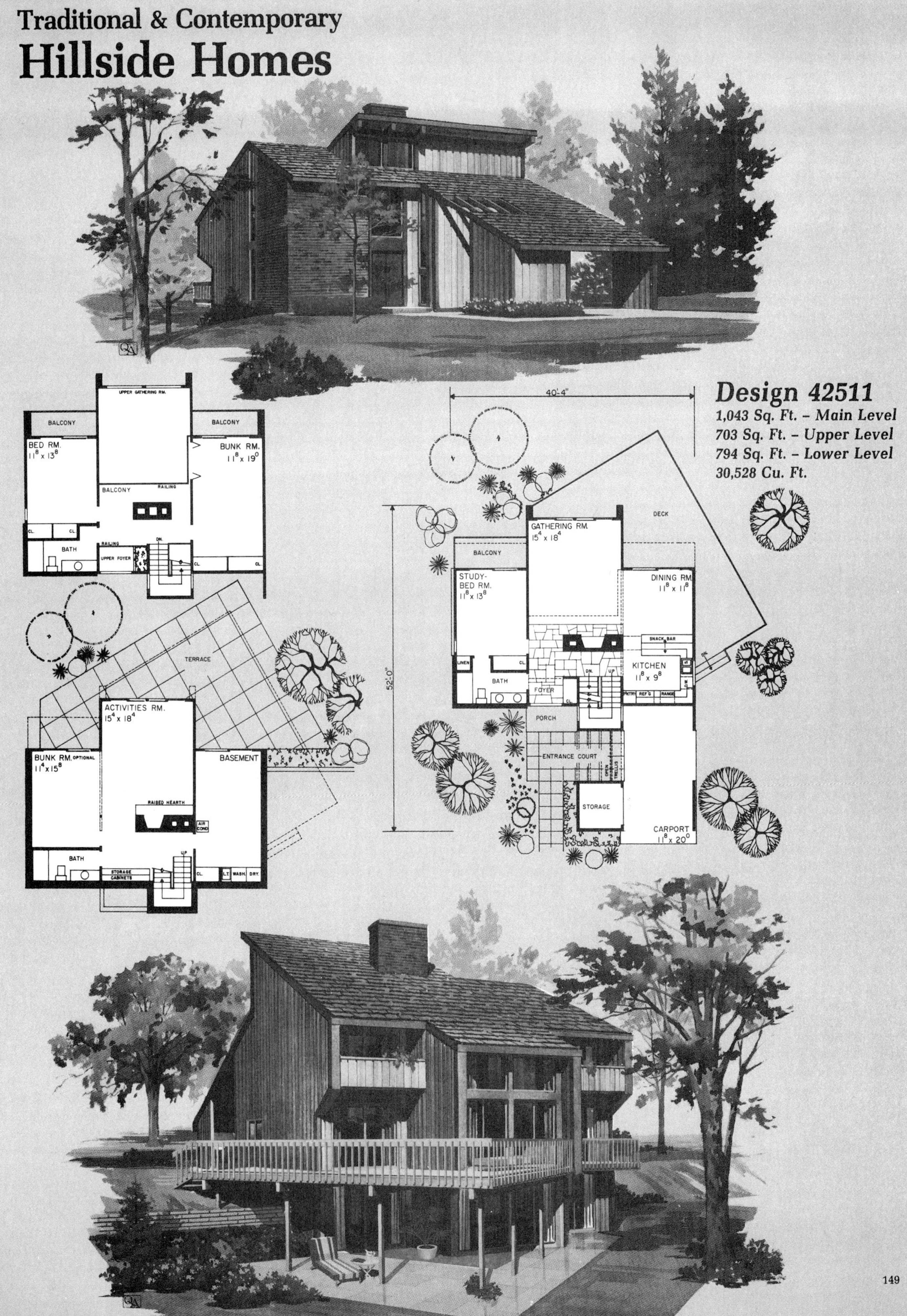

Design 42511

1,043 Sq. Ft. – Main Level
703 Sq. Ft. – Upper Level
794 Sq. Ft. – Lower Level
30,528 Cu. Ft.

Design 41974 *1,680 Sq. Ft. - Main Level; 1,344 Sq. Ft. - Lower Level; 34,186 Cu. Ft.*

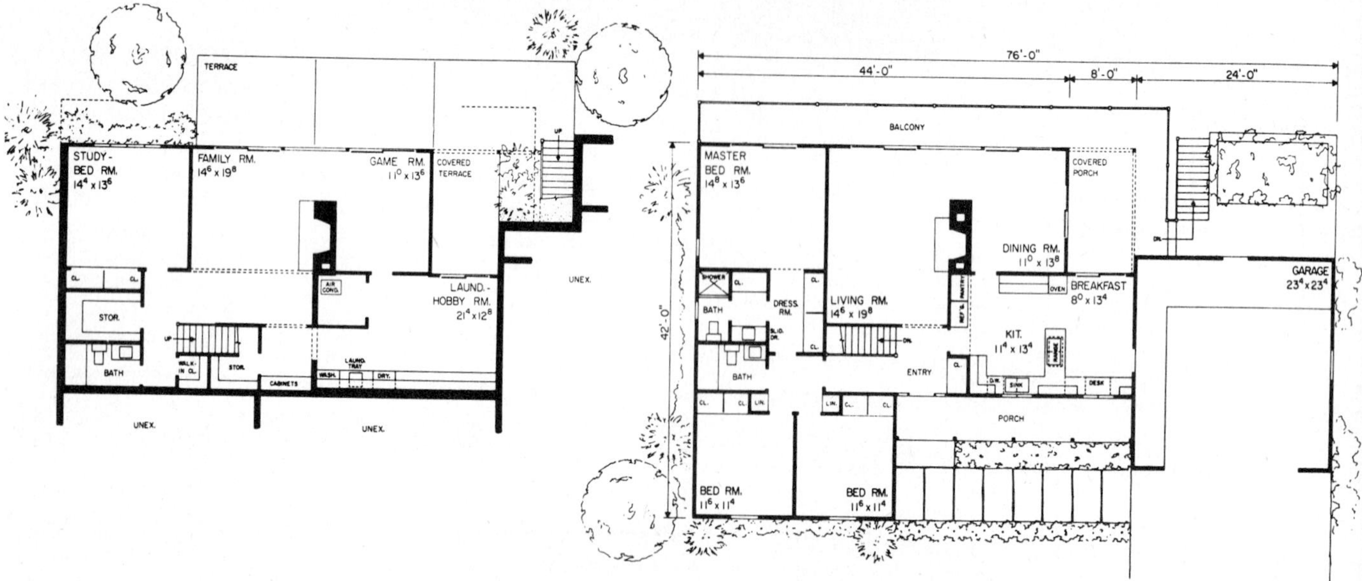

● You would never guess from looking at the front of this traditional design that it possessed such a strikingly different rear. From the front, you would guess that all of its livability is on one floor. Yet, just imagine the tremendous amount of livability that is added to the plan as a result of exposing the lower level - 1,344 square feet of it. Living in this hillside house will mean fun. Obviously, the most popular spot will be the balcony. Then again, maybe it could be the terrace adjacent to the family room. Both the terrace and the balcony have a covered area to provide protection against unfavorable weather. The interior of the plan also will serve the family with ease.

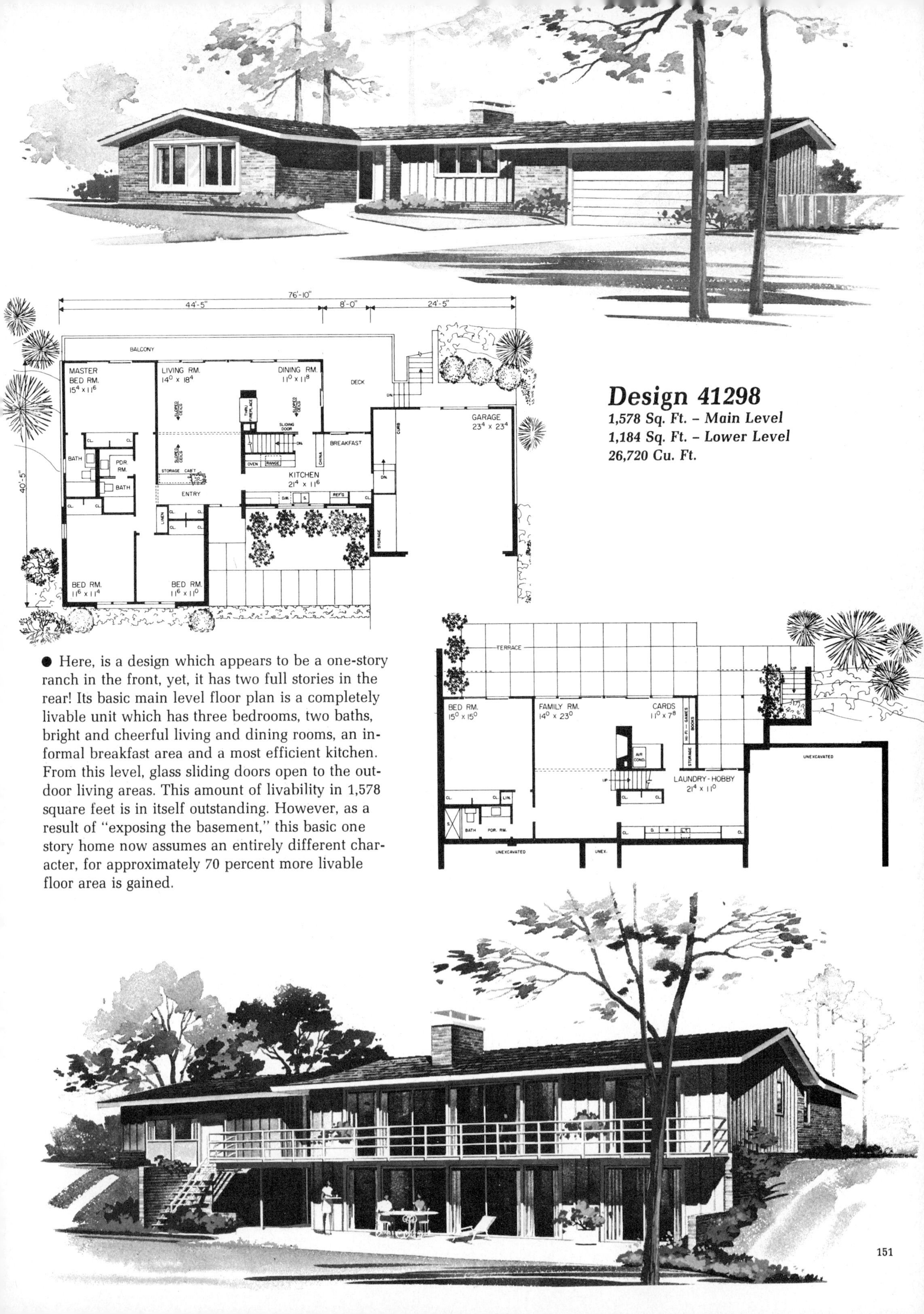

Design 41298

1,578 Sq. Ft. – Main Level
1,184 Sq. Ft. – Lower Level
26,720 Cu. Ft.

● Here, is a design which appears to be a one-story ranch in the front, yet, it has two full stories in the rear! Its basic main level floor plan is a completely livable unit which has three bedrooms, two baths, bright and cheerful living and dining rooms, an informal breakfast area and a most efficient kitchen. From this level, glass sliding doors open to the outdoor living areas. This amount of livability in 1,578 square feet is in itself outstanding. However, as a result of "exposing the basement," this basic one story home now assumes an entirely different character, for approximately 70 percent more livable floor area is gained.

Elegance And Grandeur On Two Levels

● The above illustrations of the front and rear views of this hillside contemporary design are impressive. And indeed rightly so! For the varied design features are so numerous and they are so delightfully incorporated under the wide overhanging roofs that the result is almost breathtaking at first glance. Consider the basic L-shape of the house and garage. Note how it lends itself to a large drive court. Observe the simplicity of the masses of brick and vertical character of the glass areas. Notice the inviting recessed double front doors. Around to the rear, the architectural interest is indeed extremely exciting. The glass areas are as dramatic as is the wood deck. The covered porch and the two covered terraces complete the facilities for gracious outdoor living fun.

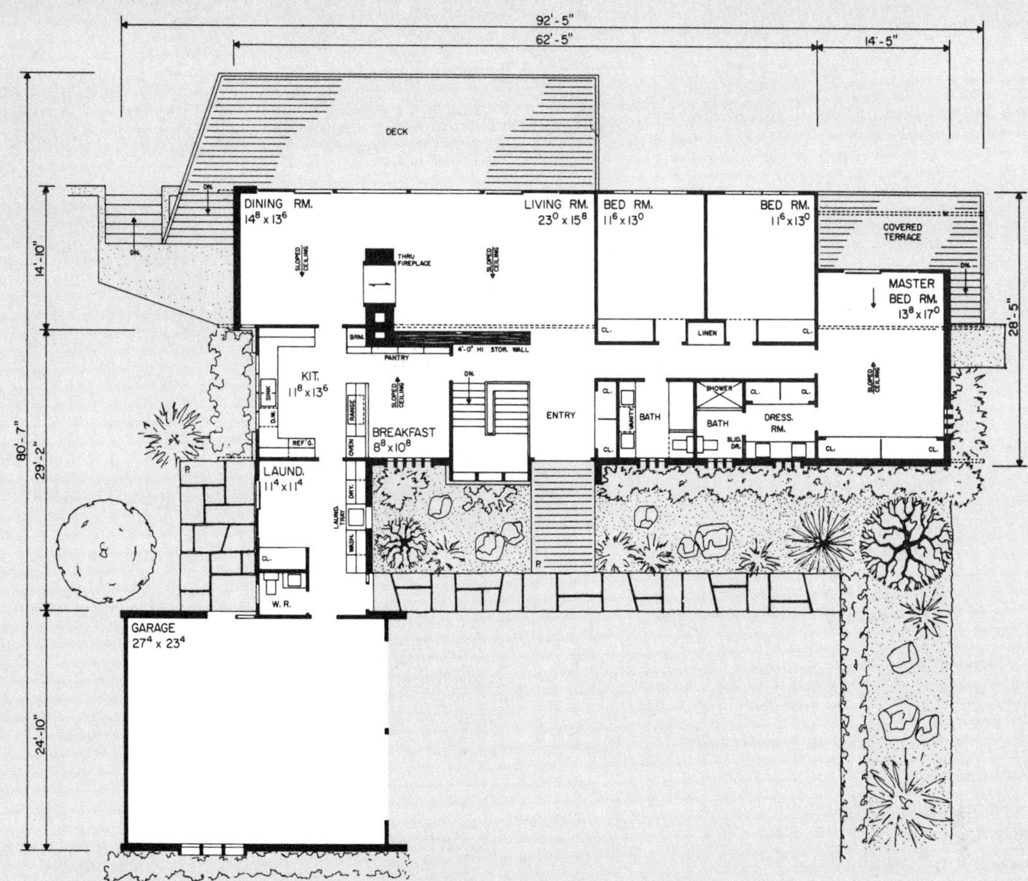

Design 41963 *2,248 Sq. Ft. – Main Level; 1,948 Sq. Ft. – Lower Level; 42,422 Cu. Ft.*

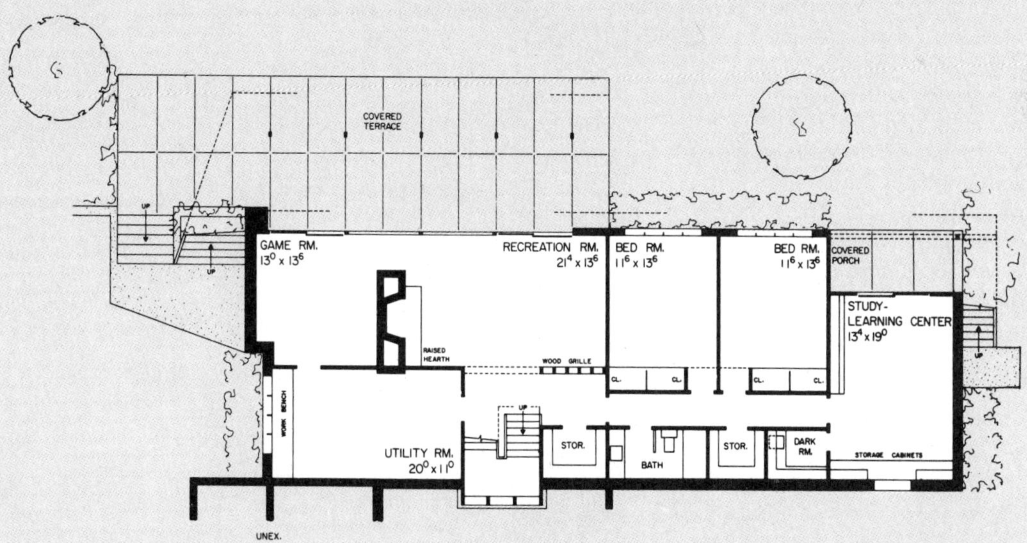

● Over four thousand square feet are available for use by the young, active family. And use them, they certainly will! There are five bedrooms to sleep a large active crew. The living areas are varied and numerous. In addition to the conventional formal living and dining rooms, there is a study/learning center. This is where all the mechanical paraphenalia like the tape recorder, film and slide projectors, phonograph, radio and television will be kept. A great way to keep all the equipment together. Note adjacent dark room. Then there is the recreation room with raised hearth fireplace. The game room will house the pool table, while the utility room will cater to the hobbyists. There are three full baths, plus an extra wash room and laundry adjacent to the kitchen.

Design 42719

2,363 Sq. Ft. – Main Level
1,523 Sq. Ft. – Lower Level; 47,915 Cu. Ft.

● If you have a flair for something different and useful at the same time, then expose the basement for hillside living. This design offers three large living areas: gathering room, family room and all-purpose activity room. Note the features in each of the three: balcony, sloping ceiling and thru-fireplace in the gathering room; deck and eating area in the family room; terrace and raised hearth fireplace in the activities room. The staircase to the lower level is delightfully open which adds to the spacious appeal of the entry hall. Cabinets and shelves are also a delightful feature of this area. Three bedrooms, the master bedroom suite on the main level with the other two on the lower level. An efficient U-shaped kitchen to easily serve the eating area of the family room and the formal dining room. The laundry is just a step away. The front projection of the two-car garage reduces the size of the lot required to build this exciting contemporary home.

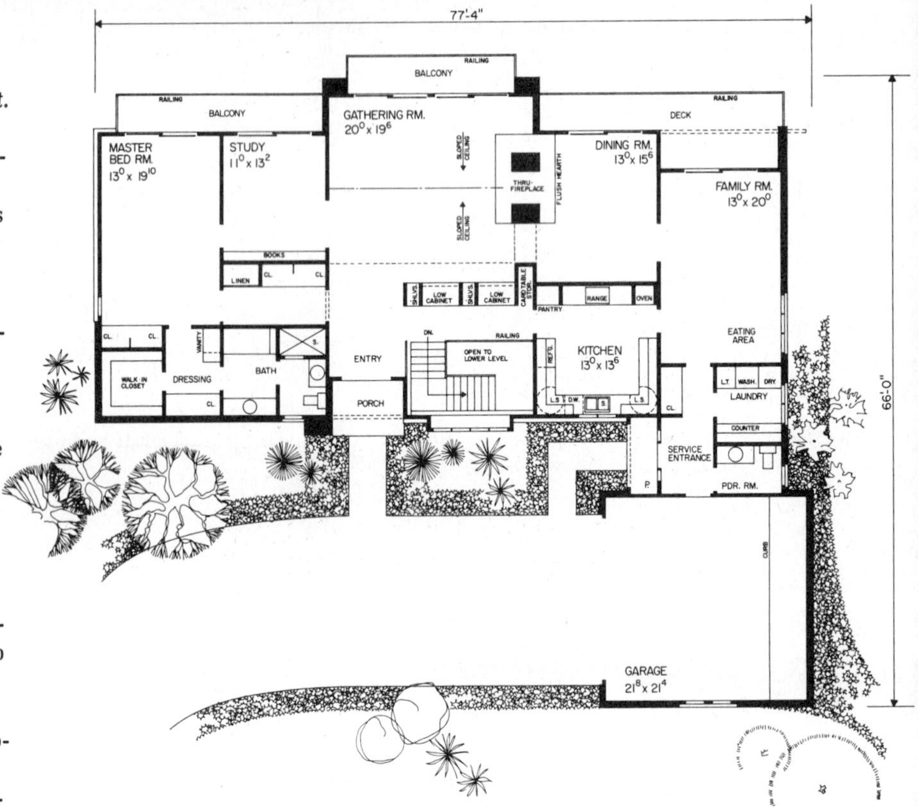

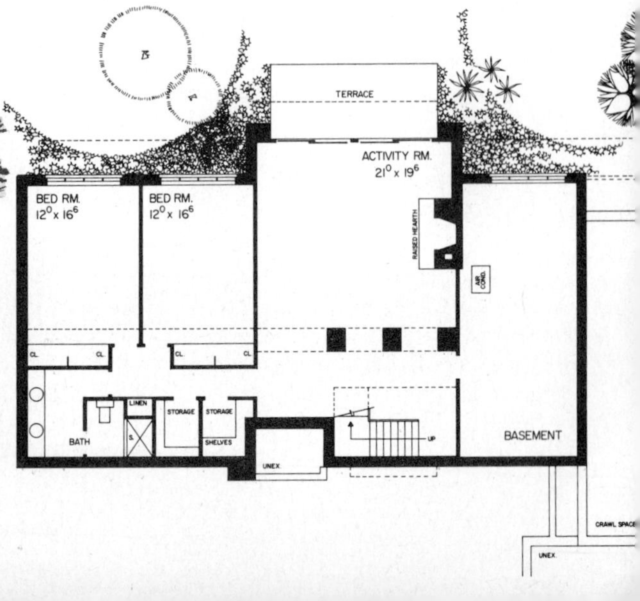

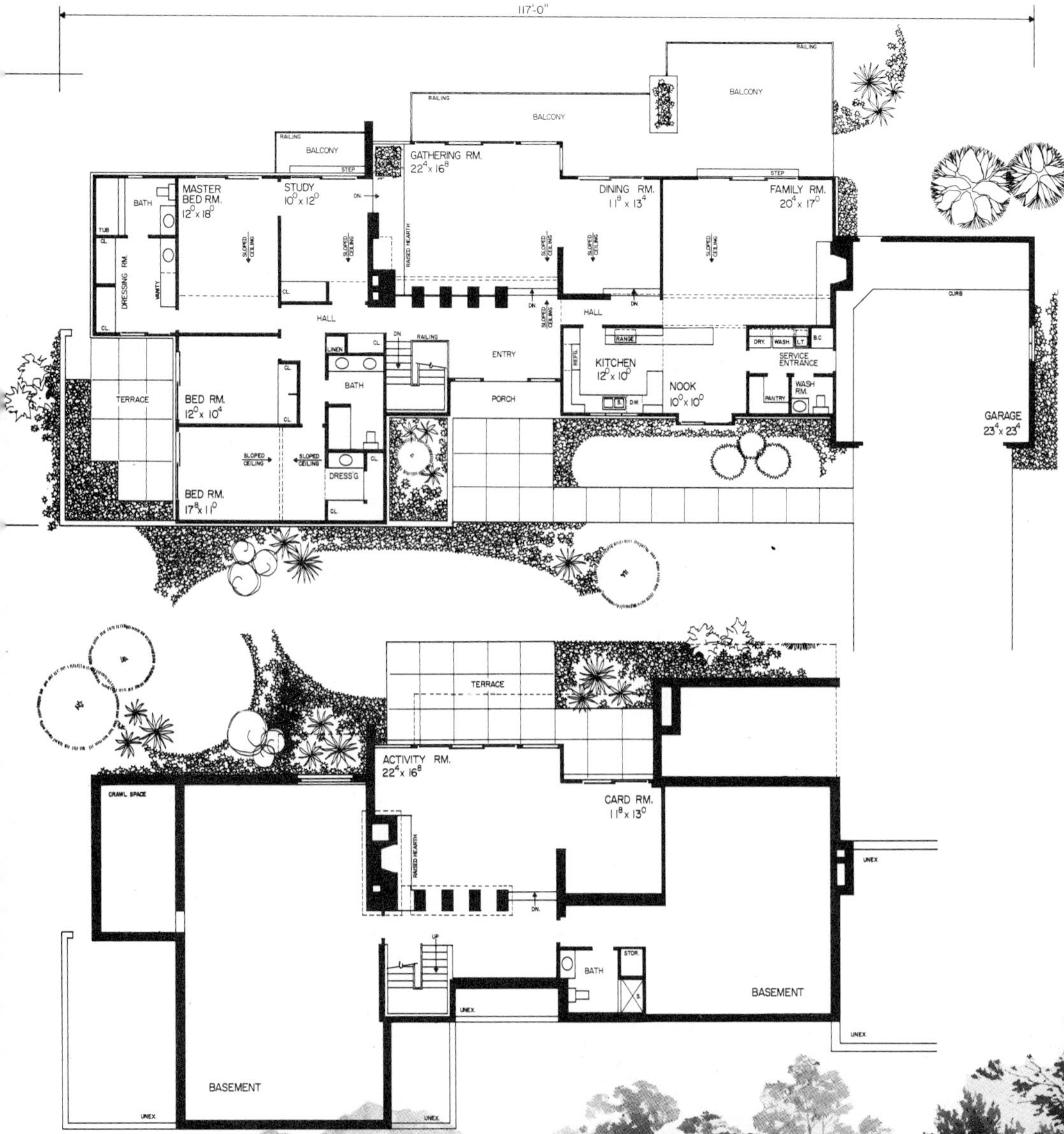

Design 42578

2,877 Sq. Ft. – Main Level
1,011 Sq. Ft. – Lower Level
47,525 Cu. Ft.

● How about three fireplaces in a delightfully livable floor plan! This outstanding home offers fireplaces in the gathering room, the family room and the activity room. That could make you look forward to winter. Also, the warm weather will be enjoyed on the rear terrace and balcony. The exposed lower level contributes an abundance of space which will enjoy natural light. Bulk items will be easily stored in the two basement areas. The country-size kitchen has an efficient work space and a separate breakfast nook. A first floor laundry is adjacent to the nook. Three bedrooms, including a complete master bedroom.

Design 41853 **1,274 Sq. Ft. – Main Level; 784 Sq. Ft. – Upper Level; 792 Sq. Ft. – Lower Level; 34,982 Cu. Ft.**

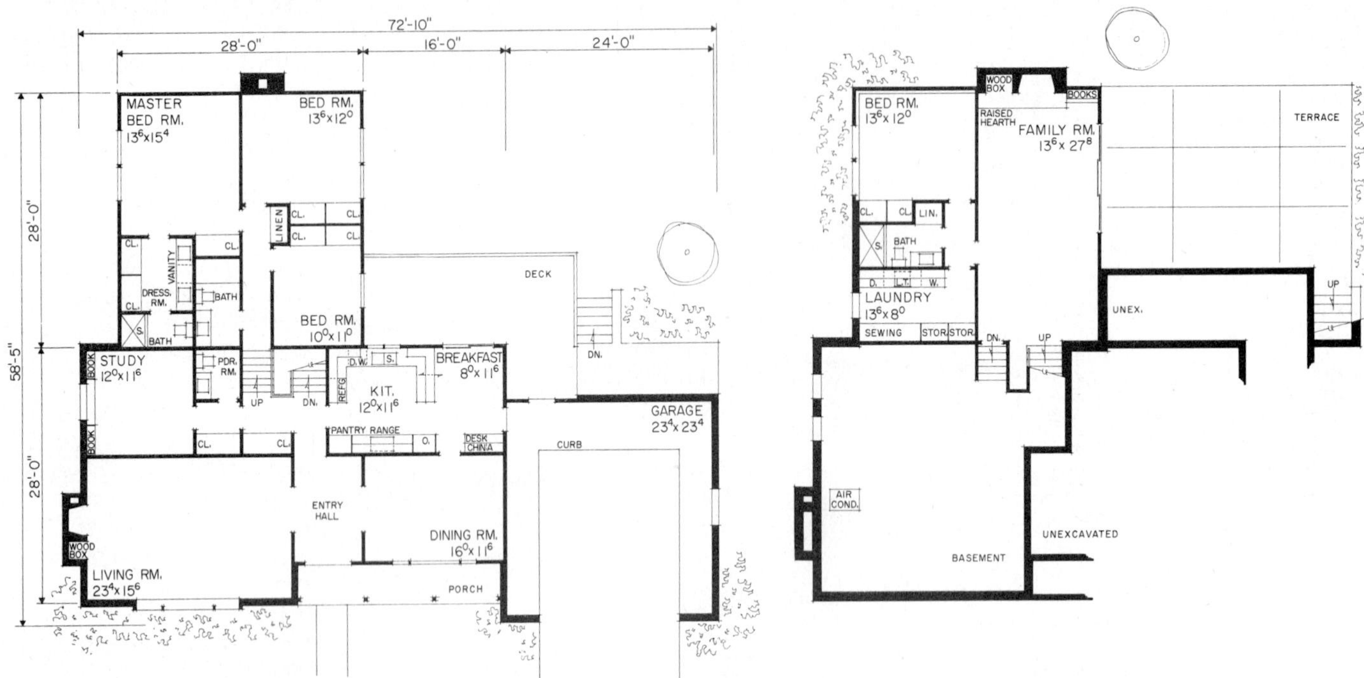

● Here, on these two pages, are two traditional homes designed specifically with the hillside site in mind. From the street they appear to be cozy one-story homes. From the rear, however, their appearance is pleasingly different. The lower level is exposed and opens onto the outdoor terrace area. The decks provide the main levels with outdoor living facilities which will be everyone's favorite area for relaxation. Study the floor plans with care. The convenient living potential of each design is outstanding. Which house satisfies your family living requirements?

Design 41739 *1,281 Sq. Ft. – Main Level; 857 Sq. Ft. – Sleeping Level; 687 Sq. Ft. – Lower Level; 37,624 Cu. Ft.*

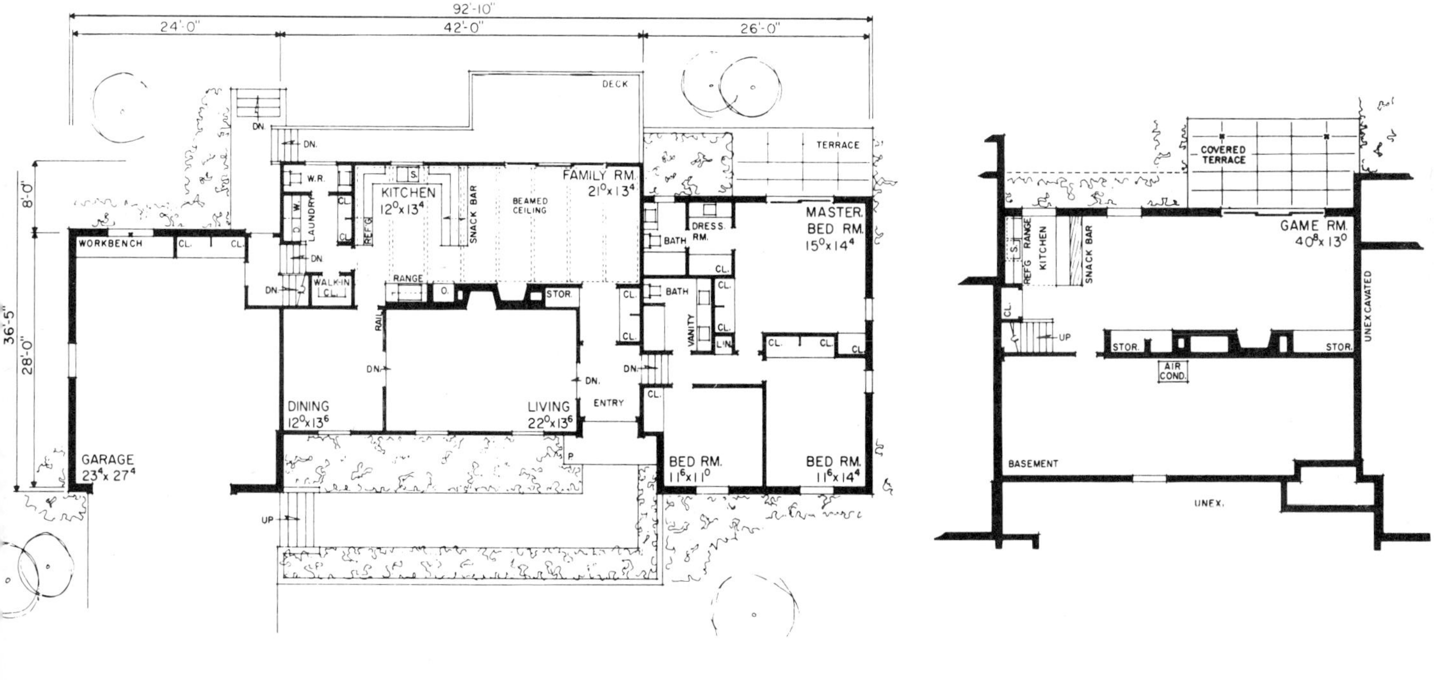

Design 42546 *1,143 Sq. Ft. – Main Level; 746 Sq. Ft. – Upper Level 1,143 Sq. Ft. – Lower Level; 31,128 Cu. Ft.*

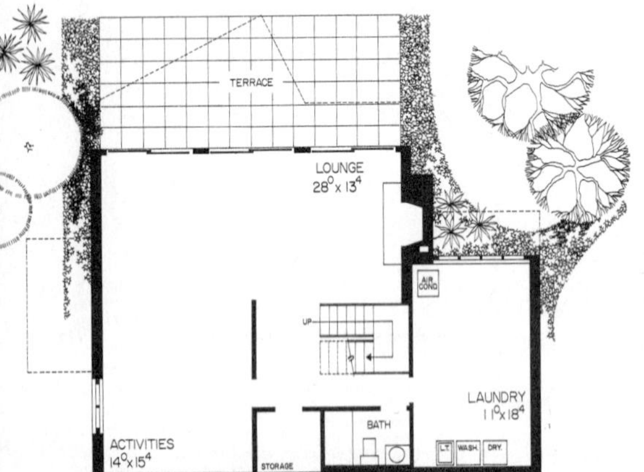

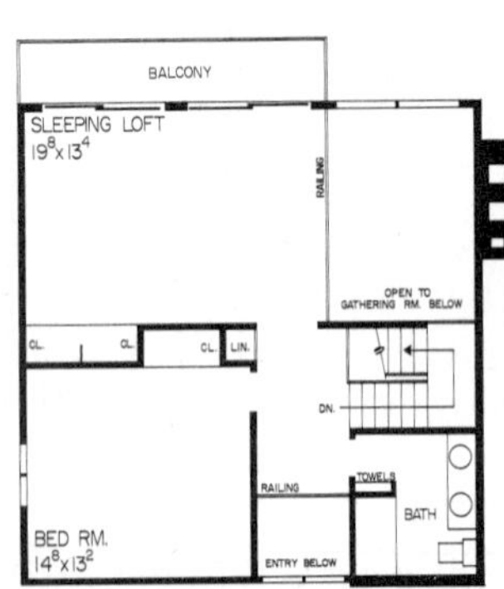

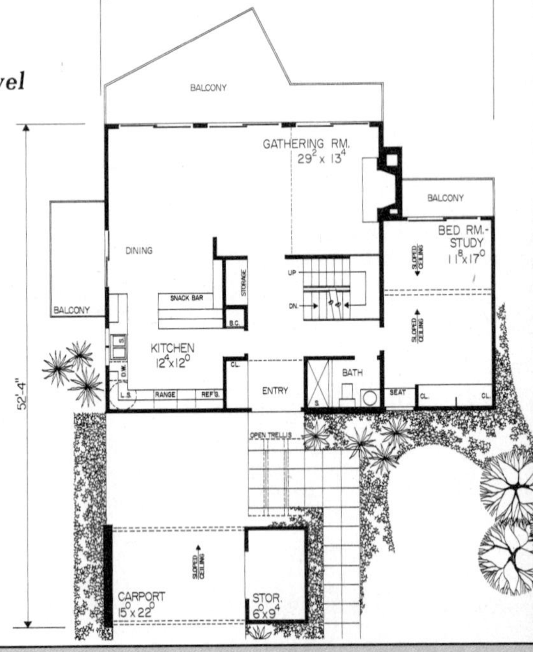

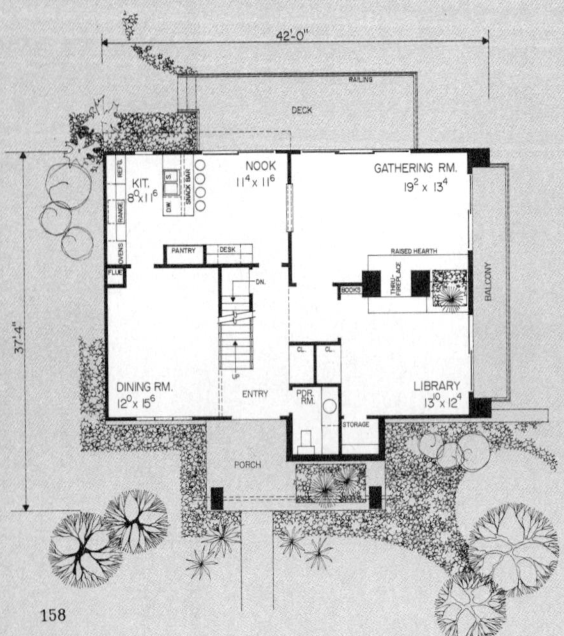

Design 42770
1,182 Sq. Ft. – Main Level
998 Sq. Ft. – Upper Level
25,830 Cu. Ft.

● If you are looking for a home with loads of livability, then consider these two-story contemporary homes which have an exposed lower level.

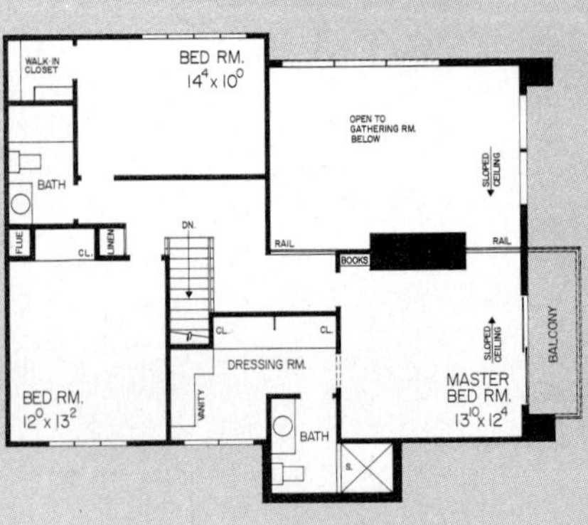

Design 42548 *1,109 Sq. Ft. – Main Level; 739 Sq. Ft. – Upper Level*
869 Sq. Ft. – Lower Level; 31,370 Cu. Ft.

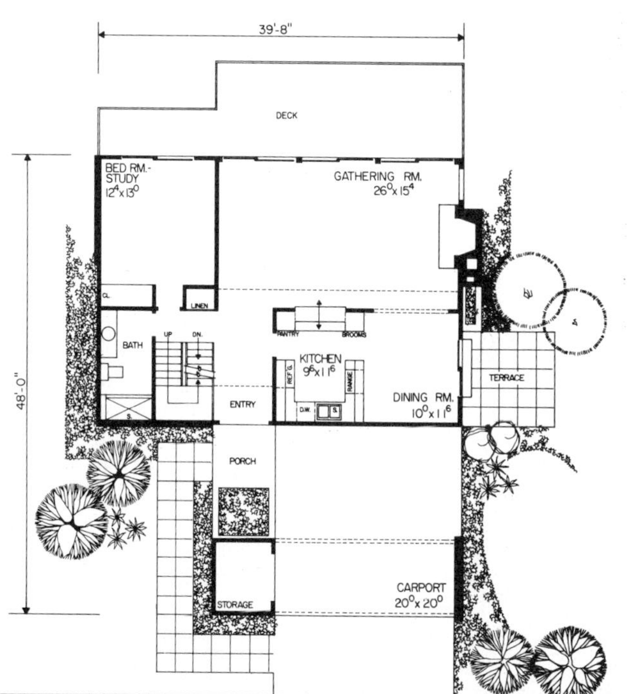

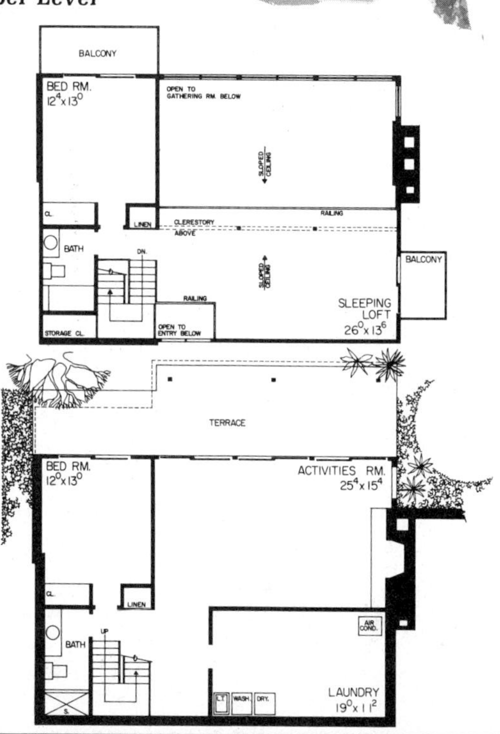

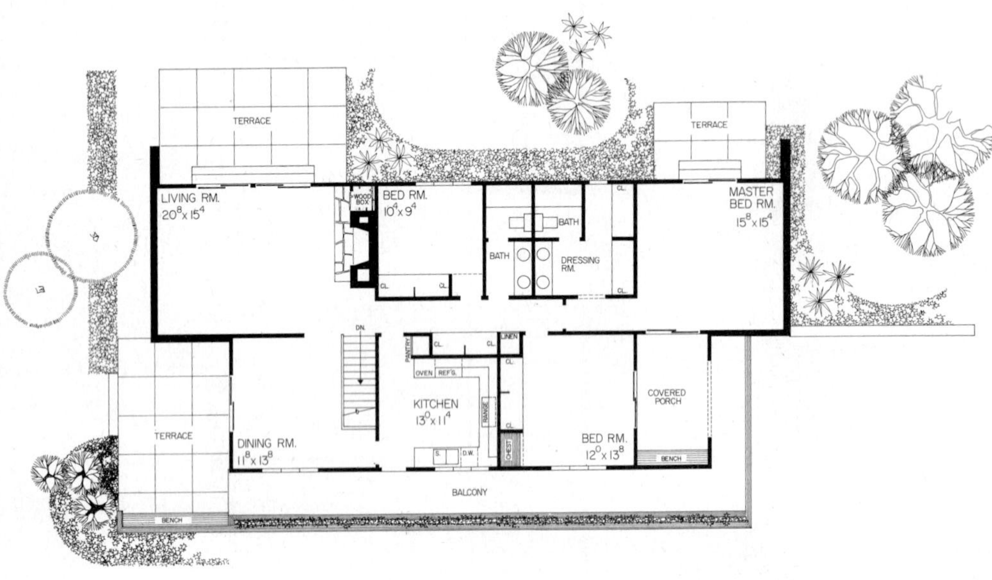

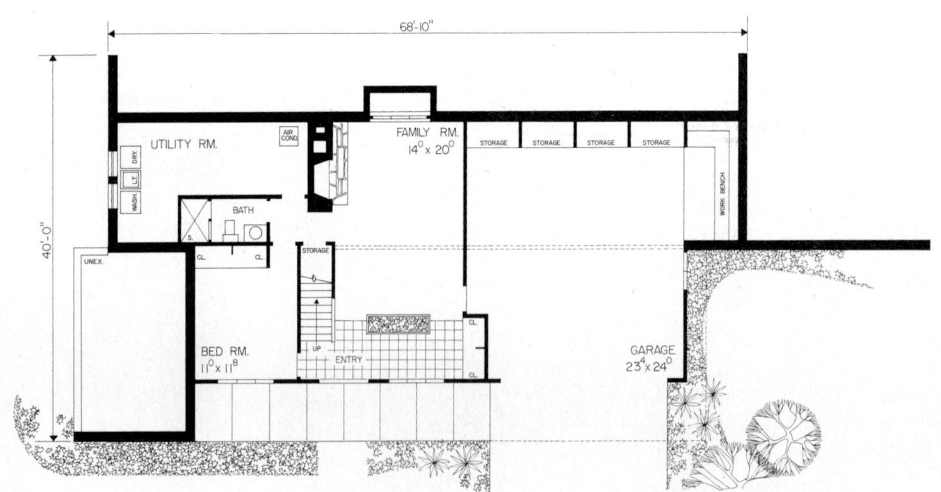

Design 42601

1,724 Sq. Ft. – Upper Level
986 Sq. Ft. – Lower Level
28,240 Cu. Ft.

● Surely it would be difficult to improve upon the refreshing appeal of this contemporary bi-level design. Built into a sloping site, its lower level has direct grade access at the rear. With four bedrooms and three full baths, the large family will be well-served. Each of the living areas has a fireplace. The family dining room is but a step or two from the side terrace for convenient, outdoor dining. The kitchen is handy to the balcony for ease of caring for those potted plants. Worthy of particular note, is the covered porch which functions with two of the upper level bedrooms. Don't miss the oversized garage with all those storage facilities and the work bench. While the blueprints call for an exterior of stone and stucco, you may wish to substitute your own choice of materials.

Design 42714

2,207 Sq. Ft. - Upper Level
1,226 Sq. Ft. - Lower Level
45,060 Cu. Ft.

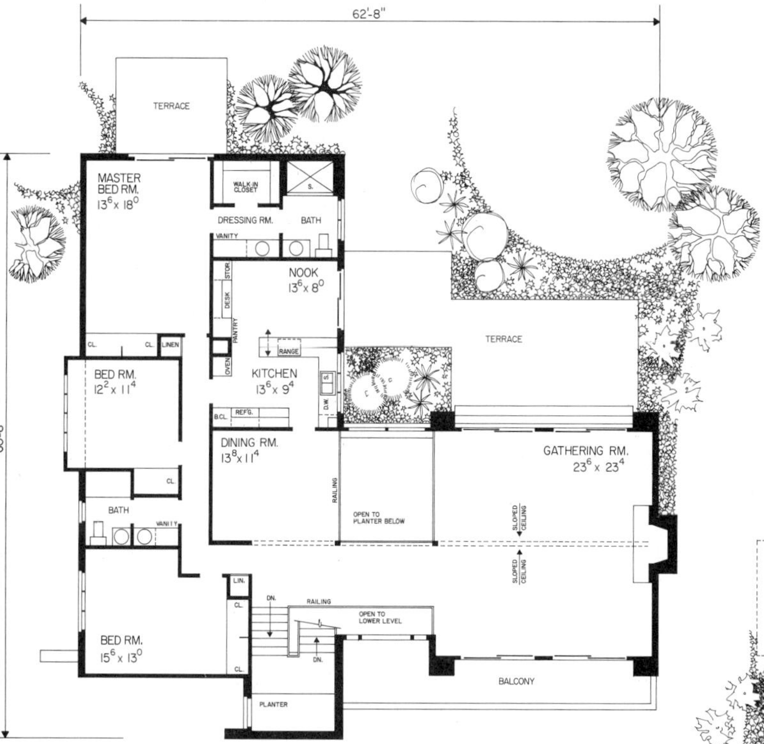

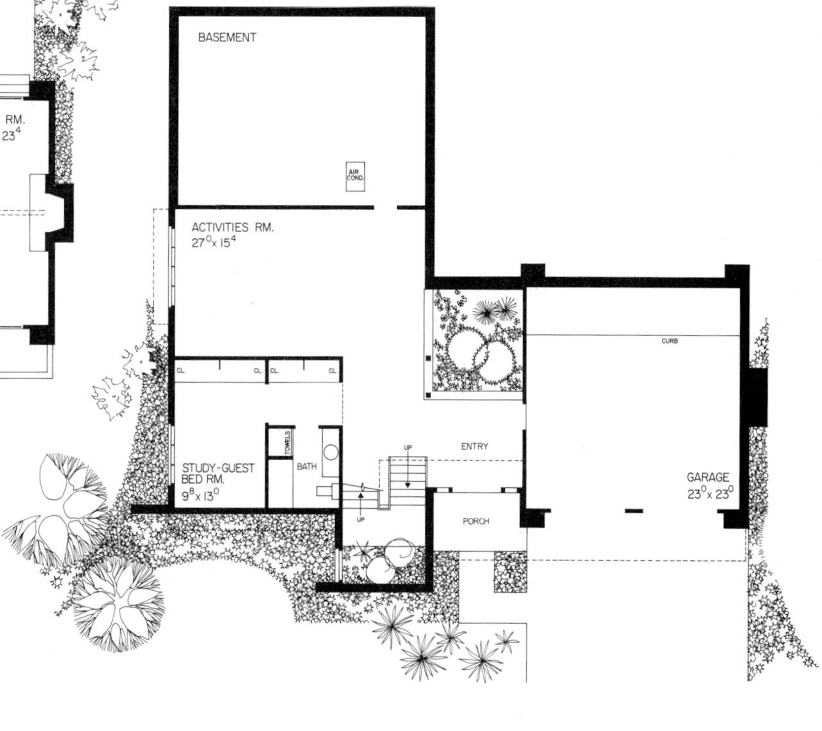

62'-8"
60'-8"
TERRACE
MASTER BED RM. $13^6 \times 18^0$
WALK-IN CLOSET
DRESSING RM.
BATH
VANITY
NOOK $13^6 \times 8^0$
KITCHEN $13^6 \times 9^4$
RANGE
PANTRY
DESK
OVEN
REFG.
B.CL.
D.W.
CL.
LINEN
BED RM. $12^2 \times 11^4$
DINING RM. $13^8 \times 11^4$
TERRACE
GATHERING RM. $23^6 \times 23^4$
OPEN TO PLANTER BELOW
RAILING
SLOPED CEILING
BATH
VANITY
LIN.
DN.
RAILING
OPEN TO LOWER LEVEL
BED RM. $15^6 \times 13^0$
PLANTER
BALCONY

BASEMENT
AIR COND.
ACTIVITIES RM. $27^0 \times 15^4$
CURB
CL.
STUDY-GUEST BED RM. $9^8 \times 13^0$
TOWELS
BATH
UP
ENTRY
PORCH
GARAGE $23^0 \times 23^0$

● Start with the dining and gathering room, both overlooking an indoor garden on the lower level. Then note how the gathering room opens to a terrace on one side, to a balcony on the other. Greenery everywhere you look! Sloped ceilings and a fireplace, too. Plus four large bedrooms, including a master suite with a private terrace which is ideal for quiet moments alone! A spacious kitchen with pantry, built-in desk and two other storage closets. Three full baths. The activity room on the lower level will serve the family's every informal need. Basement is included in the plan for bulk storage. This home makes the most use of that natural or graded sloping site.

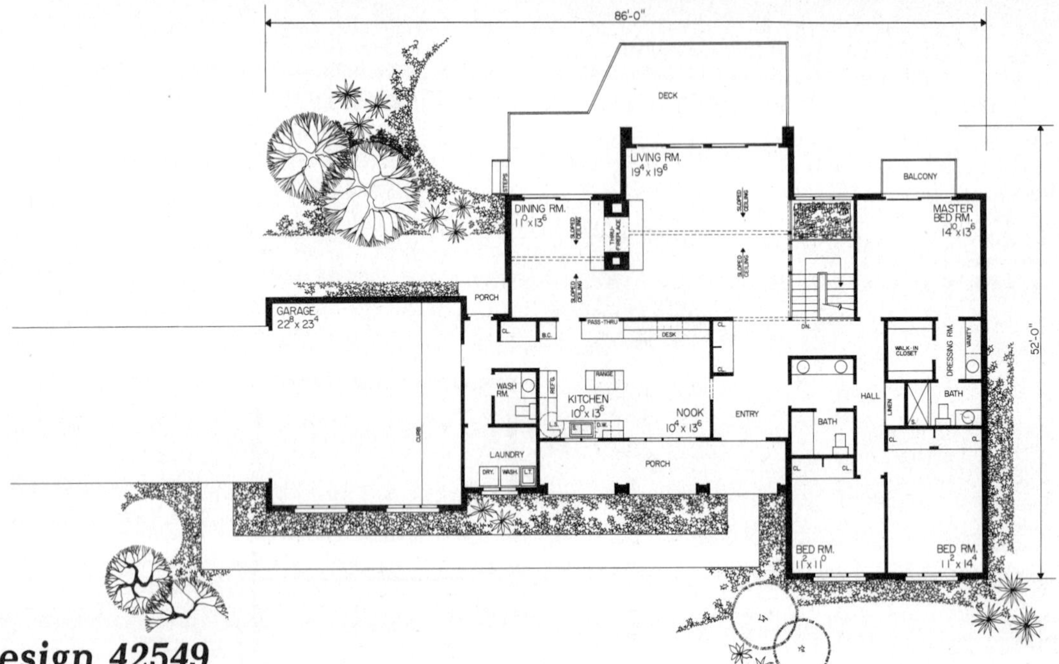

Design 42549

2,260 Sq. Ft. – Main Level
1,406 Sq. Ft. – Lower Level; 51,857 Cu. Ft.

● This hillside home gives all the appearances of being a one-story ranch home; and what a delightful one at that! Should the contours of your property slope to the rear, this plan permits the exposing of the lower level. This results in the activities room and bedroom/study gaining direct access to outdoor living. Certainly a most desirable aspect for active, outdoor family living. The large and growing family will be admirably served with five bedrooms and three baths. An extra wash room and separate laundry add to the convenient living potential. The kitchen is spacious with its island cooking surface and breakfast nook. Sloping ceilings and sliding glass doors make for a spacious living area. The open stairway to the lower level is dramatic and features a well-lighted planter area. Observe the deck, the balcony and the covered front porch. Notice the two separate basement areas.

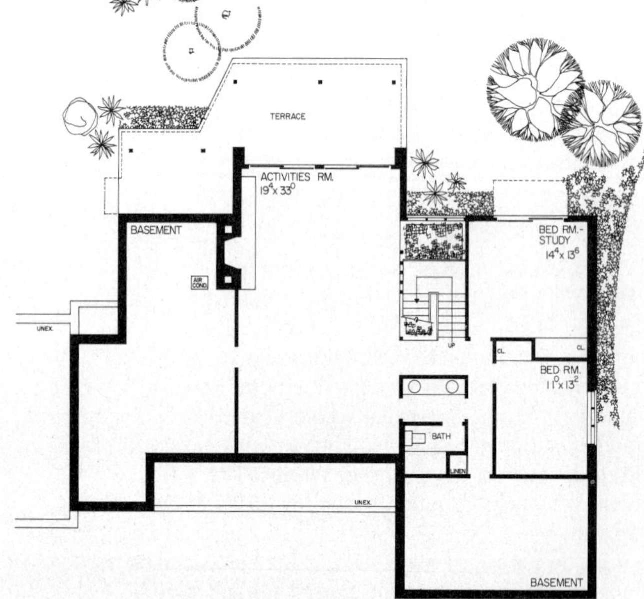

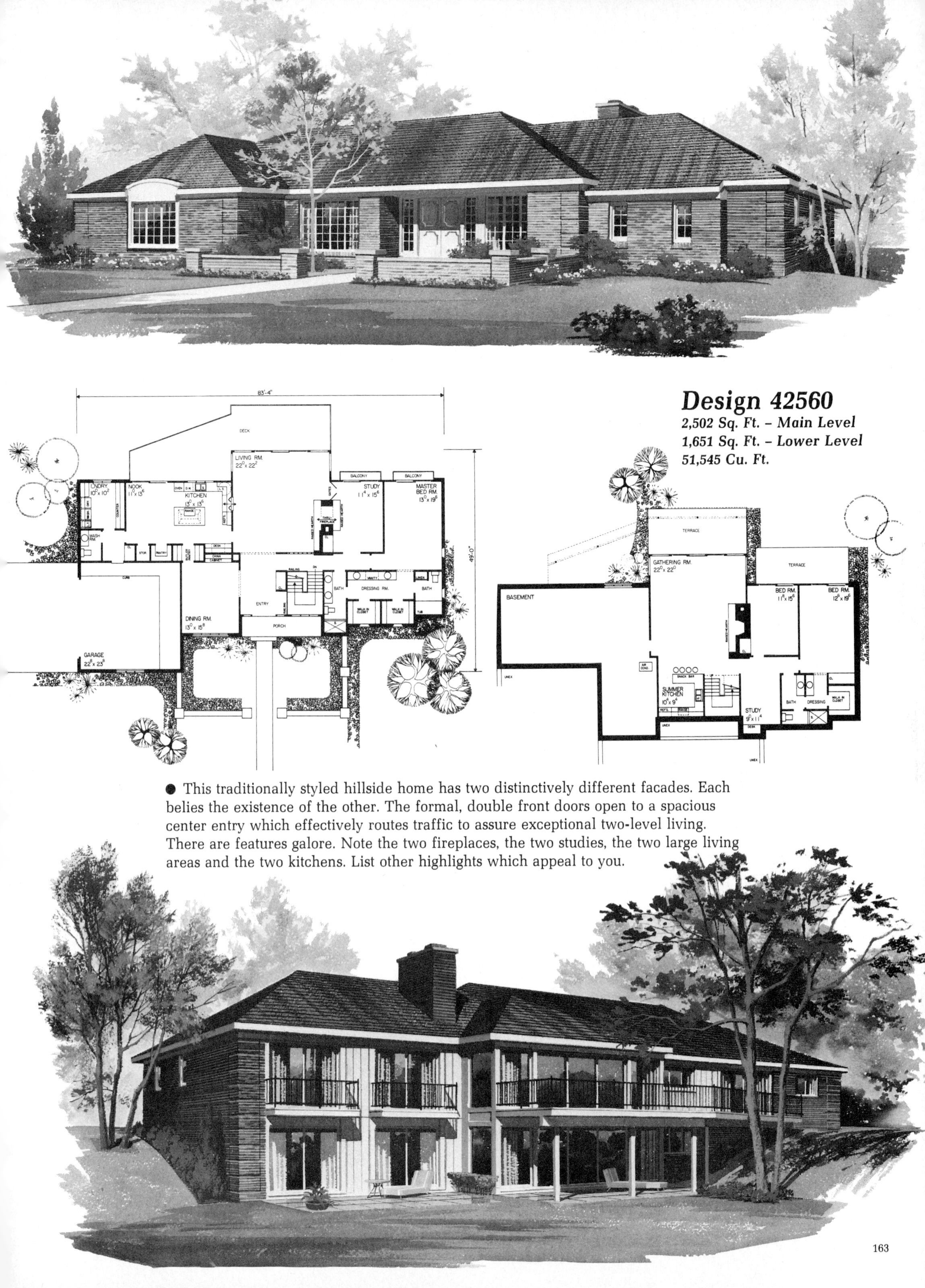

Design 42560

2,502 Sq. Ft. – Main Level
1,651 Sq. Ft. – Lower Level
51,545 Cu. Ft.

● This traditionally styled hillside home has two distinctively different facades. Each belies the existence of the other. The formal, double front doors open to a spacious center entry which effectively routes traffic to assure exceptional two-level living. There are features galore. Note the two fireplaces, the two studies, the two large living areas and the two kitchens. List other highlights which appeal to you.

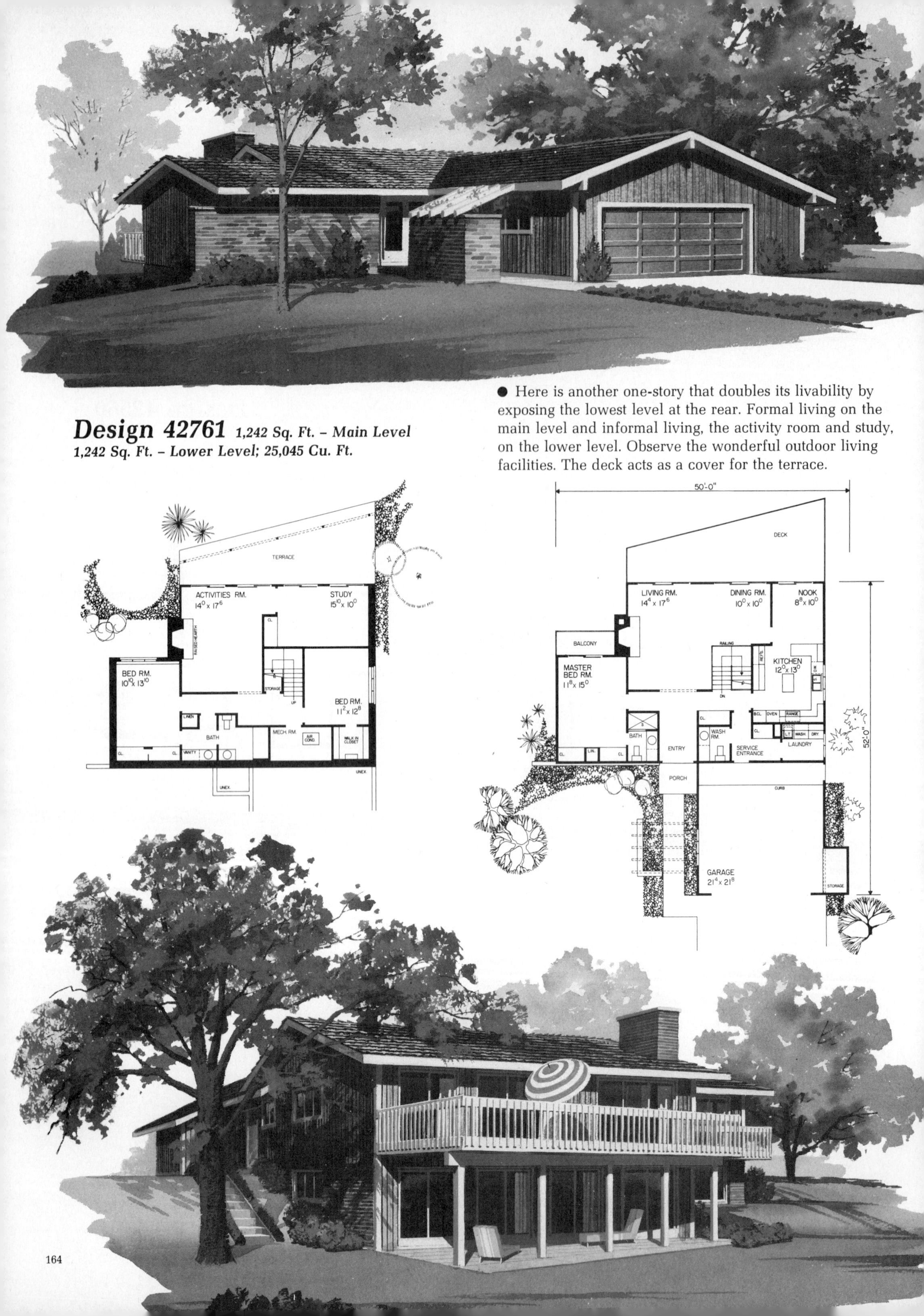

Design 42761

1,242 Sq. Ft. - Main Level
1,242 Sq. Ft. - Lower Level; 25,045 Cu. Ft.

● Here is another one-story that doubles its livability by exposing the lowest level at the rear. Formal living on the main level and informal living, the activity room and study, on the lower level. Observe the wonderful outdoor living facilities. The deck acts as a cover for the terrace.

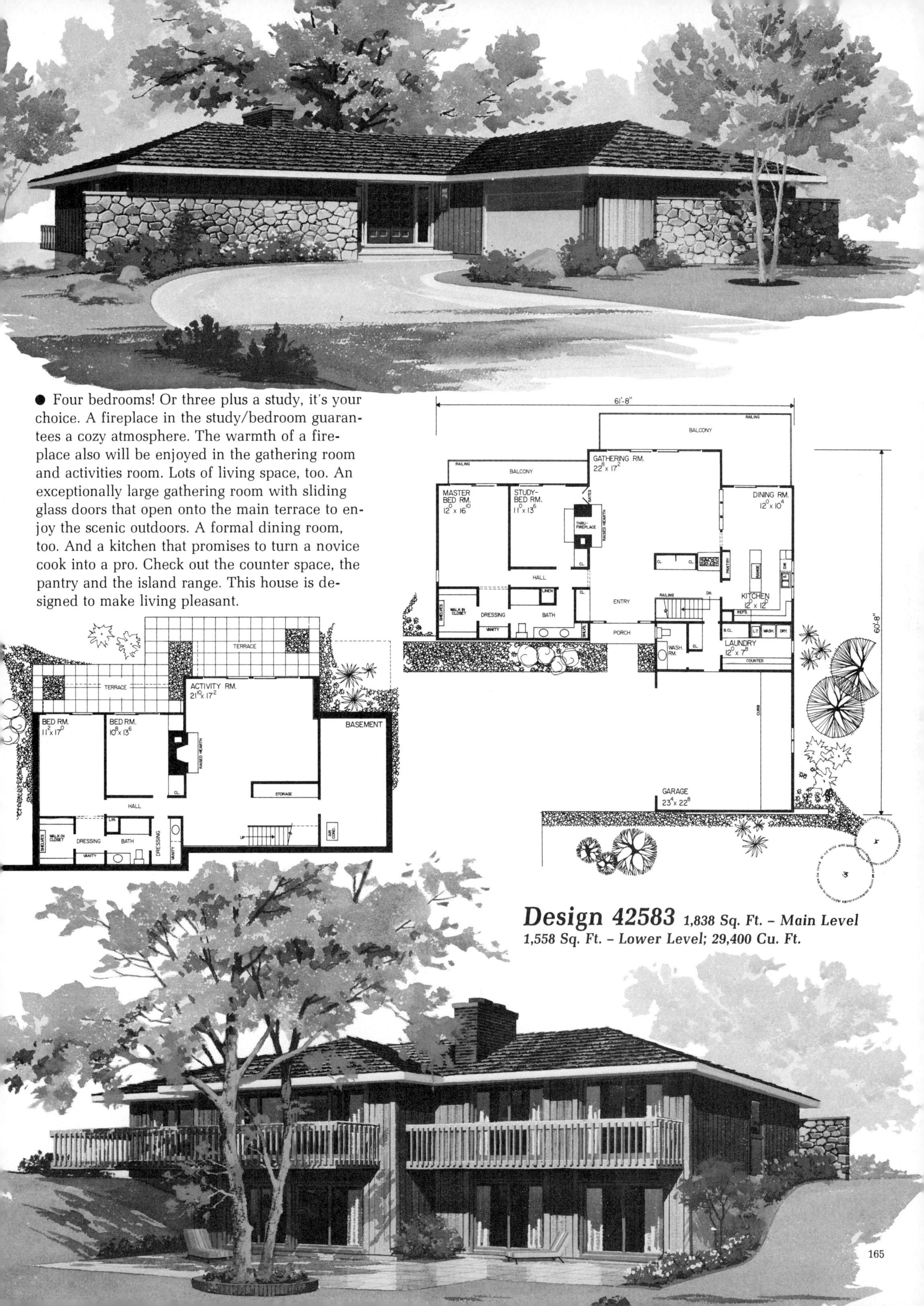

● Four bedrooms! Or three plus a study, it's your choice. A fireplace in the study/bedroom guarantees a cozy atmosphere. The warmth of a fireplace also will be enjoyed in the gathering room and activities room. Lots of living space, too. An exceptionally large gathering room with sliding glass doors that open onto the main terrace to enjoy the scenic outdoors. A formal dining room, too. And a kitchen that promises to turn a novice cook into a pro. Check out the counter space, the pantry and the island range. This house is designed to make living pleasant.

Design 42583 *1,838 Sq. Ft. - Main Level*
1,558 Sq. Ft. - Lower Level; 29,400 Cu. Ft.

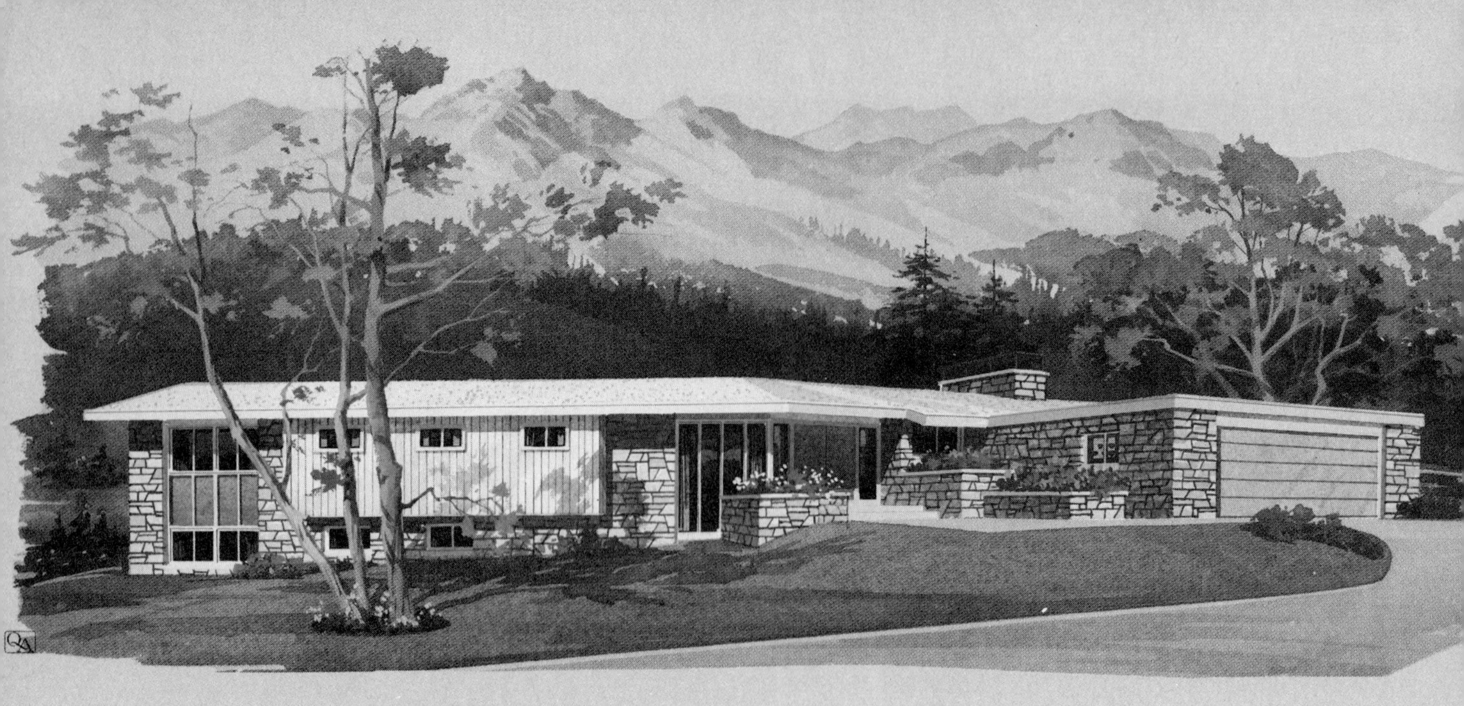

Rear Living Enjoys Maximum View

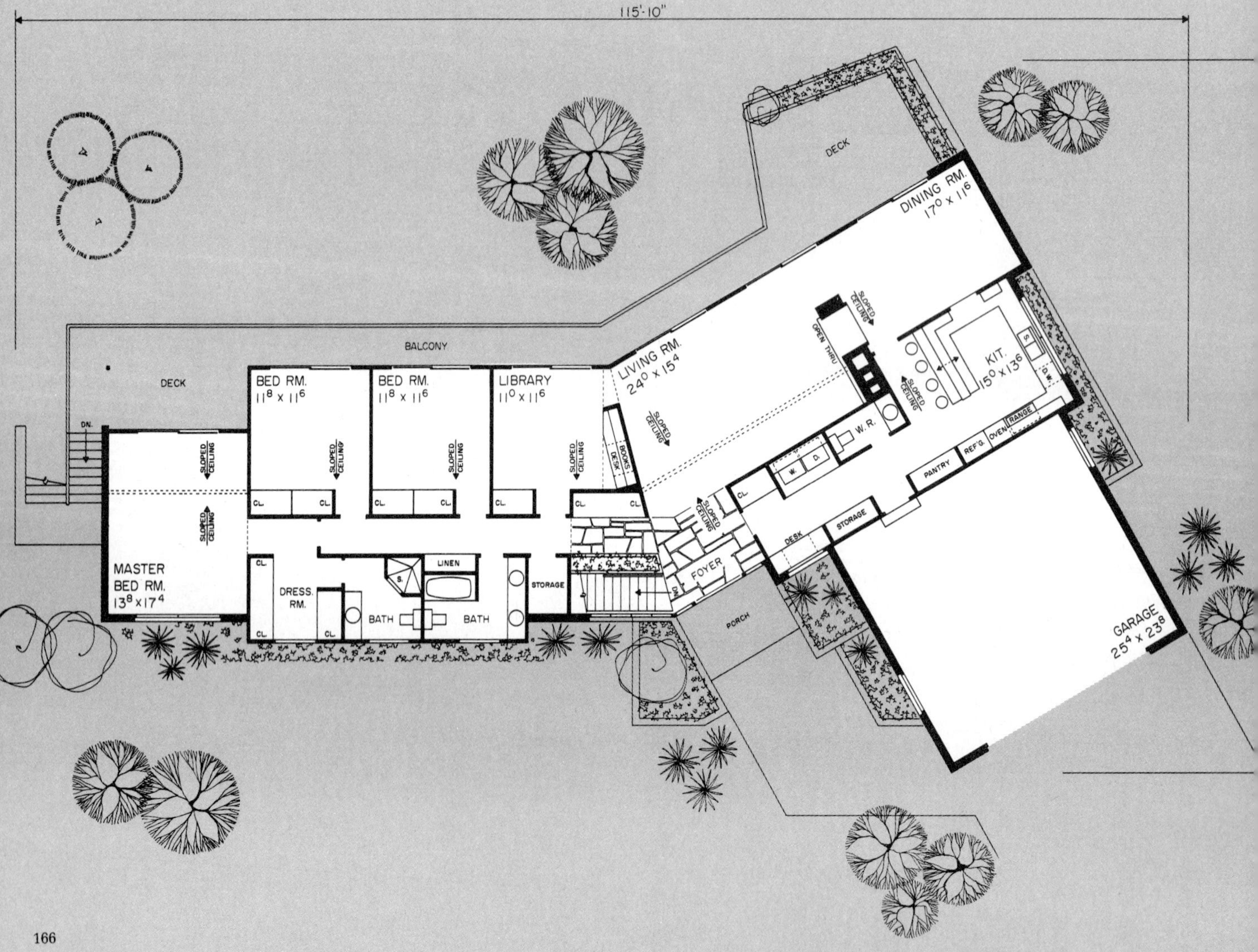

115'-10"
DECK
DINING RM. $17^0 \times 11^6$
SLOPED CEILING
OPEN THRU
LIVING RM. $24^0 \times 15^4$
KIT. $15^0 \times 13^6$
BALCONY
DECK
DN.
BED RM. $11^8 \times 11^6$
BED RM. $11^8 \times 11^6$
LIBRARY $11^0 \times 11^6$
DESK
BOOKS
W.R.
W. D.
PANTRY
REF'G.
OVEN
RANGE
CL.
MASTER BED RM. $13^8 \times 17^4$
DRESS. RM.
LINEN
BATH
BATH
STORAGE
FOYER
DESK
STORAGE
PORCH
GARAGE $25^4 \times 23^8$

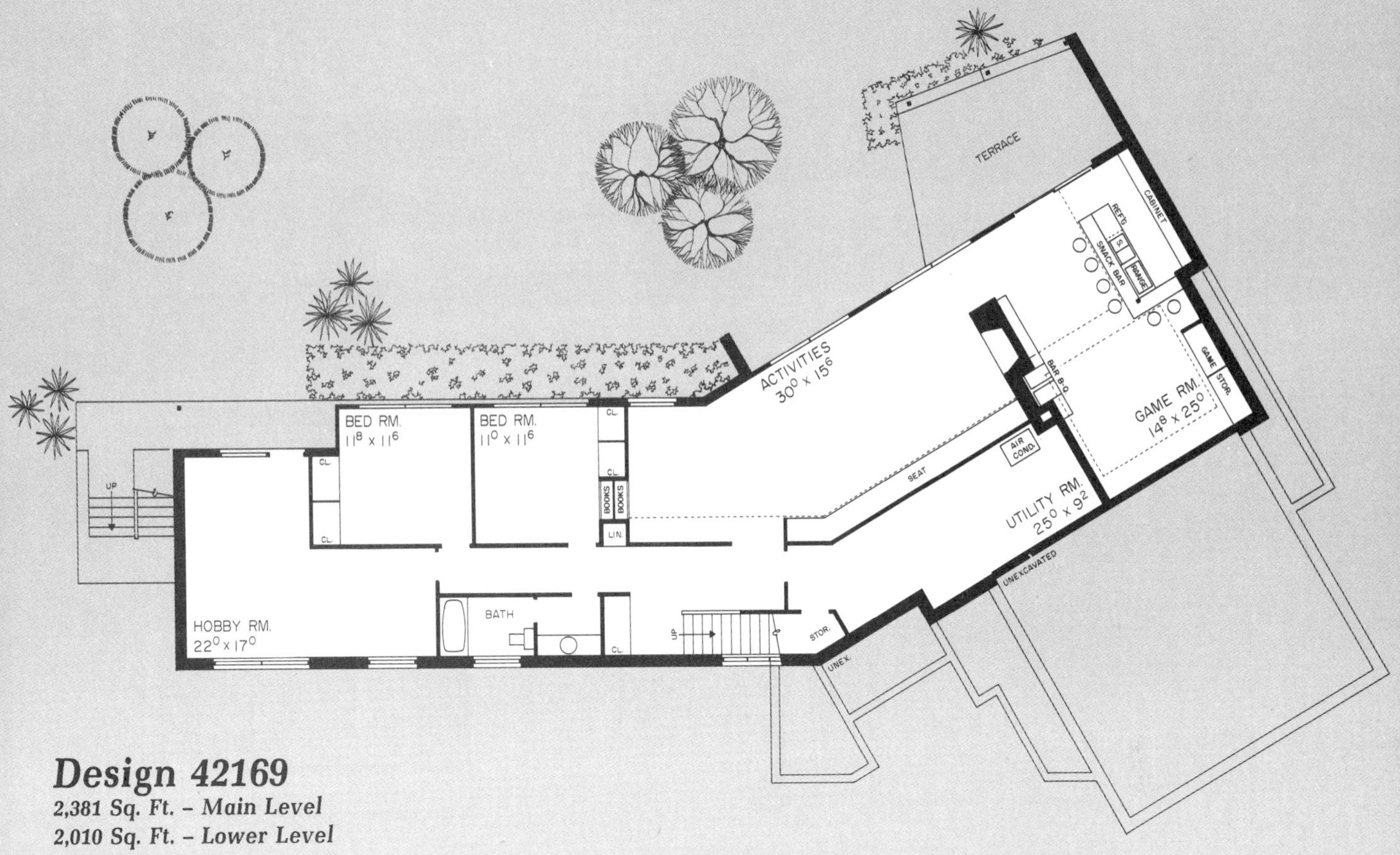

TERRACE
CABINET
REF'G.
SNACK BAR
S.
RANGE
GAME STOR.
ACTIVITIES $30^{0} \times 15^{6}$
BAR B-Q.
GAME RM. $14^{8} \times 25^{0}$
BED RM. $11^{8} \times 11^{6}$
BED RM. $11^{0} \times 11^{6}$
CL.
BOOKS
BOOKS
LIN.
SEAT
AIR COND.
UTILITY RM. $25^{0} \times 9^{2}$
UNEXCAVATED
UP
BATH
HOBBY RM. $22^{0} \times 17^{0}$
STOR.
UNEX.

Design 42169

2,381 Sq. Ft. – Main Level
2,010 Sq. Ft. – Lower Level
44,000 Cu. Ft.

● Behold, the view! If, when looking toward the rear of your site, nature's scene is breathtaking or in any way inspiring, you may wish to maximize your enjoyment by orienting your living areas to the rear of your plan. In addition to greater enjoyment of the landscape, such floor planning will provide extra privacy from the street. The angular configuration can enhance the enjoyment of a particular scene, plus it adds appeal to the exterior of the design. A study of both levels reveals that the major living areas look out upon the rear yard. Further, the upper level rooms have direct access to the decks and balcony. The kitchen with its large window over the sink is not without its view. With five bedrooms, plus a library, a game, activities and hobby room, the active family will have an abundance of space to enjoy individualized pursuits. Can't you envision your family living in this house?

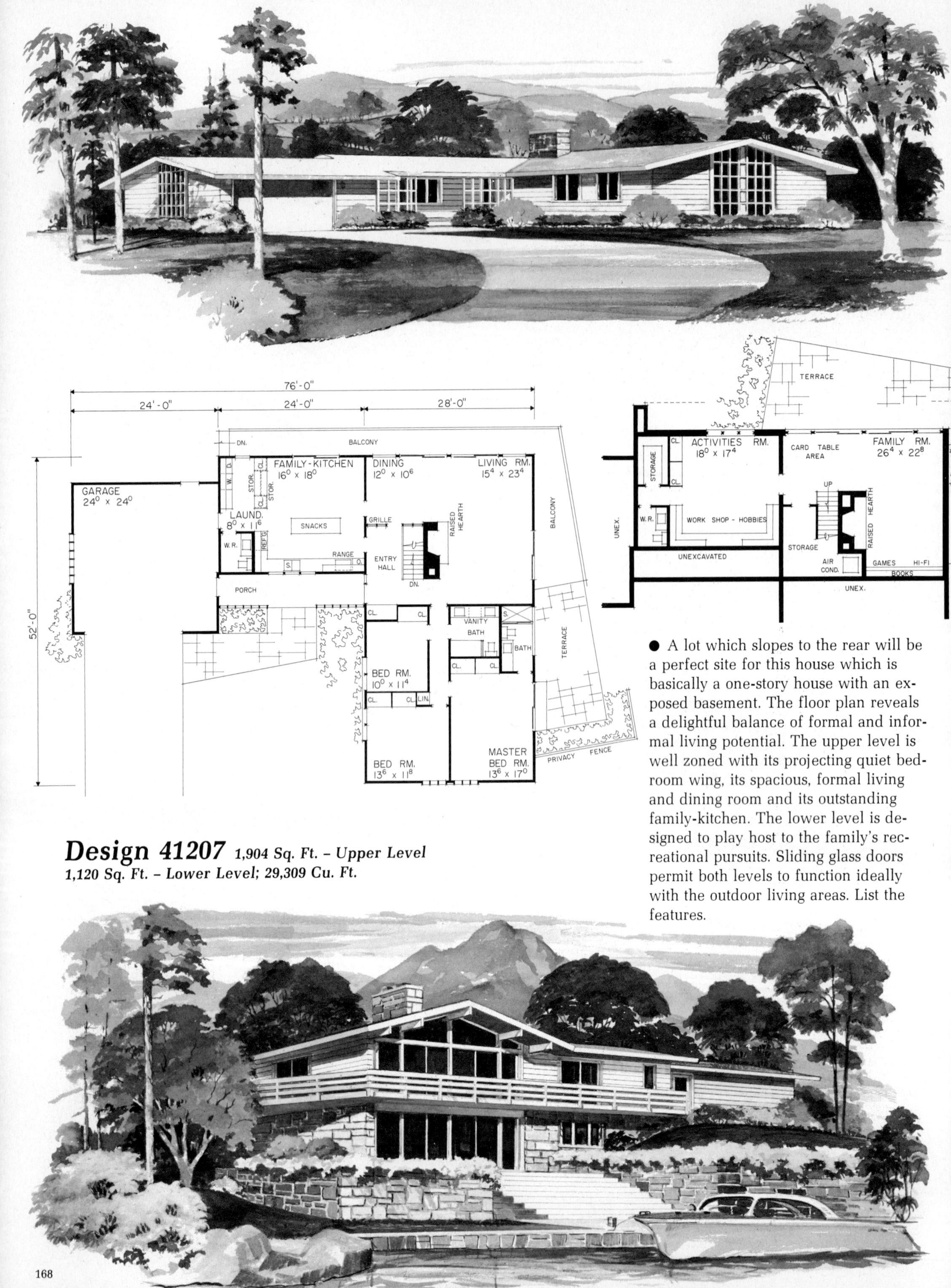

Design 41207 **1,904 Sq. Ft. - Upper Level**
1,120 Sq. Ft. - Lower Level; 29,309 Cu. Ft.

● A lot which slopes to the rear will be a perfect site for this house which is basically a one-story house with an exposed basement. The floor plan reveals a delightful balance of formal and informal living potential. The upper level is well zoned with its projecting quiet bedroom wing, its spacious, formal living and dining room and its outstanding family-kitchen. The lower level is designed to play host to the family's recreational pursuits. Sliding glass doors permit both levels to function ideally with the outdoor living areas. List the features.

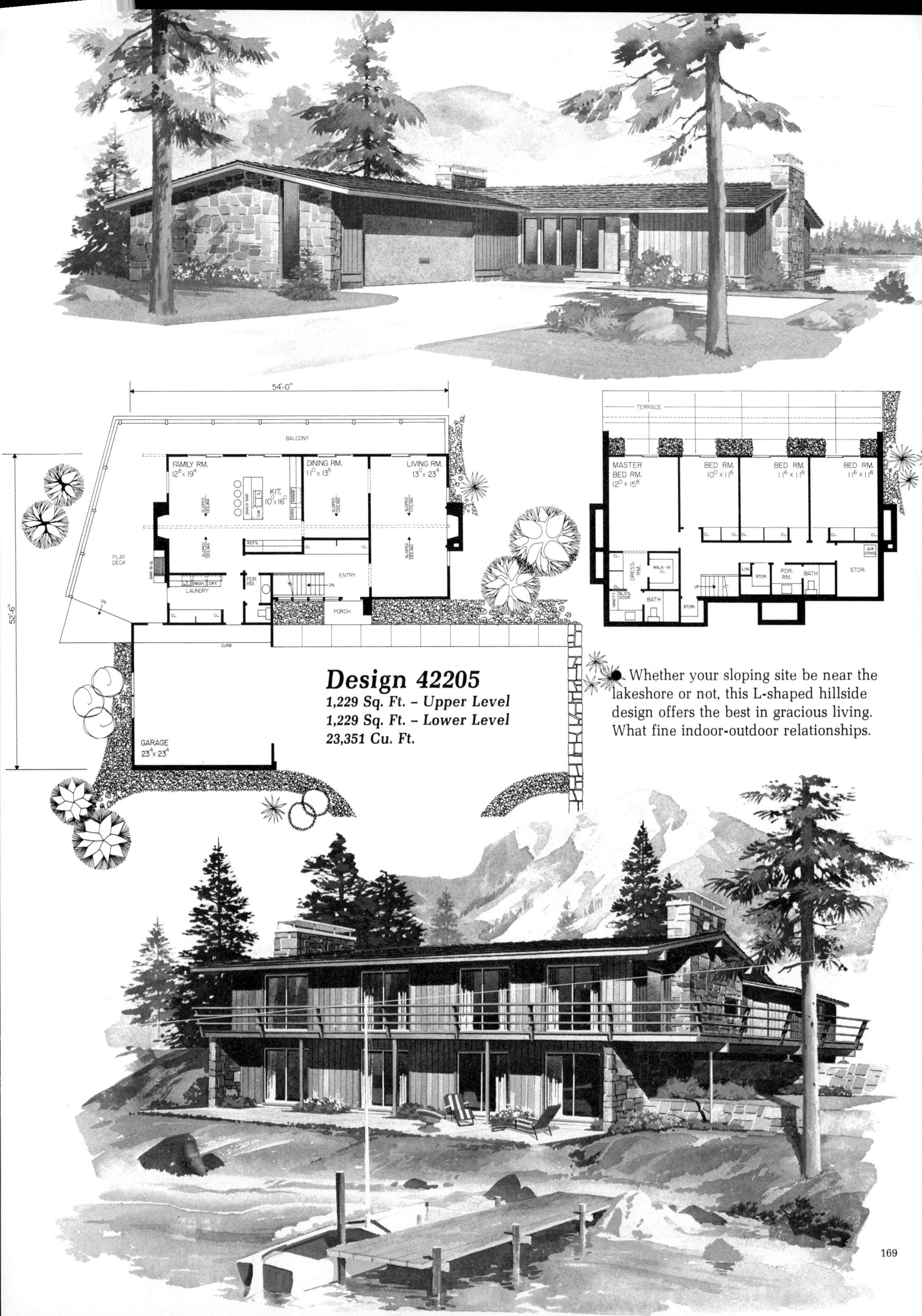

Design 42205

1,229 Sq. Ft. – Upper Level
1,229 Sq. Ft. – Lower Level
23,351 Cu. Ft.

Whether your sloping site be near the lakeshore or not, this L-shaped hillside design offers the best in gracious living. What fine indoor-outdoor relationships.

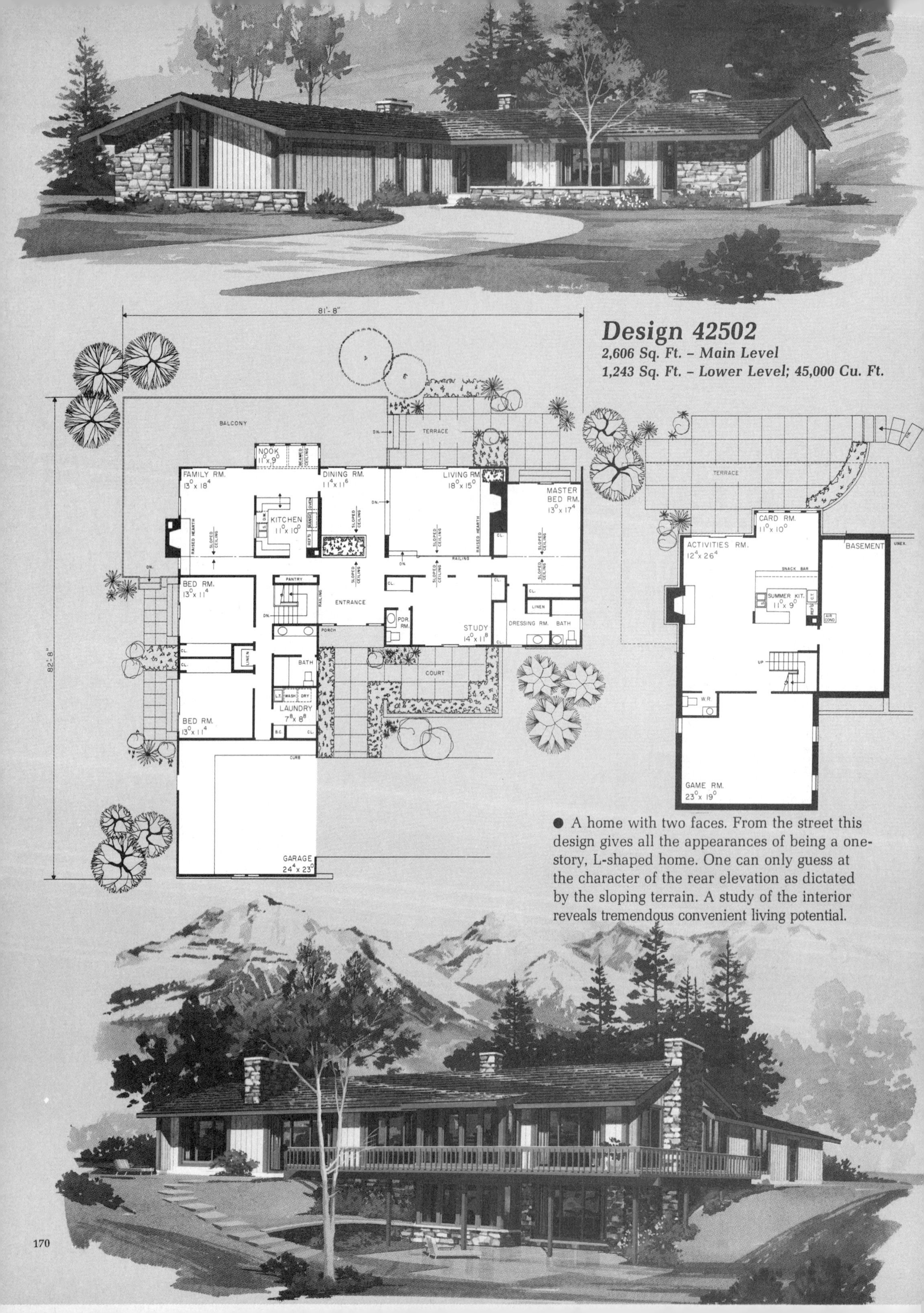

Design 42502

2,606 Sq. Ft. – Main Level
1,243 Sq. Ft. – Lower Level; 45,000 Cu. Ft.

● A home with two faces. From the street this design gives all the appearances of being a one-story, L-shaped home. One can only guess at the character of the rear elevation as dictated by the sloping terrain. A study of the interior reveals tremendous convenient living potential.

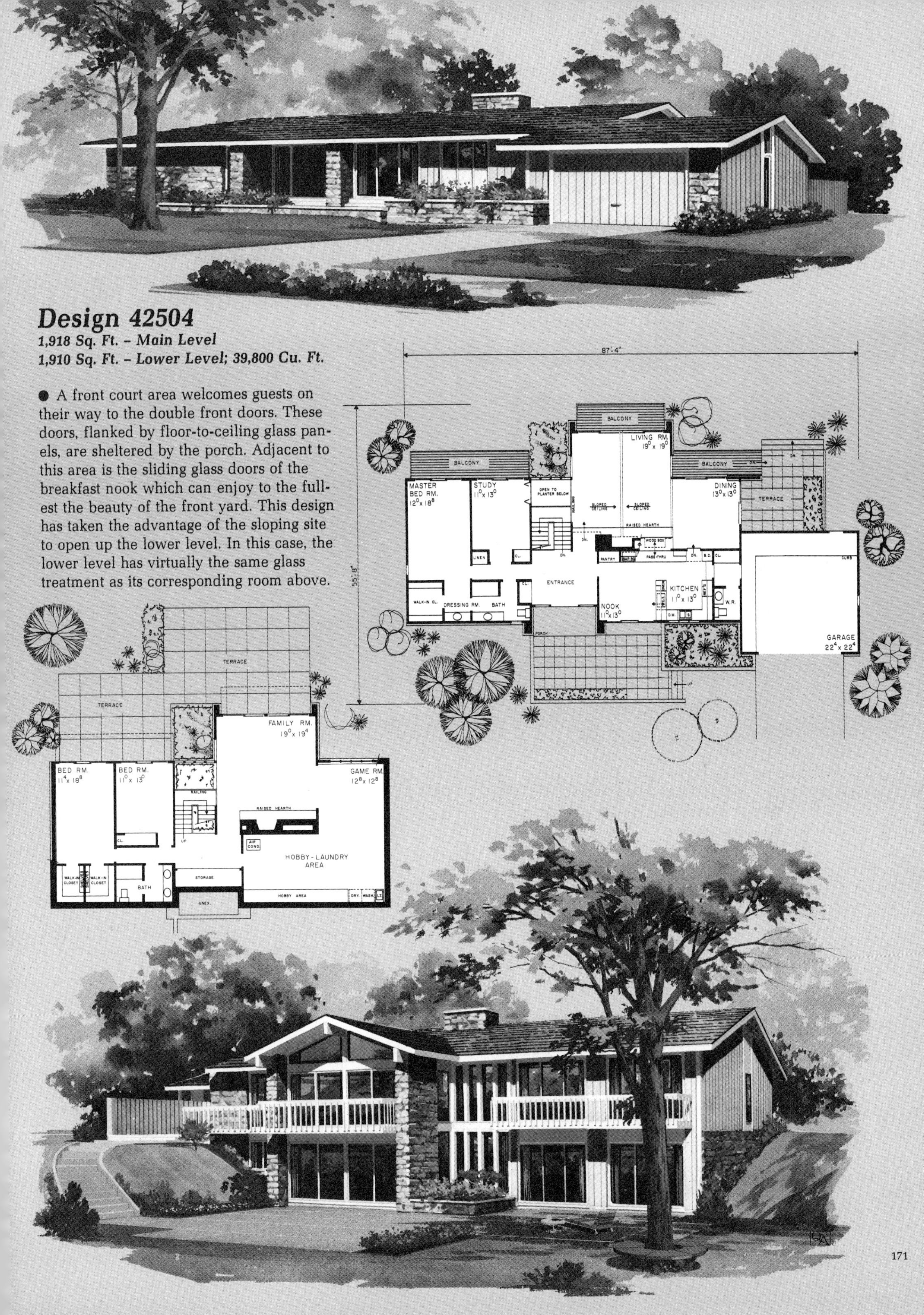

Design 42504

1,918 Sq. Ft. – Main Level
1,910 Sq. Ft. – Lower Level; 39,800 Cu. Ft.

● A front court area welcomes guests on their way to the double front doors. These doors, flanked by floor-to-ceiling glass panels, are sheltered by the porch. Adjacent to this area is the sliding glass doors of the breakfast nook which can enjoy to the fullest the beauty of the front yard. This design has taken the advantage of the sloping site to open up the lower level. In this case, the lower level has virtually the same glass treatment as its corresponding room above.

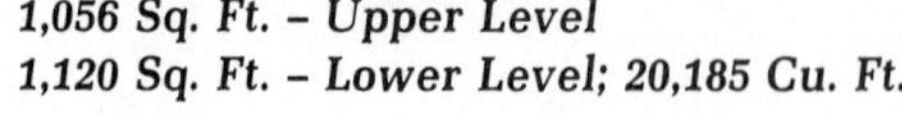

Design 41203

1,056 Sq. Ft. - Upper Level
1,120 Sq. Ft. - Lower Level; 20,185 Cu. Ft.

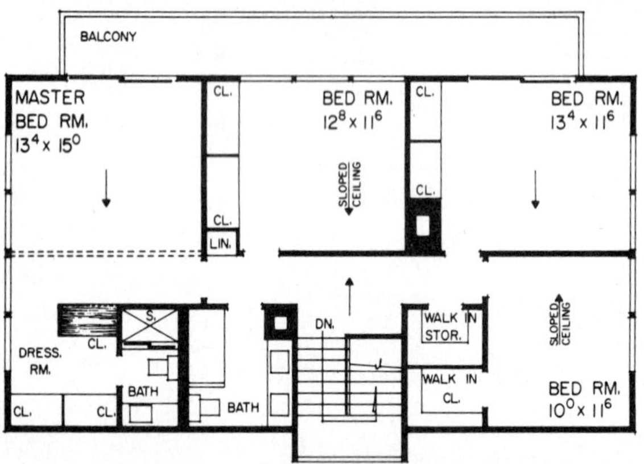

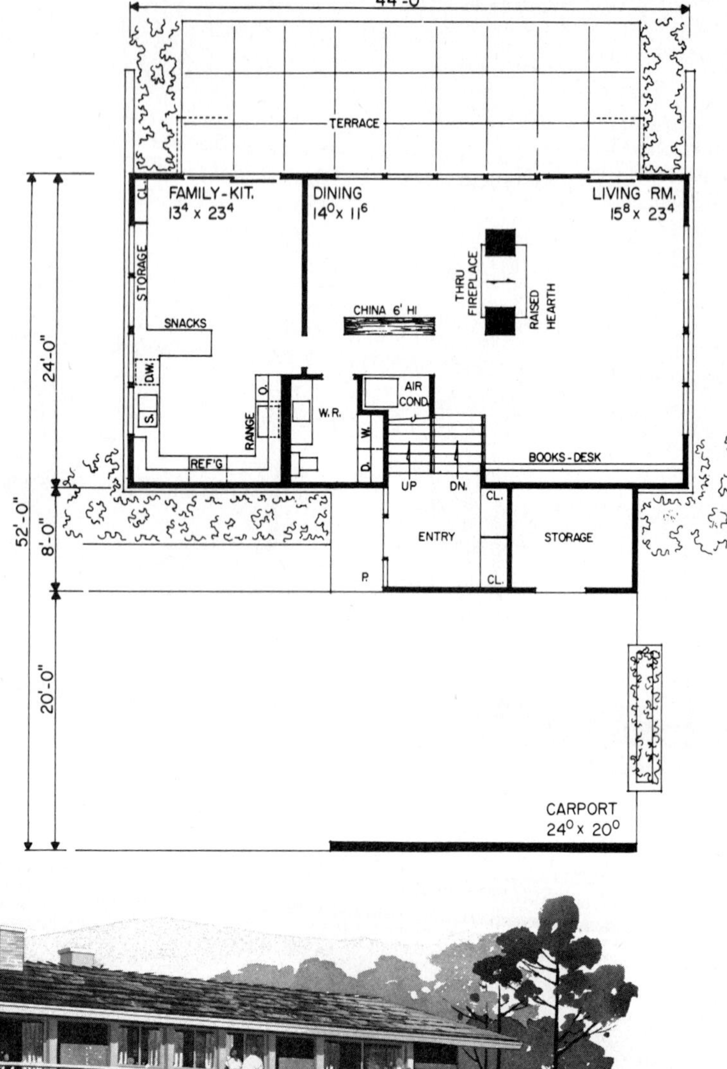

● On a gently sloping site this bi-level house makes excellent sense by separating living and sleeping areas. This type of two-level living is hard to beat for compactness and economy. While privacy from the street is assured on the one hand, a delightful awareness of the outdoors is experienced from the large glass areas on the other.

Design 42213

1,671 Sq. Ft. - Upper Level
1,033 Sq. Ft. - Lower Level; 27,249 Cu. Ft.

● Whether you locate this contemporary bi-level home on a sloping or flat site, it will certainly command its share of attention and provide the family with wonderful living patterns. The front entry is a separate level with stairs leading directly to the lower and the upper levels.

The most captivating feature of this home may very well be the spacious living and dining areas. An exposed beam is the apex of sloped ceilings. The projecting, glass-gabled end allows for a full measure of natural light. Two pairs of sliding glass doors open onto the balcony. The living balcony wraps around both front and rear to provide appealing planting areas. The kitchen is an efficient one in which to work, while the breakfast nook is but a step away. The sleeping zone has three bedrooms plus two full baths. Don't overlook the fireplace with its wood box.

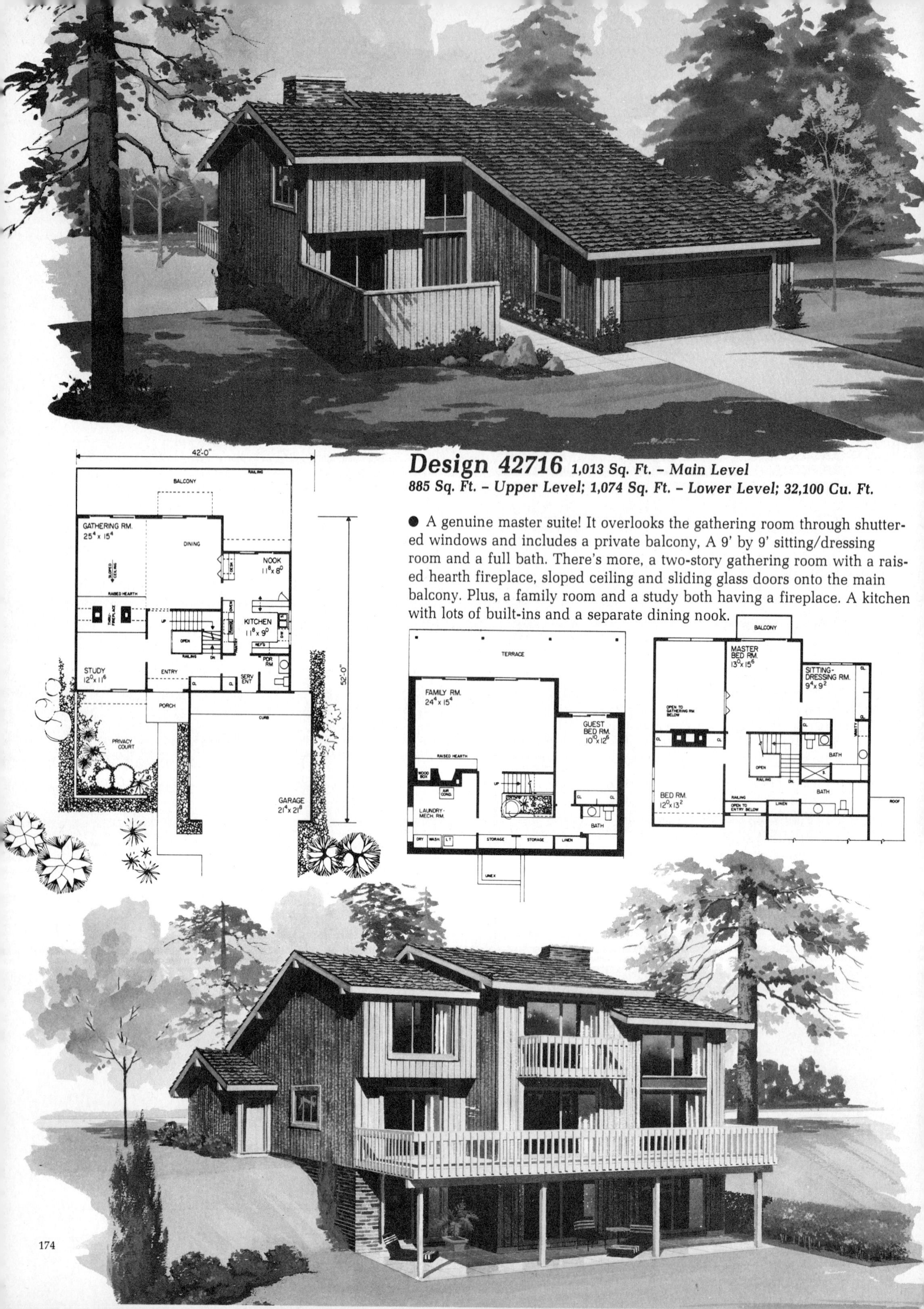

Design 42716 *1,013 Sq. Ft. - Main Level*

885 Sq. Ft. - Upper Level; 1,074 Sq. Ft. - Lower Level; 32,100 Cu. Ft.

● A genuine master suite! It overlooks the gathering room through shuttered windows and includes a private balcony, A 9' by 9' sitting/dressing room and a full bath. There's more, a two-story gathering room with a raised hearth fireplace, sloped ceiling and sliding glass doors onto the main balcony. Plus, a family room and a study both having a fireplace. A kitchen with lots of built-ins and a separate dining nook.

Design 42552 *1,437 Sq. Ft. - Main Level; 1,158 Sq. Ft. - Upper Level; 1,056 Sq. Ft. - Lower Level; 43,(*

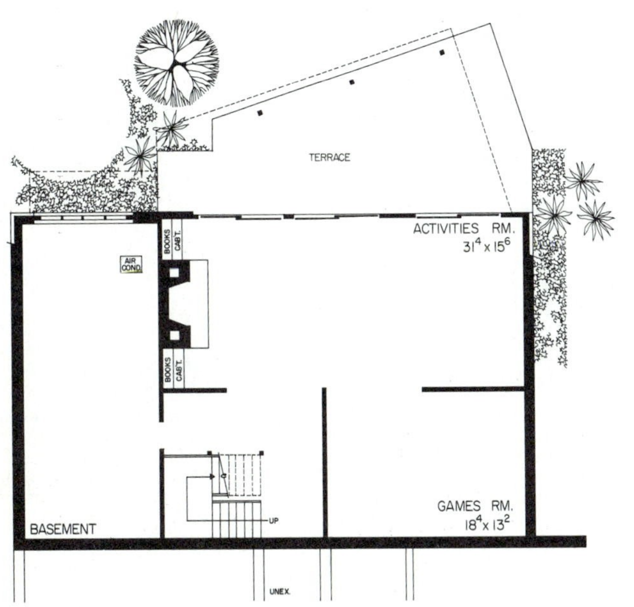

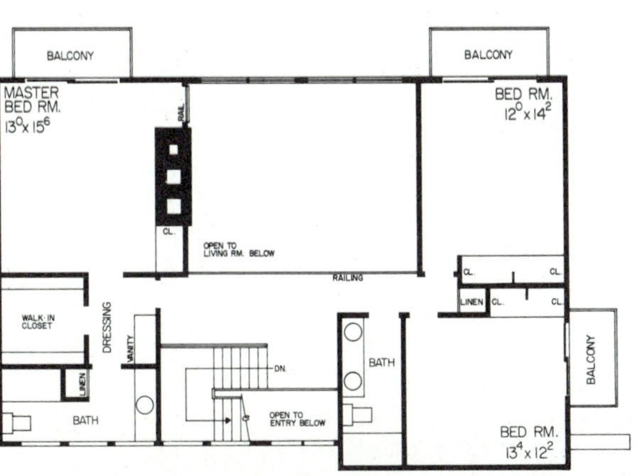

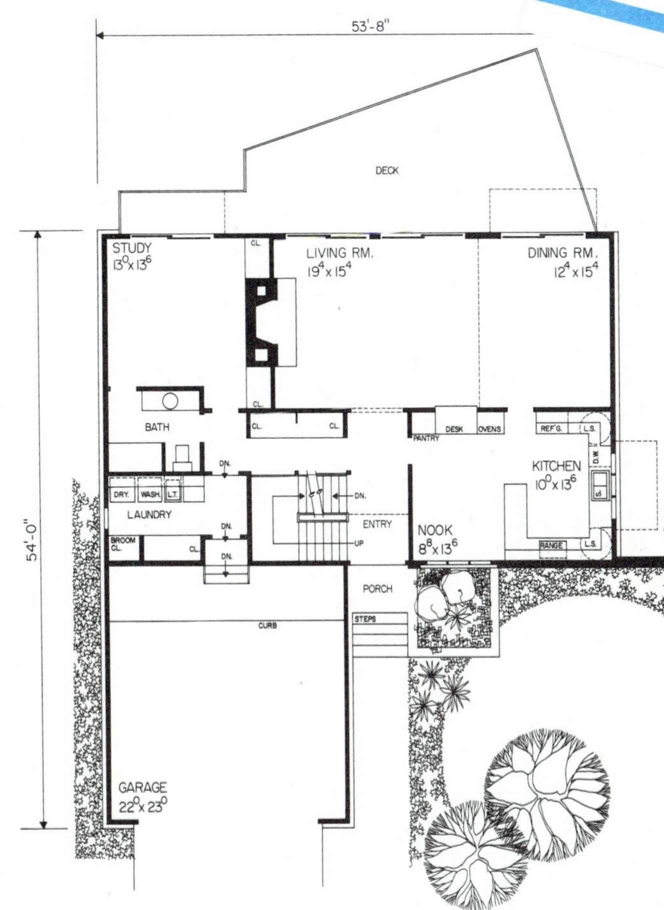

● Whatever you call this design - a hillside home or a two-story with an exposed basement - it will deliver an abundance of family livability. Study the three levels carefully. Notice how the upper level hall and master bedroom look down into the living room. Observe the study with access to a full bath.

Design 41812

1,726 Sq. Ft. – Main Level
1,320 Sq. Ft. – Lower Level
30,142 Cu. Ft.

● A home with two faces! The street view of this contemporary design presents a most pleasingly formal facade. The wide overhanging roof, the projecting masonry piers, the attractive glass treatment and the recessed front entrance all go together to form a perfectly delightful image. The rear terrace view is a picture of informality. The upper level outdoor balcony, overhanging the sweeping lower level terrace, will make summer entertaining an occasion to look forward to. The balcony gives way to the spacious deck which will be an ideal spot on which to sunbathe or dine. The plan is interesting, indeed. A two-way fireplace separates the big living/dining area. Observe the work center; study the lower level layout. Be sure to note the extra bedroom.

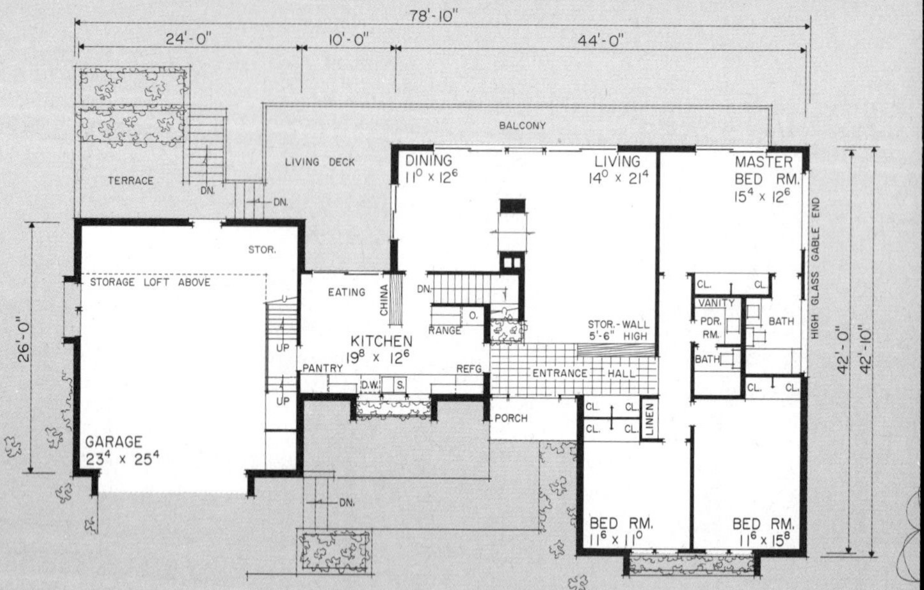

78'-10"
24'-0" 10'-0" 44'-0"
TERRACE
DN.
LIVING DECK
DN.
BALCONY
DINING $11^{0} \times 12^{6}$
LIVING $14^{0} \times 21^{4}$
MASTER BED RM. $15^{4} \times 12^{6}$
HIGH GLASS GABLE END
STOR.
STORAGE LOFT ABOVE
EATING
CHINA
DN.
O.
RANGE
KITCHEN $19^{8} \times 12^{6}$
UP
PANTRY
D.W. S.
REFG.
UP
STOR.-WALL 5'-6" HIGH
ENTRANCE HALL
CL. CL.
VANITY
PDR. RM.
BATH
BATH
CL. CL.
CL. CL.
CL. CL.
LINEN
PORCH
26'-0"
42'-0"
42'-10"
GARAGE $23^{4} \times 25^{4}$
DN.
BED RM. $11^{6} \times 11^{0}$
BED RM. $11^{6} \times 15^{8}$

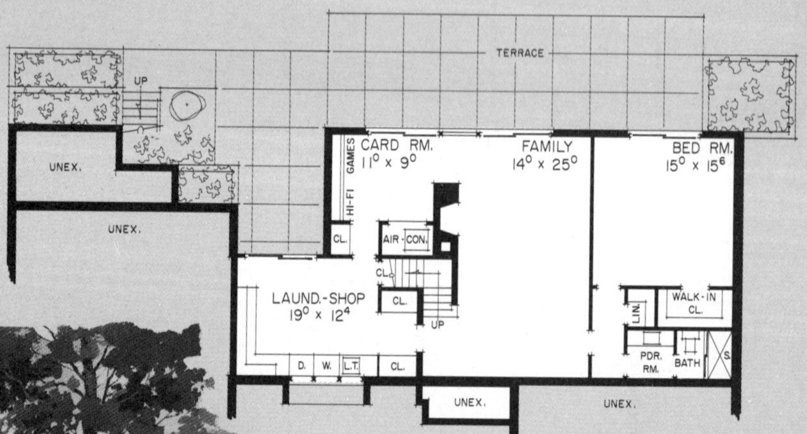

TERRACE
UP
UNEX.
UNEX.
GAMES
HI-FI
CARD RM. $11^{0} \times 9^{0}$
FAMILY $14^{0} \times 25^{0}$
BED RM. $15^{0} \times 15^{6}$
CL.
AIR-CON.
CL.
CL.
LAUND.-SHOP $19^{0} \times 12^{4}$
UP
LIN.
WALK-IN CL.
PDR. RM.
BATH
S.
D. W. L.T. CL.
UNEX.
UNEX.

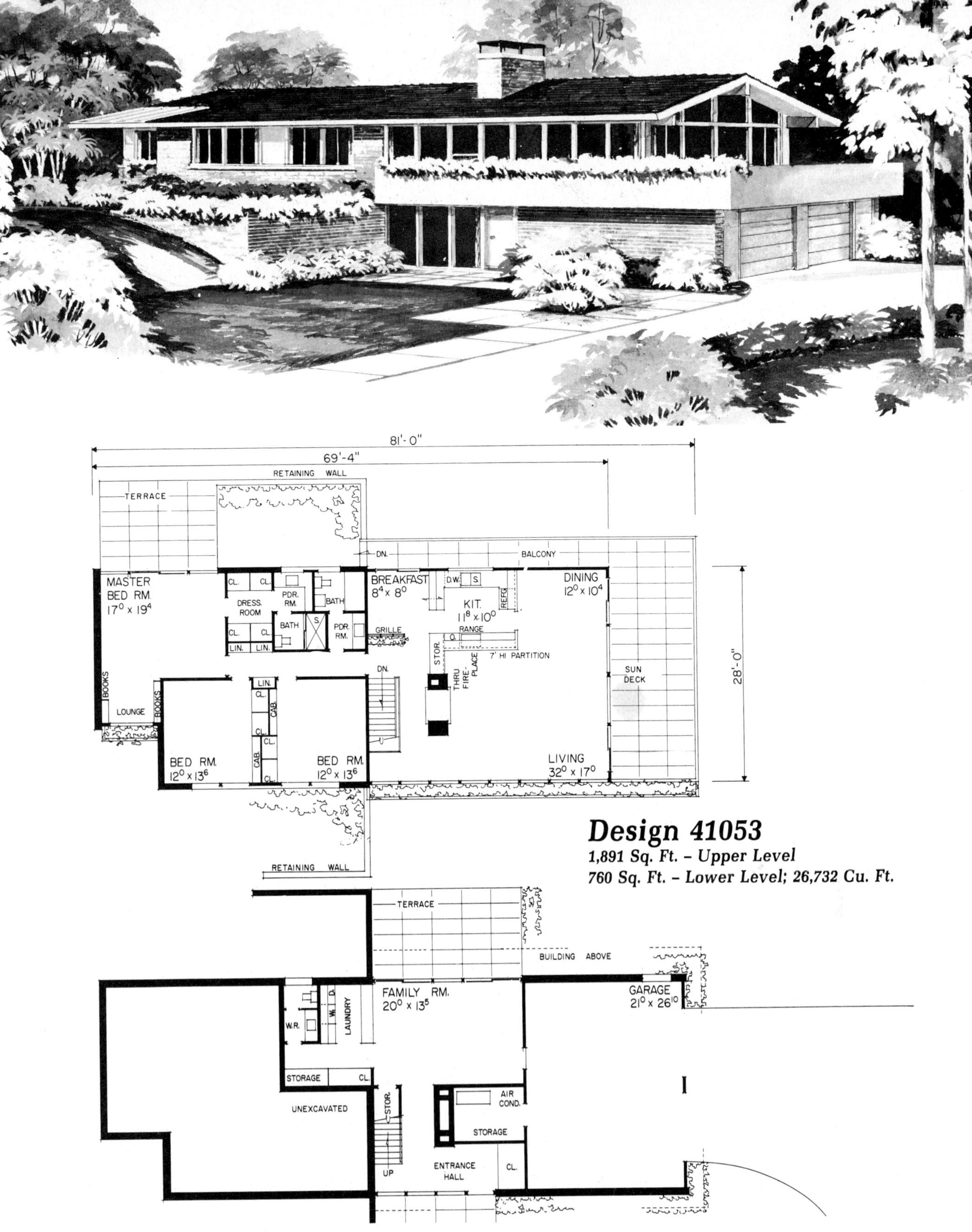

Design 41053

1,891 Sq. Ft. - Upper Level
760 Sq. Ft. - Lower Level; 26,732 Cu. Ft.

● If there is a view to be preserved and enjoyed to the fullest this two-level home will perform admirably. You enter on the lower level where the family room, laundry and extra wash room are located. Up the stairs are the living areas where spaciousness is the byword. The formal living and dining areas seem to flow through the glass wall onto the high deck with an equally as fantastic a view of the front yard. The U-shaped kitchen is convenient to both the dining room and the breakfast nook which has sliding glass doors to the rear balcony. The balcony and terrace provide outdoor living to the rear. Note the master bedroom layout and the various storage units. The other two bedrooms also have the advantage of built-in cabinets adjacent to the closet.

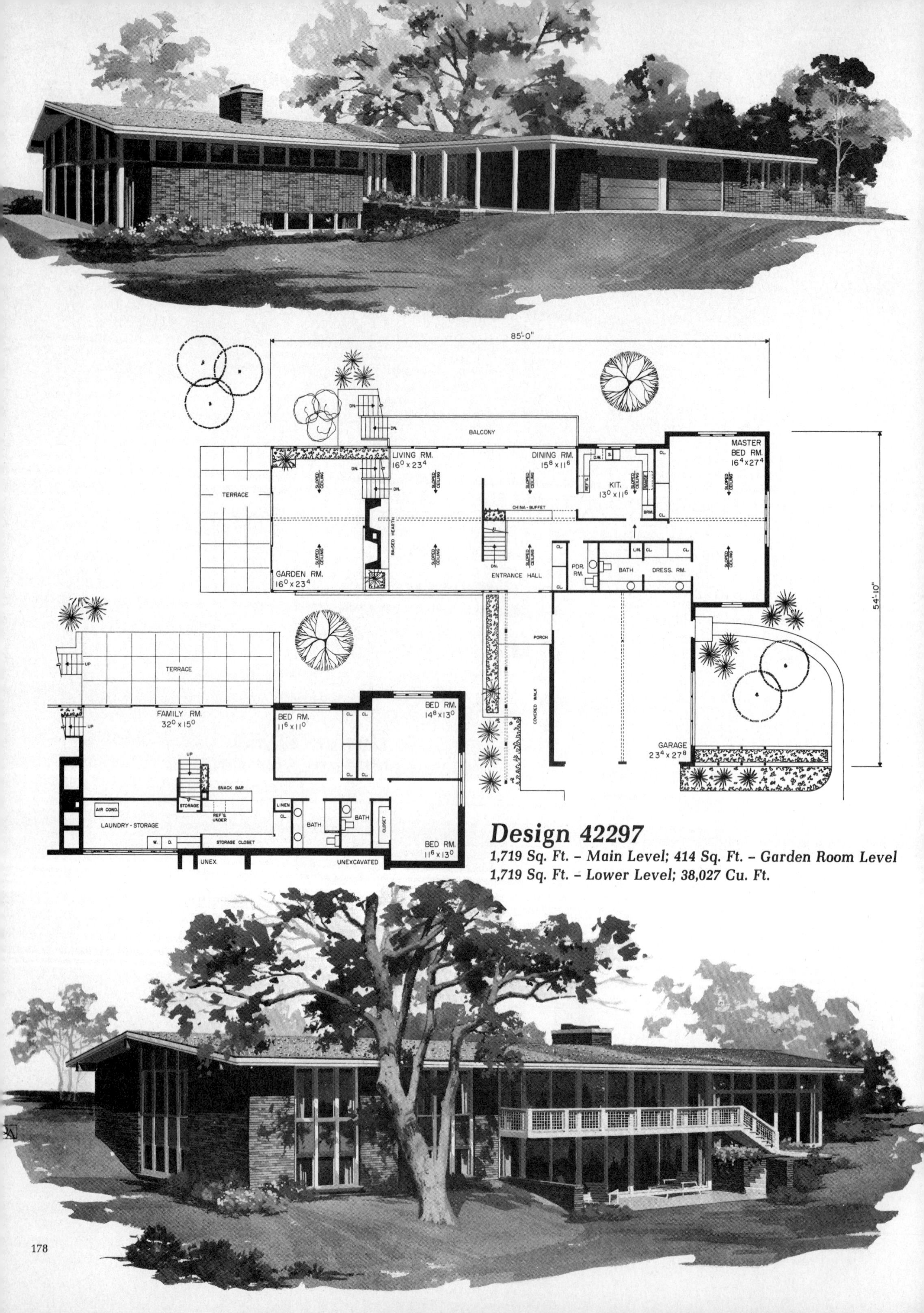

Design 42297

1,719 Sq. Ft. – Main Level; 414 Sq. Ft. – Garden Room Level
1,719 Sq. Ft. – Lower Level; 38,027 Cu. Ft.

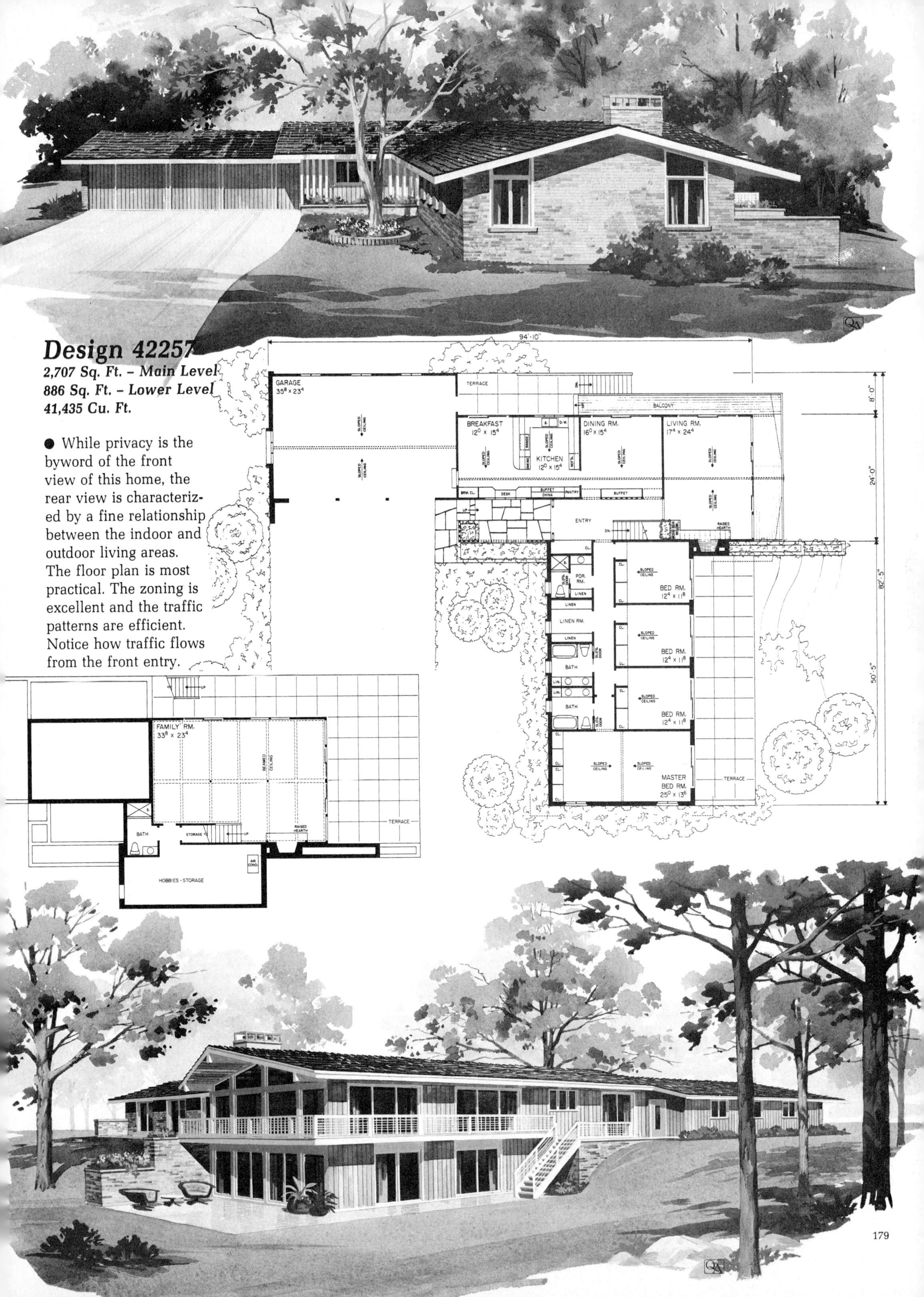

Design 42257

2,707 Sq. Ft. – Main Level
886 Sq. Ft. – Lower Level
41,435 Cu. Ft.

● While privacy is the byword of the front view of this home, the rear view is characterized by a fine relationship between the indoor and outdoor living areas. The floor plan is most practical. The zoning is excellent and the traffic patterns are efficient. Notice how traffic flows from the front entry.

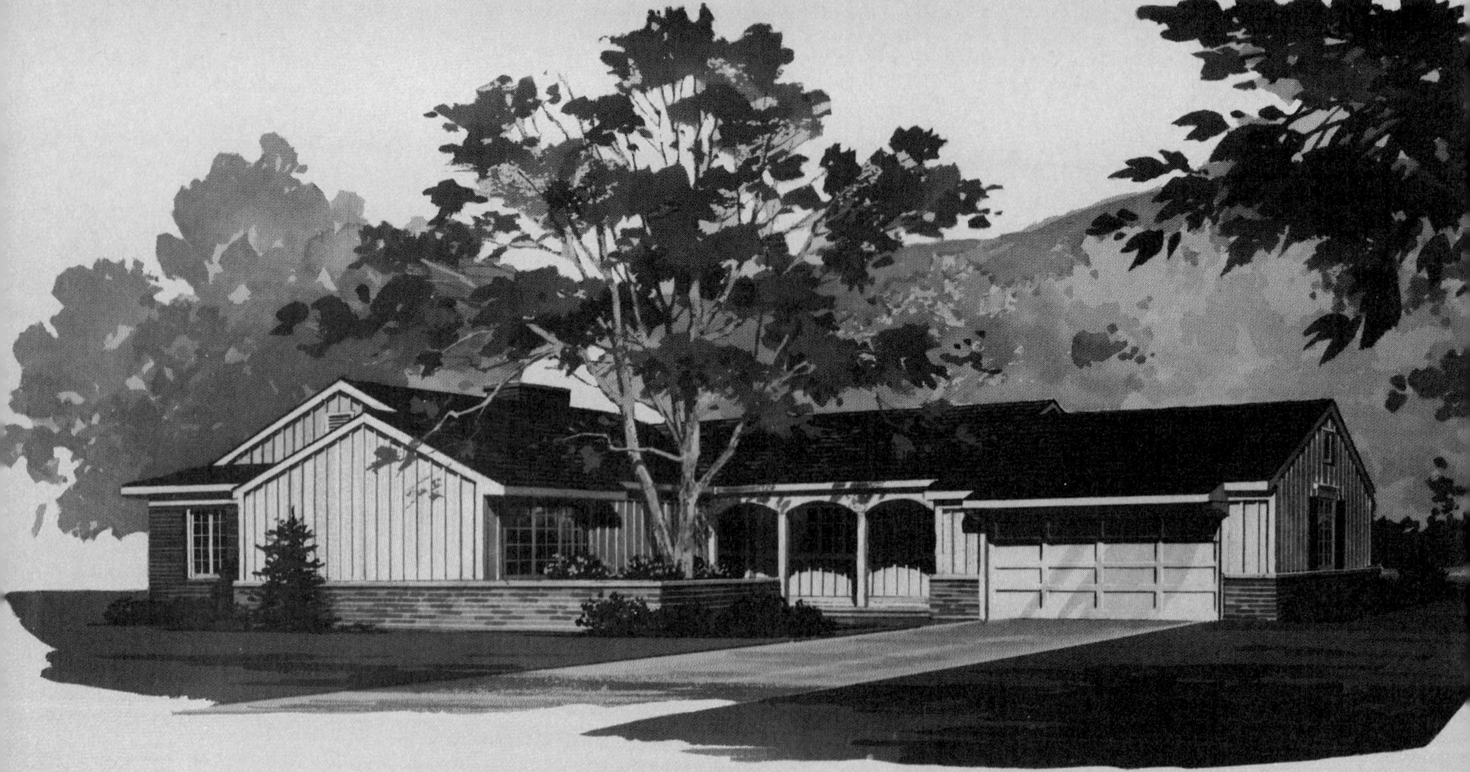

Colonial Ambience With Hillside Livability

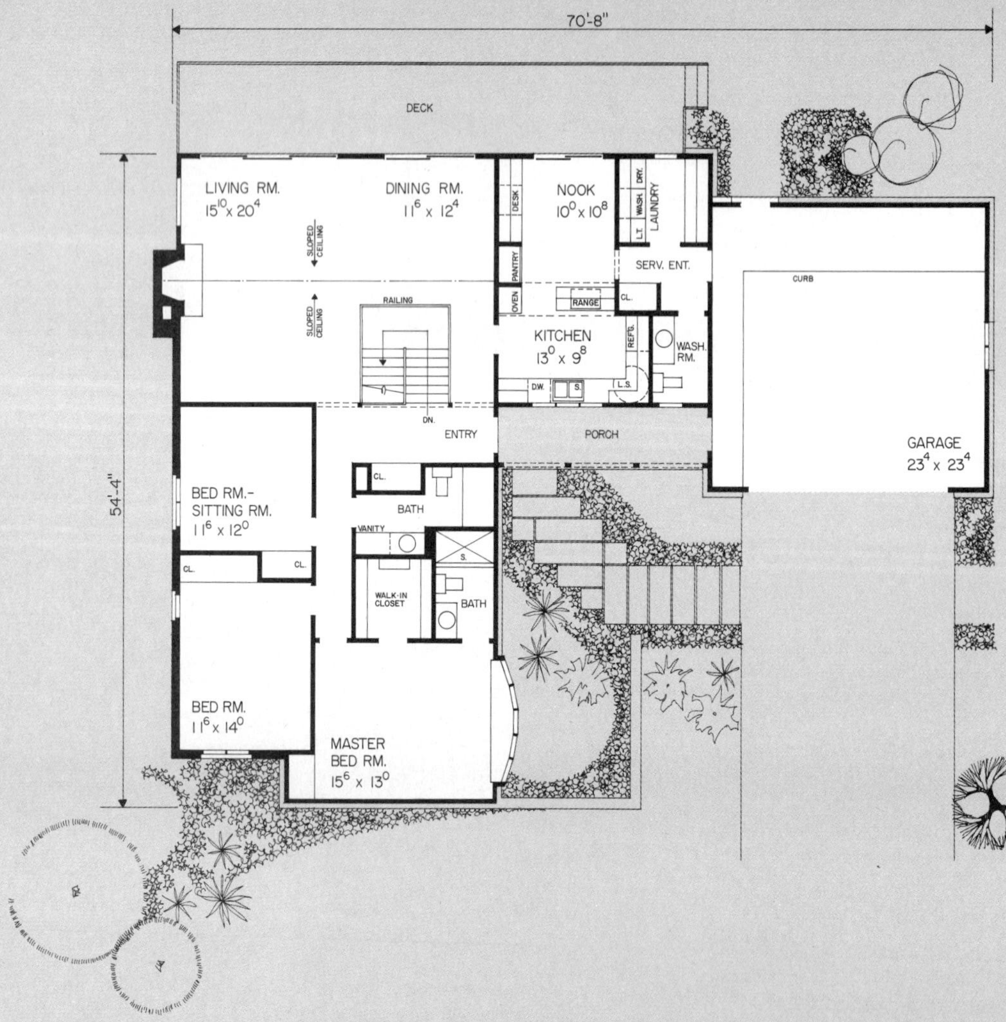

70'-8"
54'-4"
DECK
LIVING RM.
15^{10} x 20^{4}
DINING RM.
11^{6} x 12^{4}
SLOPED CEILING
SLOPED CEILING
RAILING
DN.
NOOK
10^{0} x 10^{8}
DESK
PANTRY
OVEN
RANGE
L.T.
WASH
DRY.
LAUNDRY
SERV. ENT.
CL.
KITCHEN
13^{0} x 9^{8}
REFG.
D.W.
S.
L.S.
WASH. RM.
CURB
ENTRY
PORCH
GARAGE
23^{4} x 23^{4}
BED RM.-
SITTING RM.
11^{6} x 12^{0}
CL.
CL.
CL.
BATH
VANITY
S.
WALK-IN CLOSET
BATH
BED RM.
11^{6} x 14^{0}
MASTER
BED RM.
15^{6} x 13^{0}

Design 42769

1,898 Sq. Ft. – Main Level
1,134 Sq. Ft. – Lower Level
41,910 Cu. Ft.

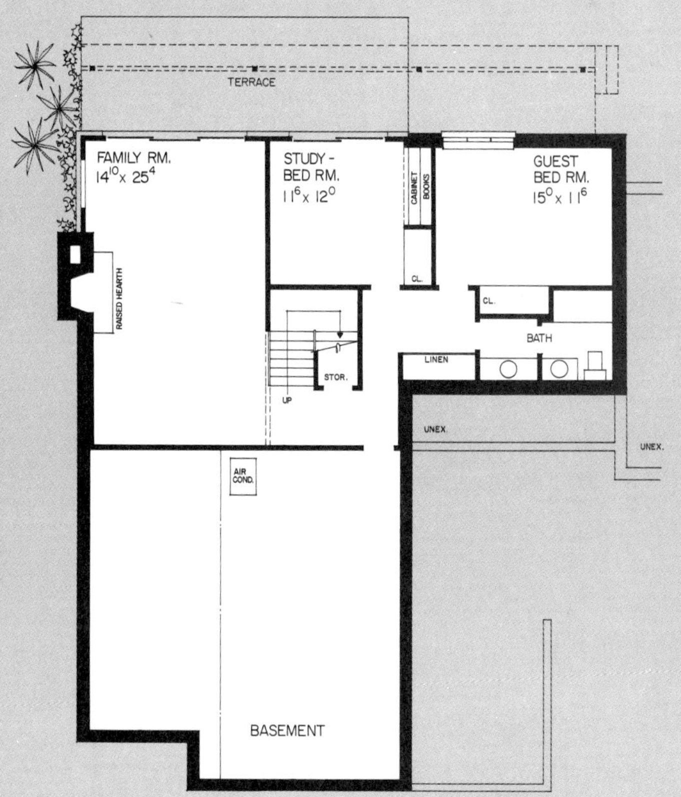

● A house design can take many forms but none will be quite as unique as this hillside design. Shielded at the front from passers-by to provide privacy, this L-shaped traditional home opens to embrace the outdoors at the rear. This outstanding design will be a welcome change to the entire family. The front one-story exterior is delightful. A picture of charm with any setting as its backdrop. Upon entrance, one will be greeted at the covered front porch and enter into the spacious entry hall. At this point the interior livability has only begun. Adding both levels, there are five good-sized bedrooms. Two of the five are noted to have flexible uses. Convert the bedroom adjacent to the living room into a sitting room. It would be a great area to enjoy the peace and quiet after a long hard day at work. The smaller bedroom on the lower level would serve ideally as a study. It already has built-in book shelves and cabinets. The master bedroom has a beautiful bay window to view the front yard, a large walk-in closet and a private bath. All formal entertaining will be enjoyed in the open-planned living/dining area which enhances the flow of light and space within, while sloping ceilings and a fireplace add a dramatic touch. Two sets of sliding glass doors provide access to the full house-wide deck. The deck also can be entered from the breakfast nook. The work center of this main level is outstanding. The kitchen, along with the nook, has many built-in features including appliances, pantry and desk. The wash room and laundry are conveniently located. Informal activities can be pursued in the family room on the lower level which, like the study/bedroom, opens up to the terrace.

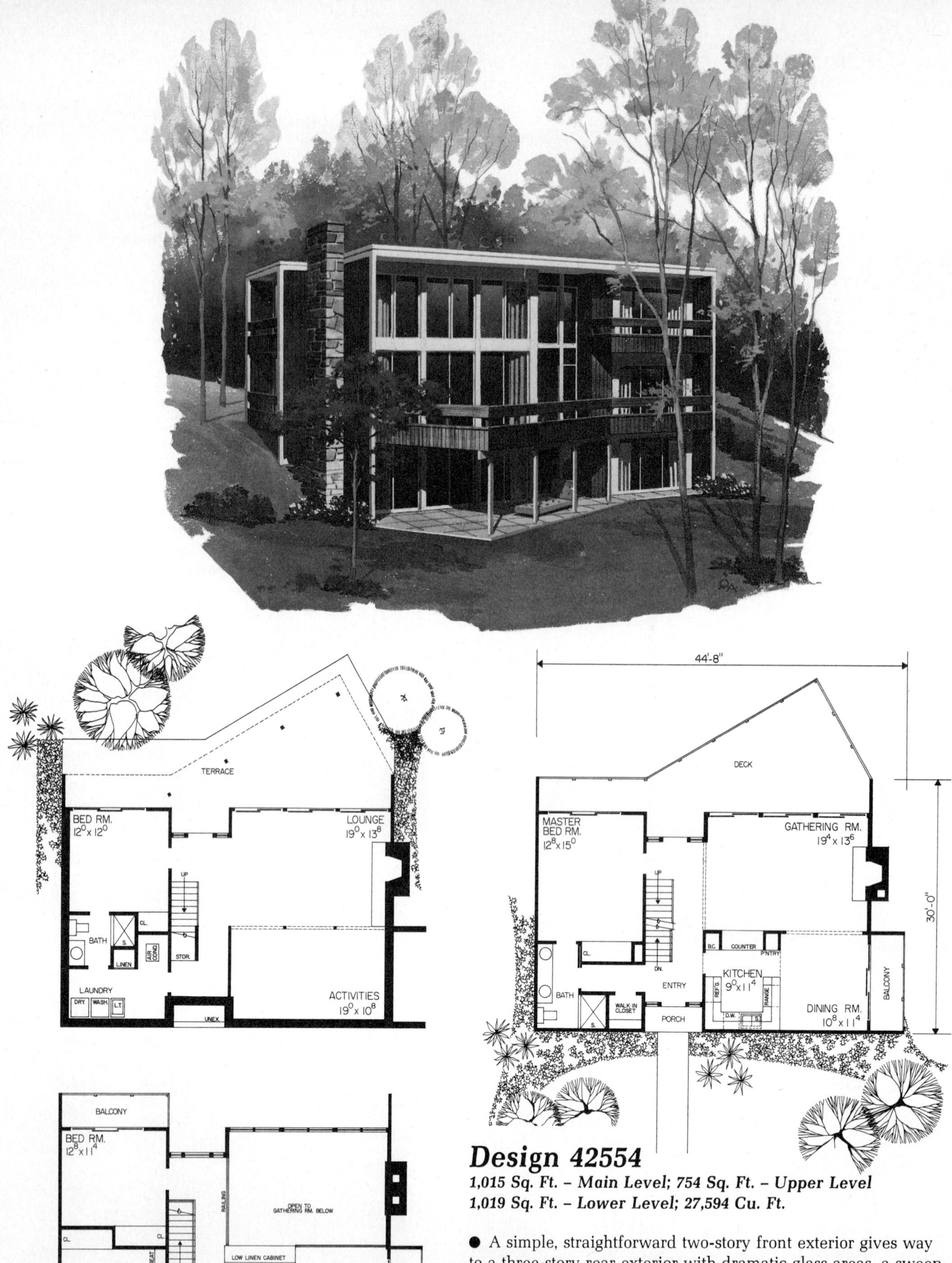

Design 42554

1,015 Sq. Ft. – Main Level; 754 Sq. Ft. – Upper Level
1,019 Sq. Ft. – Lower Level; 27,594 Cu. Ft.

● A simple, straightforward two-story front exterior gives way to a three-story rear exterior with dramatic glass areas, a sweeping deck and a high balcony. Two additional balconies serve the dining room and sleeping loft. This home with its four bedrooms and loft will sleep quite a gathering. And there is plenty of living space to accomodate everyone during waking hours, too.

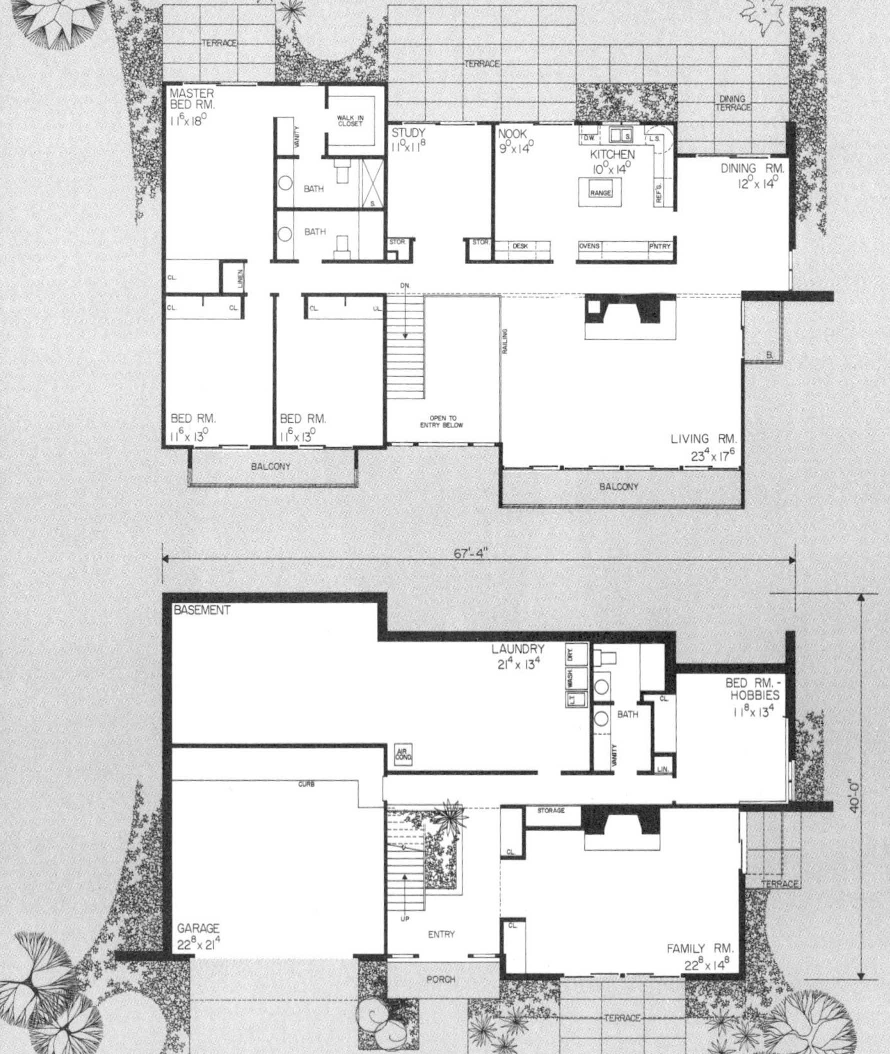

Design 42545

2,120 Sq. Ft. – Upper Level
1,094 Sq. Ft. – Lower Level
42,200 Cu. Ft.

● This contemporary bi-level seems to impart a message that funloving living patterns await the occupants. And little wonder, with all those terraces and balconies. These areas emphasize the fine indoor-outdoor living relationships this design has to offer. Built into the sloping site, the upper level walks out onto grade at the rear. The two-story entrance hall with its enchanting planter and appealing glass areas is both cheerful and dramatic. Note the open planning aspect between lower level hall and the upper level hall and living room. The breakfast nook-kitchen-dining room relationship is a nice one. If yours is a large family, this design will function well as a five bedroom home. As a three bedroom it offers a study and hobby room. In addition to the family room, there is the bonus space of the laundry-basement area. Notice the two fireplaces and three baths.

A Lifetime of Exciting, Contemporary Living Patterns

● Here is a home for those with a bold, contemporary living bent. The exciting exteriors give notice of an admirable flair for something delightfully different. The varying roof planes and textured blank wall masses are distinctive. Two sets of panelled front doors permit access to either level. The inclined ramp to the upper main level is dramatic, indeed. The rear exterior highlights a veritable battery of projecting balconies. This affords direct access to outdoor living for each of the major rooms in the house. Certainly an invaluable feature should your view be particularly noteworthy. Notice two covered outdoor balconies plus a covered terrace. Indoor-outdoor living at its greatest.

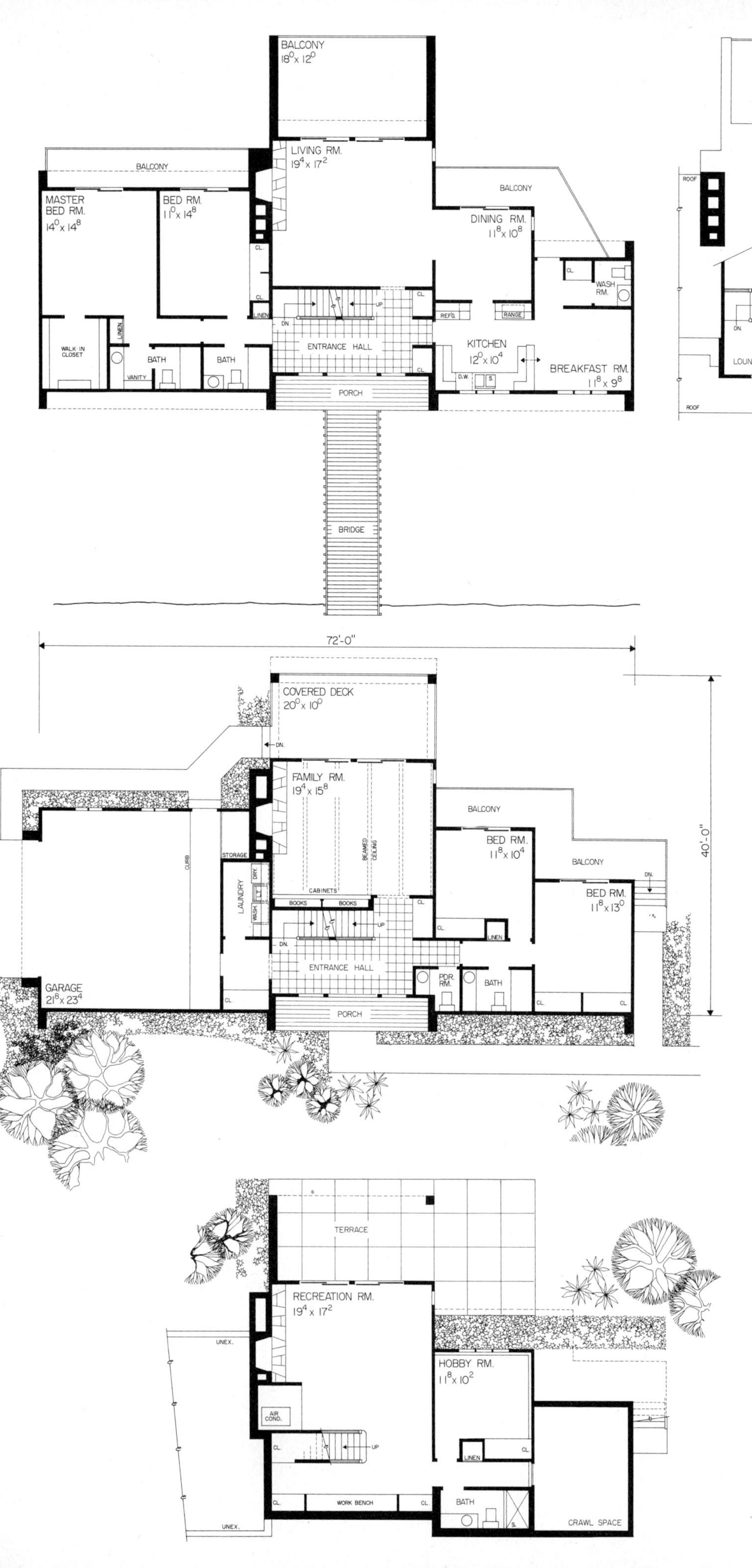

Design 42392

1,691 Sq. Ft. – Main Level
1,127 Sq. Ft. – Lower Entry Level
396 Sq. Ft. – Upper Level
844 Sq. Ft. – Lower Level
40,026 Cu. Ft.

● Try to imagine the manner in which you and your family will function in this four-level hillside design. Surely it will be an adventure in family living that will be hard to surpass. For instance, can you picture a family member painting or sewing in the upper level studio, while another is building models or developing pictures in the lower level hobby room? Or, can you visualize a group in quiet conversation in the living room, another lounging in the family room, while a third plays table tennis or pool in the recreation room? Be sure not to overlook the fireplace in each of these living areas. As for sleeping and bath facilities, your family will have plenty, four bedrooms and four baths, plus a powder room and a wash room. They also will enjoy the eating facilities with a breakfast room, a dining room and an outdoor balcony nearby. Then, too, there is the lounge of the upper level.

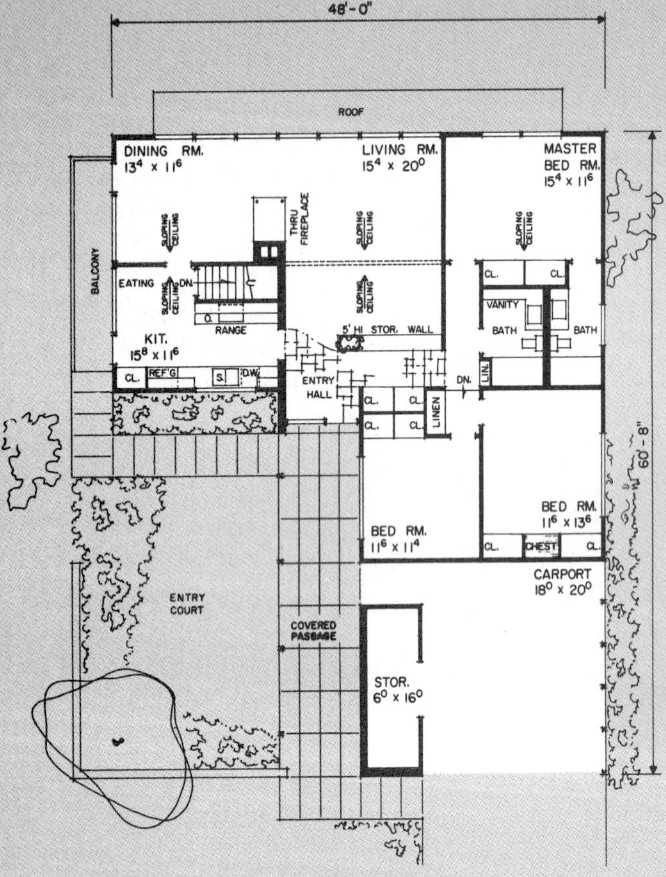

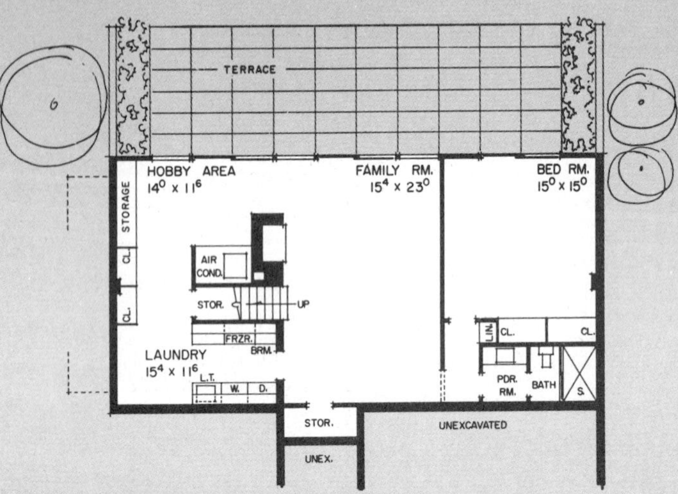

Design 41103

1,560 Sq. Ft. – Main Level
1,152 Sq. Ft. – Lower Level; 24,296 Cu. Ft.

● One-story living in the front and two-story living in the rear. Such is the story of the indoor/outdoor relationships fostered by this hillside design. One of the appealing features of this home is that it is only 48 feet wide and requires a relatively small piece of property. Its livability potential, however, is tremendous. Study the two levels carefully. With four bedrooms, three baths, large living areas and an abundance of storage facilities your family living patterns will be efficient and enjoyable.

Design 41976

1,616 Sq. Ft. – Upper Level
1,472 Sq. Ft. – Lower Level
29,909 Cu. Ft.

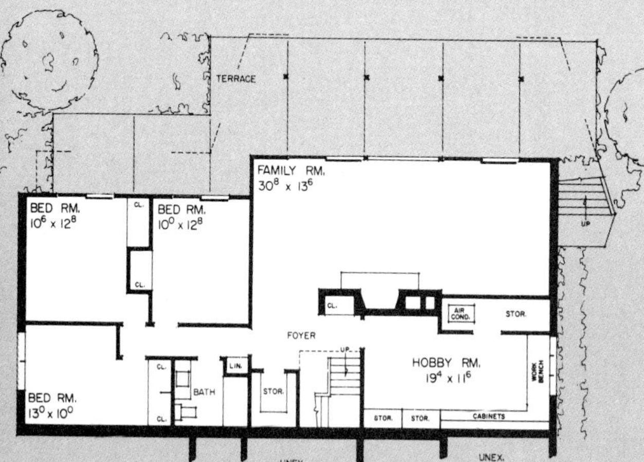

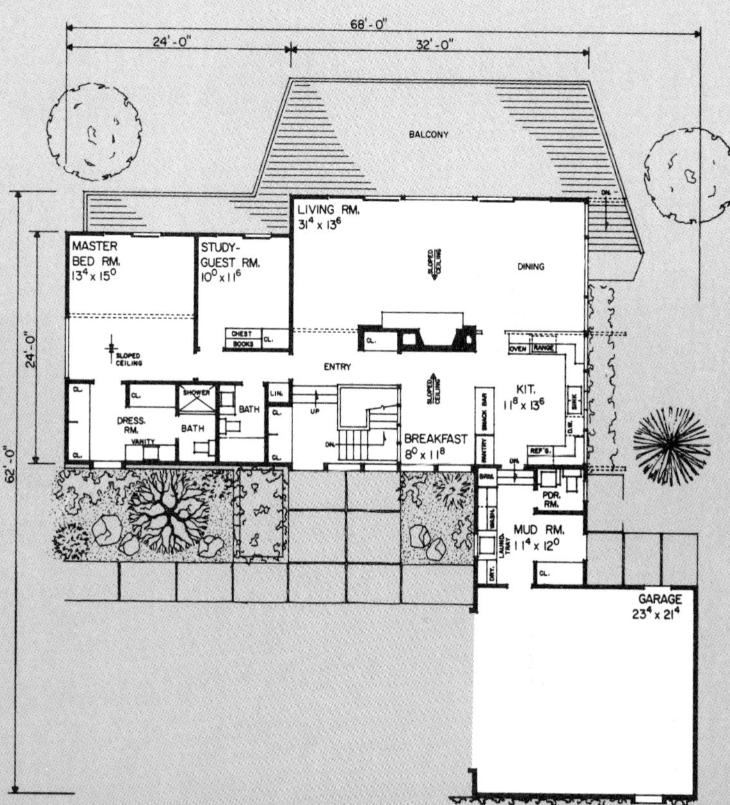

● Here's a hillside design just patterned for the large, active family. Whatever the pursuits and interests of the various members, you'd have to guess there would be more than enough space to service one and all with plenty of room to spare. If the children were teenagers, just imagine the fun they would have with their bedrooms, their family room and their hobby room on the lower level. The parents would be equally thrilled with their more formal facilities on the upper level.

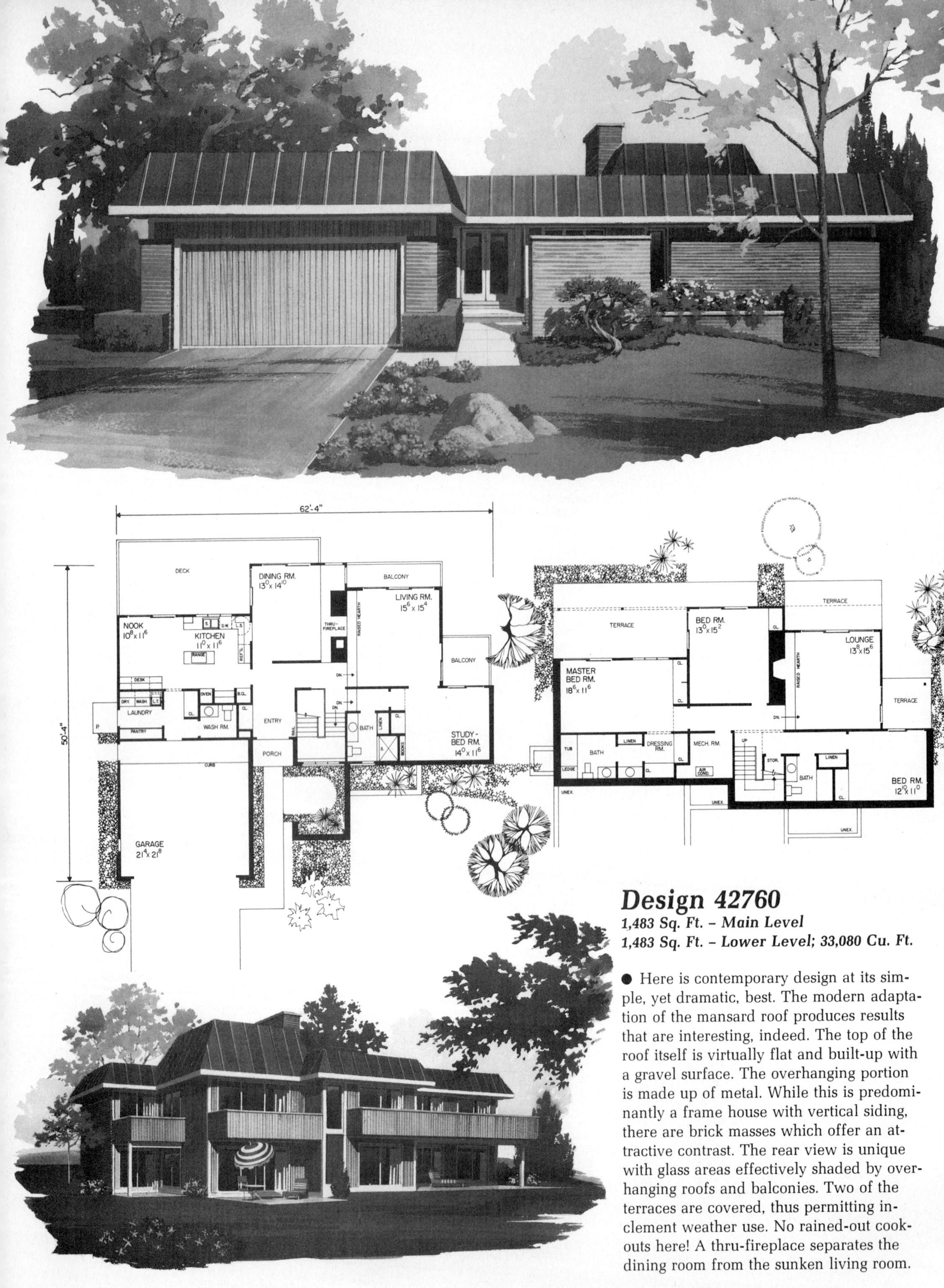

Design 42760

1,483 Sq. Ft. – Main Level
1,483 Sq. Ft. – Lower Level; 33,080 Cu. Ft.

● Here is contemporary design at its simple, yet dramatic, best. The modern adaptation of the mansard roof produces results that are interesting, indeed. The top of the roof itself is virtually flat and built-up with a gravel surface. The overhanging portion is made up of metal. While this is predominantly a frame house with vertical siding, there are brick masses which offer an attractive contrast. The rear view is unique with glass areas effectively shaded by overhanging roofs and balconies. Two of the terraces are covered, thus permitting inclement weather use. No rained-out cookouts here! A thru-fireplace separates the dining room from the sunken living room.

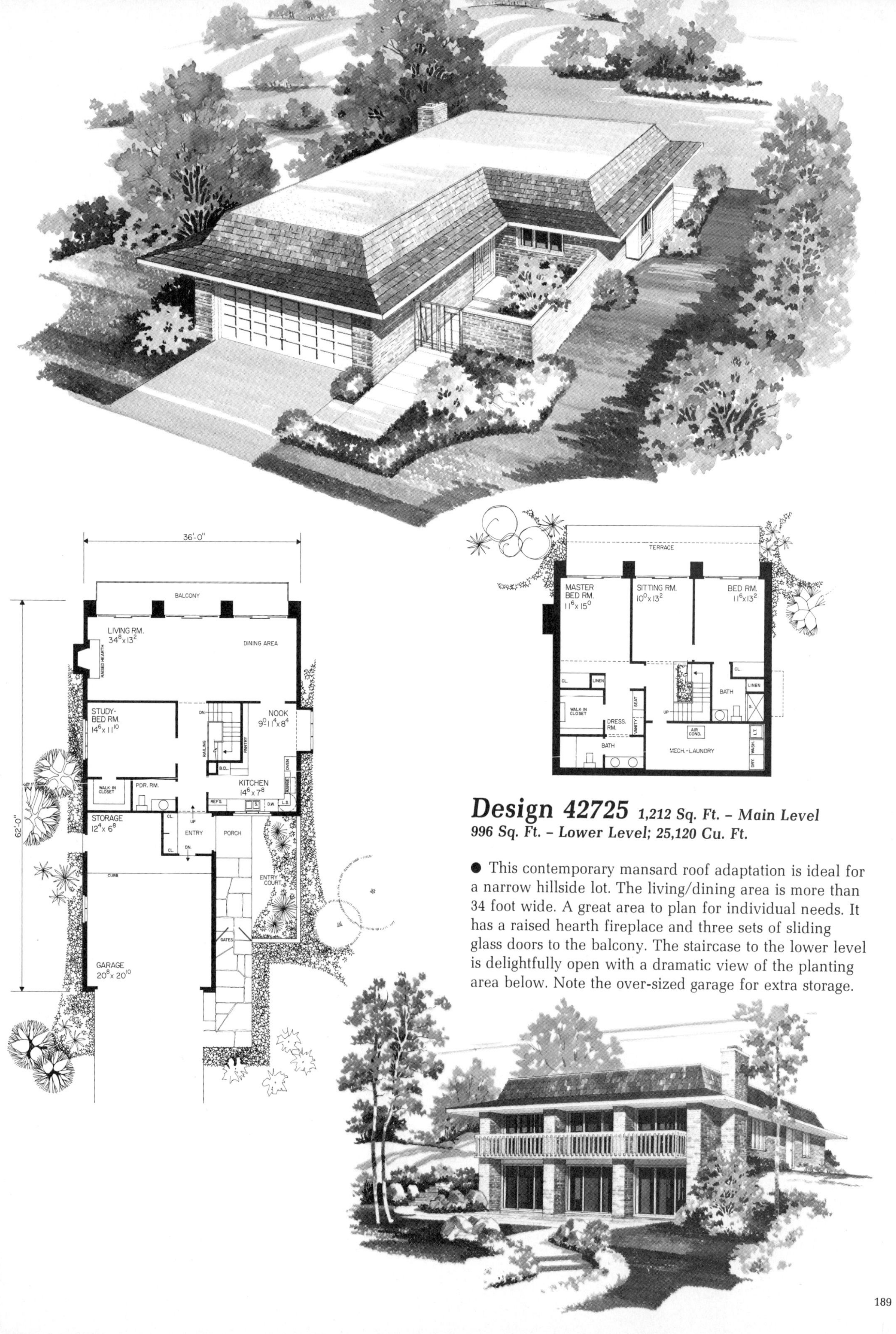

Design 42725 *1,212 Sq. Ft. - Main Level*

996 Sq. Ft. - Lower Level; 25,120 Cu. Ft.

● This contemporary mansard roof adaptation is ideal for a narrow hillside lot. The living/dining area is more than 34 foot wide. A great area to plan for individual needs. It has a raised hearth fireplace and three sets of sliding glass doors to the balcony. The staircase to the lower level is delightfully open with a dramatic view of the planting area below. Note the over-sized garage for extra storage.

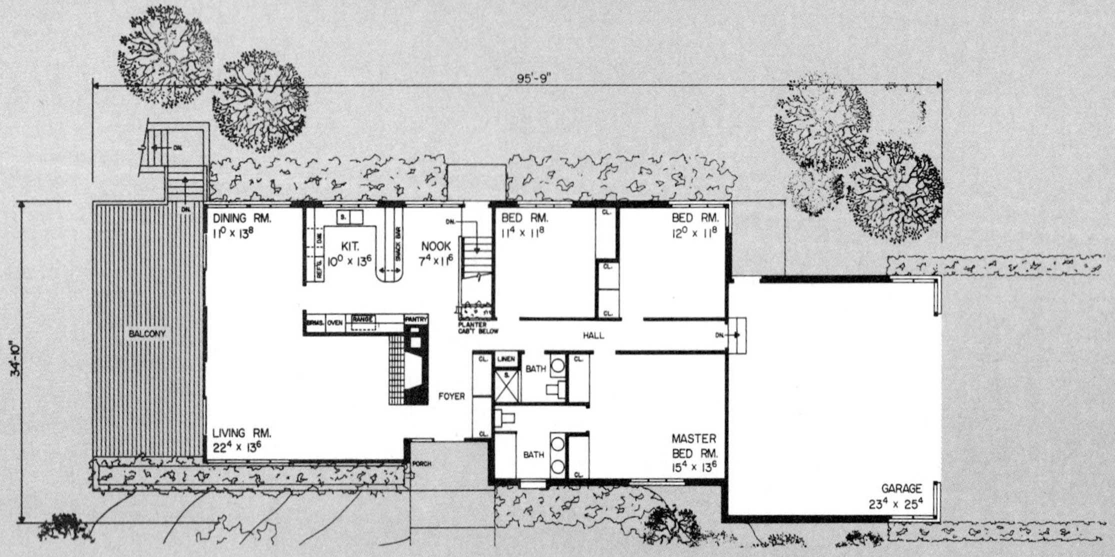

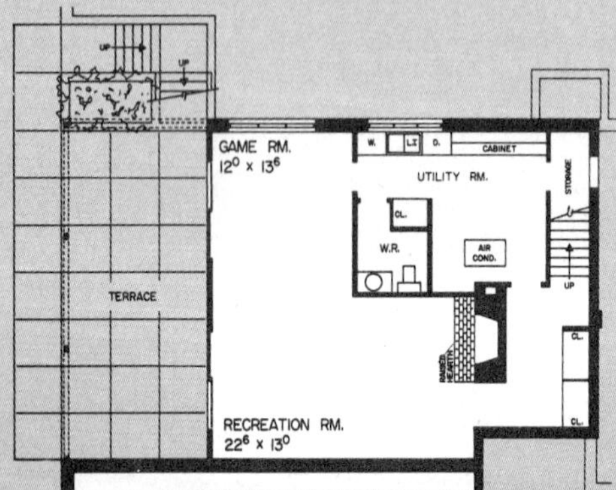

Design 42272

1,731 Sq. Ft. – Main Level
672 Sq. Ft. – Lower Level; 27,802 Cu. Ft.

● Certainly not a huge house. But one, nevertheless, that is long on livability and one that surely will be fun to live in. With its wide-overhanging hip roof, this unadorned facade is the picture of simplicity. As such, it has a quiet appeal all its own. The living-dining area is one of the focal points of the plan. It is wonderfully spacious. The large glass areas and the accessibility, through sliding glass doors, of the outdoor balcony are fine features. For recreation, there is the lower level area which opens onto a large terrace covered by the balcony above.

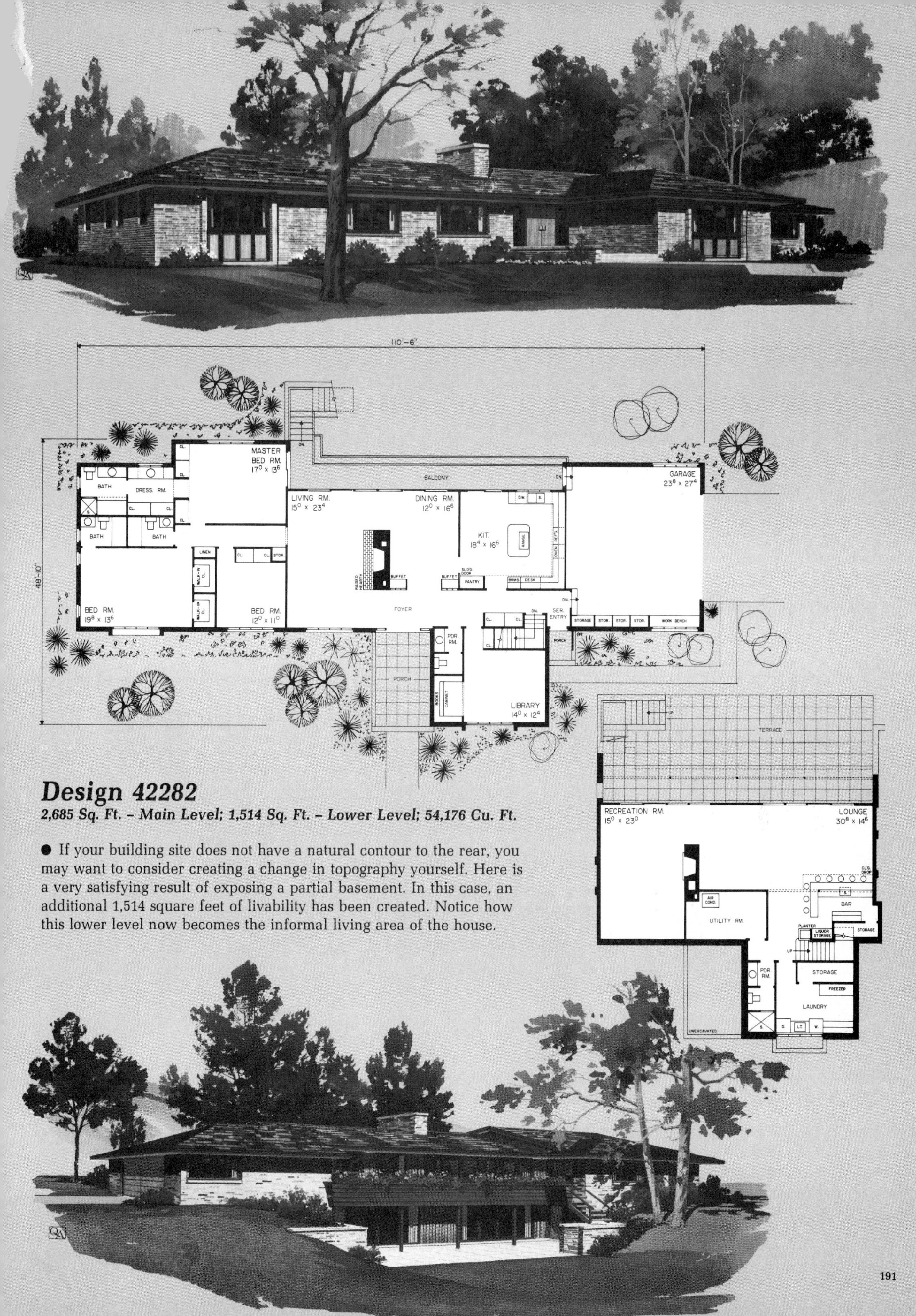

Design 42282

2,685 Sq. Ft. – Main Level; 1,514 Sq. Ft. – Lower Level; 54,176 Cu. Ft.

● If your building site does not have a natural contour to the rear, you may want to consider creating a change in topography yourself. Here is a very satisfying result of exposing a partial basement. In this case, an additional 1,514 square feet of livability has been created. Notice how this lower level now becomes the informal living area of the house.

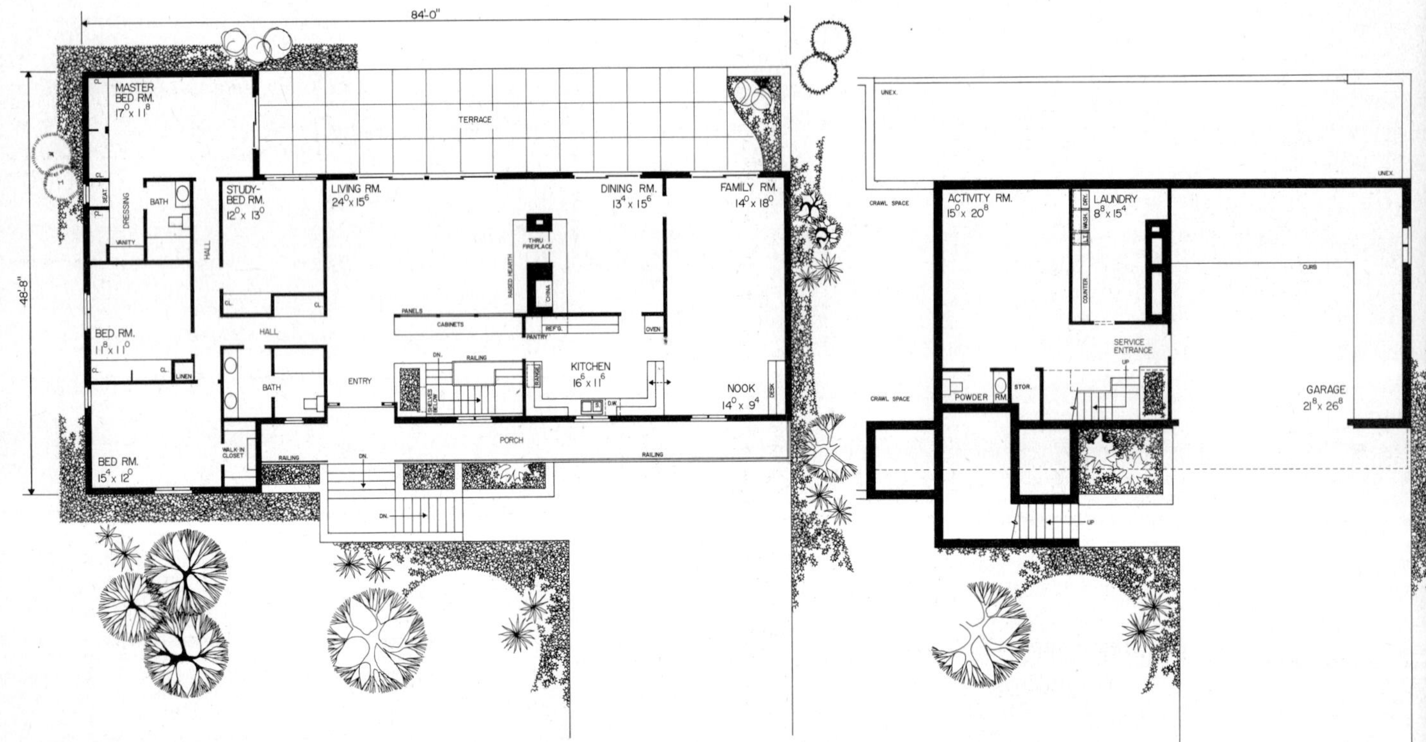

Design 42576 *2,805 Sq. Ft. – Main Level; 785 Sq. Ft. – Lower Level; 41,562 Cu. Ft.*

● This delightfully styled hip-roofed house features all the living facilities on one level with an additional lower level housing an activity room, powder room, laundry and garage. The result is a very captivating design. Formal living patterns will prevail in this house. The living room and dining room will have a delightful view of the rear yard, possess a dramatic thru-fireplace and has natural light through the openness that the sliding glass doors provide, along with access to the terrace. Two full baths serve the three bedrooms and the cozy study/bedroom. Food preparation could hardly have better facilities than those offered by this well-planned front, U-shaped kitchen. The pass-thru from the kitchen to the nook make the quick, informal meal an enjoyable one.